RENEWALS 458-4574

DATE DUE

SIERRA EAST

CALIFORNIA NATURAL HISTORY GUIDES

Arthur C. Smith, General Editor

The publisher gratefully acknowledges the generous contribution to the California Natural History Guide series provided by the following organizations:

RICHARD AND RHODA GOLDMAN FUND
UNION BANK OF CALIFORNIA

and by the General Endowment Fund of the Associates of the University of California Press

CALIFORNIA NATURAL HISTORY GUIDES, 60

SIERRA EAST

Edge of the Great Basin

EDITED BY

GENNY SMITH

With contributions by
Diana F. Tomback, Ann Howald, Mary Hill,
Harold Klieforth, Douglas Powell, Donald
Sada, Genny Smith, and Evan A. Sugden

Illustrated by
Flora Pomeroy Smith and the Authors

UNIVERSITY OF CALIFORNIA PRESS

Berkeley   Los Angeles   London

University of California Press
Berkeley and Los Angeles, California

University of California Press, Ltd.
London, England

© 2000 by
The Regents of the University of California

Library of Congress Cataloging-in-Publication Data

Sierra East : edge of the great basin / edited by Genny Smith ; with contributions by
    Diana F. Tomback . . . [et al.] ; illustrated by Flora Pomeroy Smith and the authors.
        p.   cm. — (California natural history guides ; 60)
    Includes bibliographical references (p.   ) and index.
    ISBN 0-520-08689-9 (alk. paper)
        1. Natural history—Sierra Nevada (Calif. and Nev.)   I. Smith, Genny.   II. Series.
    QH104.5.S54S55   2000
    508.794′4—dc21   99-11688
                                                                                  CIP
                                                                                  REV.

Printed in the United States of America
9 8 7 6 5 4 3 2 1

Drawings copyright Flora Pomeroy Smith

CONTENTS

CONTRIBUTORS

MARY HILL, *geologist,* who now lives in Santa Fe, New Mexico, was the editor of *California Geology* and its predecessor, *Mineral Information Service,* for many years. Her books include *Geology of the Sierra Nevada* and *California Landscape: Origin and Evolution.*

ANN HOWALD, *biology instructor* at Santa Rosa Junior College, lives in Sonoma. She holds an M.A. degree in botany from the University of California, Santa Barbara. She has spent more than twenty summers in the Eastern Sierra, teaching natural history classes and conducting plant research at the university reserves at Mammoth Lakes and Convict Creek.

HAROLD KLIEFORTH, *research meteorologist* at the Desert Research Institute and associate state climatologist for Nevada, received his M.A. degree in meteorology from the University of California, Los Angeles. He first visited the Eastern Sierra in 1937, lived in Bishop from 1951 to 1964, and has lived in both Reno and Bishop since 1965. His work includes snow surveying, weather studies, and vegetation-climate relationships.

DOUGLAS POWELL, *lecturer* in geography, teaches at Sonoma State and San Francisco State universities. His master's thesis for the Geography Department at the University of California, Berkeley, is titled "A Physical Geography of the White Mountains." He was a lecturer for many years at Deep Springs College and UC Berkeley, conducted snow surveys for thirty years, and has led at least sixty field trips in the Eastern Sierra.

DR. DONALD SADA, *aquatic ecologist,* adjunct professor at the University of Nevada, Las Vegas, and consultant, holds a doctorate in biology from the University of Nevada. Born, raised, and now living in Bishop, he has hiked and fished the Sierra most of his life. He has worked for eighteen years on research, management, and conservation biology of Great Basin fishes and mollusks.

FLORA POMEROY SMITH, *nature illustrator and wildlife artist,* has a graduate degree in natural science illustration from the University of California, Santa Cruz, and now lives in Mt. Shasta, California. She is the author and illustrator of *Granite Mountain Spring,* a nature guide to the Eastern Mojave.

GENNY SMITH, *editor and publisher* of Genny Smith Books, has a B.A. degree from Reed College and now lives in Cupertino. She first met the Eastern Sierra in the late 1940s on ski trips. Owning a cabin at Mammoth Lakes since 1957, she has edited and published *Deepest Valley, Mammoth Lakes Sierra, Mammoth Gold,* and other books on the Eastern Sierra.

DR. EVAN A. SUGDEN, *entomologist,* with a doctorate in entomology from the University of California, Davis, is a teacher, freelance biological illustrator, and technical editor living in Seattle. As a graduate student in 1978–83, he studied the ecology of bees and endangered plants in the Eastern Sierra and collected arthropods. He has investigated bees and other insects in several foreign countries.

DR. DIANA F. TOMBACK, *professor of biology,* University of Colorado at Denver, holds a doctorate in biological sciences from the University of California, Santa Barbara. From 1973 to 1977 she studied the behavioral ecology of the Clark's Nutcracker and its interaction with Whitebark Pine in the Eastern Sierra and then spent parts of three more summers pursuing further studies there.

Other illustrators: Annie Kook, scientific illustrator; Doris Marsh; Adrienne E. Morgan; Susan A. Rinehart.

Photographers: Roy Bailey, Ed Cooper, David Dunaway, Robert C. Frampton, David Hamren, Jim Hildinger, T. J. Johnston, Tom Lippert, Susan Moyer, Larry Prosor, Edwin C. Rockwell, Tom Ross, Gerhard Schumacher, John S. Shelton, D. Burton Slemmons, Beverly F. Steveson, Roland von Huene. Six of the color photographs of flowers are by J. R. Haller.

All of us have hiked, photographed, and written about the Eastern Sierra because that was part of our jobs, but also in large part because we love the magnificent Eastern Sierra and enjoy just being there.

ACKNOWLEDGMENTS

We are especially grateful to the following persons and organizations for their generous help with individual chapters.

Geologic Story: Roy A. Bailey, Paul C. Bateman, and N. King Huber, U.S. Geological Survey; Becky Purkey, Nevada Bureau of Mines and Geology; Robert Paschall.

Weather and Climate: All past and present snow surveyors, hydrologists, meteorologists, cooperative weather observers, and geographers who have provided data and insights on Sierra climatology.

Plant Communities: Mark Bagley, Gayle Dana, Dan and Leslie Dawson, Mary DeDecker, J. Robert Haller, David Leland, Sally Manning, Tim Messick, Clifford Ochs, Janet Sherwin.

Arthropods: Andy D. Anderson, University of California, Riverside; David B. Herbst, Sierra Nevada Aquatic Research Laboratory; Derham Giuliani; the late Enid Larson; Richard M. Bohart, Harry Lange, the late Robert O. Schuster, and Robbin W. Thorp, all of University of California, Davis; David Winkler, Cornell University.

Amphibians and Reptiles: Mark R. Jennings for status information and taxonomy.

Birds: Howard Cogswell, Charles Chase III, Kimball Garrett, Joseph Jehl, Sheila Mahoney, Gary Neuchterlein, Stephen Rothstein, and Robert Storer for information based on their research or field experience in the Eastern Sierra; Tom and Jo Heindel for sharing distribution data.

Mammals: Terry Russi, John Wehausen, Ken Zanzi.

Flora Pomeroy Smith, illustrator, thanks Mary DeDecker, Betty and Gil Gilchrist, Mary and Fred Pomeroy, the late Bessie Poole, and Vince and Ann Yoder.

Genny Smith, editor, thanks Alan Gubanich, University of Nevada, Reno; Jim McLaughlin and Neal Brecheisen, Bureau of Land Management, Reno; Bob McQuivey, Nevada Department of Wildlife, Reno; Terisa Vogelsang, U.S. Geological Survey, Carson City; Nevada Historical Society Research Library; Mary Ann Henry.

CHAPTER ONE

INTRODUCTION

GENNY SMITH

The beginnings of the Sierra Nevada rise gently from the Creosote Bush and Joshua Trees of the Mojave Desert. As the range trends northwest, its pale gray crest gradually rises steeper and higher to a grand crescendo of treeless peaks and ridges that soar as much as two miles above the valleys below. A little over three hundred miles from its beginnings, north of Lake Tahoe, the crest slowly descends lower and lower, and the Sierra's gray granite disappears under layers of volcanic rock near Mount Lassen.

This grand mountain range is a major geographic feature, prominent on all space photographs of Earth. Adjacent to the Sierra, a vast arid region (Stephen Trimble calls it "the sagebrush ocean") stretches east across Nevada and half of Utah. This is the Great Basin, so named because its waters never reach the sea but deadend in salty, alkaline lakes. The eastern slope of the Sierra and the western edge of the Great Basin merge in an area commonly known as the Eastern Sierra, which lies mostly in eastern California and partly in western Nevada. Much of the area is thinly populated, with scattered cattle ranches and irrigated pasture lands. The only major city is Reno; Carson City, Bishop, and Ridgecrest are small towns.

The dramatic alpine scenery, the feeling of space, the heavy snowfall in winter and the refreshing, gentle mountain climate in summer: All combine to nourish a popular vacationland here, for winter sports and summer camping. Resorts around Lake Tahoe and at Mammoth Lakes draw skiers and tourists from across the country and around the world; hikers and anglers seek out the less crowded campgrounds, lakes, and trails. While the Sierra's forests and waters are the major attractions, more and more people are finding that the desert to the east is also fascinating and beautiful—although in quite different ways.

Besides its magnificent scenery, the Eastern Sierra supports a wealth of native plants and wildlife. This book will introduce you to that wealth. From well over a thousand plants, hundreds of birds, and perhaps ten thousand arthropods in the region, we have had to choose no more than our limited space can accommodate. Consequently, we have chosen species that are common and conspicuous and, in a few cases, those that are particularly interesting or unusual. Every chapter, treated comprehensively, would require a book larger than this one. So if any subject piques your interest and you would like to know more, you will find some recommended readings at the end of each chapter. Local guidebooks and well-written, popular books on all these subjects are for sale in bookstores, museums, and visitor centers in the Eastern Sierra.

SIERRA EAST

The eastern slope of the massive Sierra Nevada is a world apart from lands to the west (map 1). In contrast to its 40- to 60-mile-wide western slope that sweeps leisurely upward from the Great Central Valley, the eastern slope drops abruptly. From the ridges above Tioga Pass and Mount Rose Summit, for example, the eastern slope falls more than a mile in less than 5 horizontal miles. From Mount Whitney the slope drops a precipitous 2 miles in less than 6 horizontal miles. Mary Hill, in chapter 3, explains the geologic processes that formed this massive range long ago and that continue to operate today.

The changes in climate and vegetation are just as abrupt as the change in topography. In contrast to the west slope's dense pine-fir forests and many large rivers that reach the sea, the sparse forests of the east slope have little underbrush; below 7000 feet there may be no forest at all, only Sagebrush Scrub. The small streams of the Eastern Sierra never reach the sea, ponding instead in large alkaline lakes, such as Mono or Pyramid, or in shallow, seasonal lakes.

EDGE OF THE GREAT BASIN

The enormous mountain barrier blocks many storms from ever reaching the far side of the Sierra. Thus deprived of moisture, the lands east of the Sierra are desert, with little vegetation softening their angular outlines. However, the *rain shadow effect,* as it is called, is more complicated than commonly believed. As Doug Powell and Harold Klieforth explain in chapter 4, air masses pour over and down the eastern slope, and the warming of that air makes clouds evaporate and precipitation decline.

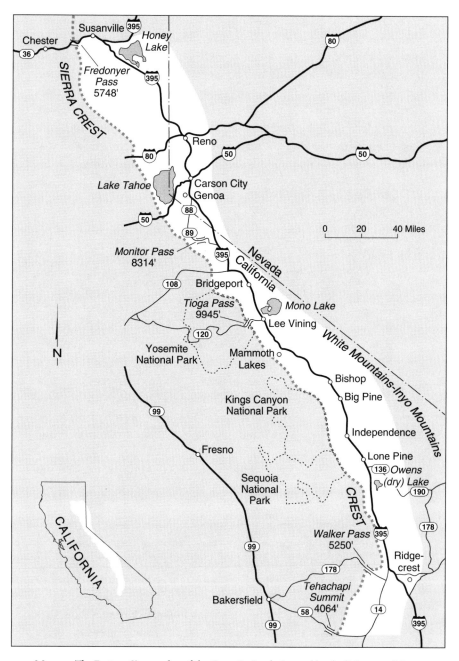

Map 1. The Eastern Sierra, edge of the Great Basin, designated by the light area. (Map by Annie Kook)

Aside from the valleys greened by irrigation, such as the Carson and Bridgeport valleys, much of the Eastern Sierra is muted tan and gray: grasses bleached to straw color, tan sand, blue-gray haze, gray-green pinyon, dead-gray shrubs, pale gray boulders. Except at dawn and dusk. Then colors seldom seen in forested lands magically appear. After the sun-scorched midday glare disappears, cliffs, peaks, shrubs, and grasses come alive with gold, salmon, pinks, and reds, and later, blues and purples. To truly understand this region, one must look eastward, for it is kin to the sagebrush lands that stretch eastward for hundreds of miles, the region known as the Great Basin.

BIOGEOGRAPHIC REGIONS
OF THE EASTERN SIERRA

The map of the Eastern Sierra's biogeographic regions (map 2) helps to explain a very great deal about the Eastern Sierra. It identifies the three remarkable regions that here adjoin and influence each other. Each of the three—the Sierra Nevada, the Great Basin Desert, and the Mojave Desert—has a unique character, a distinctive climate, and characteristic plants and animals. The Sierra has long, below-freezing winters with 10 or more feet of snow covering the ground for as long as six months. In the Great Basin Desert long, cold winters alternate with long, hot, dry summers; precipitation ranges between 5 and 15 inches, most of it coming in winter as snow. The lower, warmer Mojave Desert has even longer, hotter, and drier summers, with rainfall seldom exceeding 5 inches a year. Consequently, the number and kinds of species, both plant and animal, vary tremendously from region to region.

Picturing the Eastern Sierra as the juncture of these three quite different regions helps us understand why it is the way it is. Calling the regions *biogeographic* merely means that they are defined on the basis of their organisms as well as their geography (where plants and animals live). It should not be surprising that the regional boundaries shown in map 2 agree rather well with the geomorphic provinces mapped by geologists, the drainage systems mapped by hydrographers, and the floristic provinces mapped by botanists—though the boundaries and names may differ slightly—because all are interconnected. The Sierra Nevada is a powerful biological boundary as well as a structural landmark. Its height and bulk determine how much rain falls on lands to the east. Temperature and water in turn largely dictate which plant communities flourish where. For our purposes, since our book encompasses many aspects of the Eastern Sierra, defining our area by its biogeographic regions seems most appropriate.

None of the regions, however, has the precise boundaries that map 2 delineates; rather, the regions blend into each other. And that is the key to

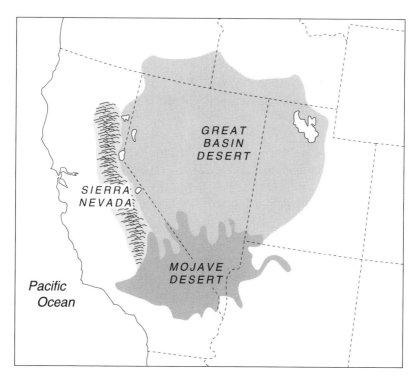

Map 2.   Biogeographic regions of the Eastern Sierra. (Based on MacMahon [1979], *North American Deserts;* map by Annie Kook in *Deepest Valley* [1995])

understanding this land, for as the plants and animals of the regions inter-mingle, they bestow upon the Eastern Sierra an amazing diversity. Desert species creep upward to mingle with montane flora and fauna; a few desert species even occur as high as subalpine elevations in the southern Sierra. Riparian woodlands follow their streams out of the canyons and into the sagebrush. The green pine-fir forests of the Sierra merge with the gray shrubs of the Great Basin Desert, and both intermingle with the Creosote Bush and other widely spaced shrubs of the Mojave Desert.

PLANT COMMUNITIES: DIVERSITY

In addition to the major differences among the large biogeographic regions, the precipitous drop of the eastern Sierra slope from its towering crest to the western edge of the Great Basin has created differing environments or habi-tats. The result of changes in elevation, moisture, soil, and topography, these varied habitats account for the rich and varied flora and fauna in the Eastern Sierra. In each habitat lives a distinct assemblage or *community* of plant and animal species. In chapter 5 Ann Howald describes the Eastern Sierra's plant communities, from the lowest valley bottoms to the highest ridges. She also discusses plants' ingenious adaptations to extreme conditions. Many animals and some plants occur in several communities, while other species are restricted to—and define—one community.

Plant communities are not only diverse but so closely spaced on the Sierra's steep slope that you may drive through four or five different communities within half an hour, as in Bishop Creek or Rock Creek canyons. In the desert valleys, by June most birds have finished nesting, the plants have set seed, and long, hot days are the norm. Yet just twenty minutes' drive west may bring you to snowbanks blocking mountain roads. In the Eastern Sierra, you are always aware of both the arid valleys and the snow-streaked mountains. Nowhere along Highway 395 can you escape the mountain crests looming thousands of feet above you, and nowhere in the mountains can you avoid looking down on arid valleys to the east. One caveat: Boundaries and classifications and plant communities are helpful concepts when we try to understand the complex world we live in, as long as we remember that plants and animals live where *they* find it suitable, which may not always agree with the artificial categories we create for them.

SURVIVAL

Throughout the world, how living things, plant and animal, manage to survive extreme heat, cold, dryness, or lack of food is one of the great wonders of life. In the Eastern Sierra, how do they survive blazing heat, freezing cold, months of no water, and nothing to eat? We describe some of the fascinating, unusual behavioral and physiological adaptations that enable creatures to survive here. Unusual as the number and diversity of species are, far more impressive are the ways Eastern Sierra plants and animals exploit their habitats and cope with the harsh mountain and desert environments.

Although precipitation varies enormously from year to year, and occurs at unpredictable times and in unpredictable quantities, most of the time the lands of the Eastern Sierra have little water. In addition to an uncertain water supply, the Eastern Sierra has intense sun, dry air, high evaporation rates, alkaline soils, long, hot summer days, and months of cold in winter. Temperatures, too, can be extreme, varying significantly between winter and summer and between daytime and nighttime. All these factors make life very difficult for native plants and animals.

Mammals have some general ways of dealing with extreme temperatures and lack of food. Some burrow under the ground in winter to avoid the cold (hibernate), and in summer some desert mammals go underground for months to avoid the heat (estivate). Some, such as deer and Bighorn Sheep, migrate from the high mountains to the lower valleys. Birds seem to have it best of all: They can fly several miles to water and food, or to higher or lower altitudes to cool off or warm up. After nesting, many fly hundreds or thousands of miles to areas where food is abundant. Animals have many other ways to cope. In chapters 8–11 Diana Tomback describes a wide spectrum of behavioral and physiological adaptations that reflects the varied habitats and harsh conditions that mammals, birds, reptiles, and amphibians encounter. Don Sada, in chapter 7, reports on native fishes, for which scarce food in winter, and not cold, is the chief problem. In chapter 6 Evan Sugden explains the adaptations that enable arthropods to survive in rigorous environments.

Unlike animals, plants can't migrate with the seasons or move downslope for the winter. And yet, when you think about it, do plants really just sit there and endure? Cannot their offspring migrate, given enough time? What about the seeds that birds or the wind carry a few feet or a few miles away, to new localities that may be favorable? Seeds can also be thought of as a plant's remarkable way of hibernating—avoiding harsh conditions and coming to life only when moisture and temperature are just right. And although perennials may look dead aboveground, in reality are their roots not hibernating, waiting for milder times before bursting into new growth?

THE EASTERN SIERRA DEFINED

Essentially, we cover the eastern slope of the entire Sierra Nevada and the High Desert valleys immediately adjacent on the east, from Tehachapi Pass (Highway 58) in Kern County north to the northern boundary of Plumas County (Highway 36). The Sierra Crest and most of the adjacent valleys lie in California; a few of the valleys are in Nevada. U.S. Highway 395 is an easy marker to generally define the area we cover, for it hugs the base of the Sierra most of the way from Mojave to Susanville, traversing the desert valleys at the western edge of the Great Basin.

But as we studied the plants and animals for this book, we found ourselves in trouble if we defined the eastern boundary rigidly as "the desert valleys immediately adjacent to the eastern Sierra slope." For example, just east of Owens Valley, atop the White Mountains, 4000-year-old Bristlecone Pines—one of the wonders of the world—have lived continuously since 2000 B.C. How could we *not* mention them, when they are so close? And in discussing native fishes, how could we *not* mention the Death Valley and Lahontan drainage systems far to the east? For they explain why native fishes are where they are today.

Another trouble cropped up as we worked north toward Lake Tahoe and Reno. From Mojave to Reno, the Sierra Crest is easily recognized as it juts skyward west of Highway 395 and divides the waters flowing east into the desert from those flowing west to the Pacific Ocean. However, north of Reno the Sierra Crest is not so easily defined. It drops to altitudes of 6000 and

7000 feet, the Feather River's Middle and North forks cut through it, and Highway 395 and the High Desert valleys lie 20–40 miles away, east of the Grizzly and Diamond mountains. So why didn't we define the crest here as the divide between east-flowing and west-flowing waters? Why didn't we choose as our western boundary the crest of the Diamond Mountains, which is the drainage divide? Mainly because the Sierra Nevada, one of the world's great mountain ranges, has always been identified as the 400-mile-long block of massive granites that extends from Tehachapi Pass nearly to Mount Lassen, where volcanic outpourings bury the bedrock granite and define its northern boundary. The Grizzly and Diamond mountains have a geologic history that is quite unlike that of the Sierra; they are composed of younger rocks and have been subjected to complex faulting. Since this book is basically a guide to the overall ecology and physical setting of the region, our boundaries had to follow where its geology took us.

THE EASTERN SLOPE

Except for the discussion of plants in chapter 5, we use the terms *Eastern Sierra* and *eastern slope* interchangeably to mean the steep slope from the Sierra Nevada's crest to its eastern base *and the adjacent valleys,* whose floors range from 3000 feet to 7000 feet in altitude. By *western slope,* we mean the long, gentle slope from the crest to the western foothills *and the adjacent Central Valley* (the Sacramento and San Joaquin valleys), which is close to sea level. Because many plants and plant communities are restricted to either the mountain slope or the adjacent valley, in chapter 5 the term *eastern slope* refers only to the slope and not to the valley. To emphasize that the Eastern Sierra is quite a different world from the land lying west of the Sierra, we have identified those plant and animal species that do not occur on the west slope with an asterisk (*). Few species are found only on the east slope, since most are able to survive in many habitats. Some plants, however, have adapted to specific Eastern Sierra environments and cannot tolerate different conditions. Most of the plants discussed in this book extend their range west of the crest, although the greater part of their distribution lies to the east.

With all its variety, the Eastern Sierra is an ideal location for university field stations where students can learn the techniques of field work and scientists can conduct long-term studies on subjects as varied as human adaptation to high altitude and the feeding habits of phalaropes. Universities and their field stations include the following: University of California, Berkeley, Sagehen Creek; UC Davis, Tahoe Research Group; UC Santa Barbara, Sierra Nevada Aquatic Research Laboratory; UC, White Mountain Research Station; University of Nevada, Whittell Forest and Wildlife Area.

SPECIES, REFERENCES, STYLE, DISTRIBUTIONS

Where appropriate, vertebrate descriptions, particularly of amphibians and reptiles, are those of the subspecies (ssp.) known to occur on the eastern slope. For some species, there is little regional differentiation. Status and taxonomy of amphibians and reptiles are based on Mark R. Jennings (1996), "Status of Amphibians," in *Sierra Nevada Ecosystem Project: Final Report to Congress,* Vol. 2: *Assessments and Scientific Basis for Management Options,* Wildland Resources Center Report 37 (Davis: UC Davis, Centers for Water and Wildland Resources), and on Jennings, "An Annotated Checklist of Amphibians and Reptiles of California," 3d rev. ed. (draft). All common and scientific names for mammals are based on R. C. Banks, R. W. McDiarmid, and A. L. Gardner (1987), *Checklist of Vertebrates of the United States, the U.S. Territories, and Canada,* U.S. Department of the Interior, Fish and Wildlife Service, Resource Publication 166 (Washington, D.C.), and C. Jones et al. (1997), *Revised Checklist of North American Mammals North of Mexico,* Museum of Texas Tech University, occasional papers, no. 173. Bird names and classification follow the 1998 *Check-list of North American Birds* published by the American Ornithologists' Union. See the introductory material of other chapters for their references.

The University of California Press style, in the California Natural History Guides, is to capitalize the common names of single species and single plant communities (Desert Horned Lizard, Quaking Aspen, Sagebrush Scrub), but to use lower case when referring to several species or communities (lizards, finches, scrub communities) or when using an abbreviated or partial name (horned lizard). The distributions of many vertebrate species, especially birds, in the Eastern Sierra are not completely known. Although we consulted several sources, in many cases information was inadequate, and in a few cases the sources differed. The distributions we report are estimates, which we hope to modify in the future as more information becomes available.

UPON LONGER ACQUAINTANCE

At first acquaintance, the Eastern Sierra strikes some people as bleak, forbidding, lonely, even depressing. Feeling alone in the desert is not necessarily the same as feeling lonely—it can be exhilarating. One can be alone *with* the vastness, become a part of, not *apart from,* the space and sky. Upon longer acquaintance with the Eastern Sierra, you, like many others, may find it starkly beautiful, dramatic, peaceful, and teeming with life. It may take time—far more time than it takes to feel a kinship with green mountains and blue water—to develop a new set of values about scenery and to discover that the absence of green makes possible a vast array of other beautiful colors. It may take time to appreciate that in arid lands life is complicated and often hidden, that much goes on underground and at night. But it is all there, waiting for you. We hope this book will open doors to different worlds and new ideas and that you will choose to walk through some of those doors, to explore and discover for yourself the magic of the Eastern Sierra.

CHAPTER TWO

DISCOVERING THE EASTERN SIERRA

GENNY SMITH

You may spend weeks, months, and even years happily exploring the Eastern Sierra, yet every time will be different from the last time you were there, and you will be surprised how many surprises you will come upon. Each stream, each meadow, each season, even each day will be unique, for that is the nature of wild things. They are in *their* place at *their* time—not yours. The best you can do is to look in the most likely places (and we give you some general clues to where you may find what), but remember, wild things appear and disappear on *their* schedules. Still, if you keep searching, you will surely have some unforgettable lucky days.

Although we mention rare or unusual plants and animals, in this chapter we mostly describe how to find areas such as meadows, wetlands, riparian woodlands, and mountain forests, without repeating lists of birds and plants that you might find in each locality. By referring to the pertinent chapters, you will find the types of locality that each species is likely to inhabit. Look especially for wetlands, oases in the desert that attract insects, amphibians, birds, and mammals. You may be disappointed that you see few of the desert mammals we mention; most of them spend their summer days keeping cool—underground, under rocks, or in shaded places. Dawn and dusk are good times to watch for them. In most Eastern Sierra valleys, the bottomlands have been farmed or grazed and the marshes drained for more than a hundred years. Consequently, native plants are absent from agricultural lands, but they still dominate pastures, marshes, and shrublands, which are the natural habitats of valley floors.

To start you on your way to discovering, we list here some locations where you can find outstanding examples of the landforms, plants, and animals described in the chapters that follow. We suggest only locations that are relatively easy to get to. All of them have stories to tell: dramatic tales of birth, death, and predation, featuring multiple plots and unpredictable casts of plants and animals—tales of mystery and wonder in the midst of seemingly monotonous expanses of gray shrubs.

MAPS AND INFORMATION

Map 1 (p. 3) shows the major roads, towns, and lakes of the Eastern Sierra. In addition to this small map, you will need to inquire locally for larger, more detailed maps. The U.S. Forest Service (USFS) and Bureau of Land Management (BLM) both have excellent large local maps and also can supply information on campgrounds, roads, trails, and weather conditions.

BLM offices: Ridgecrest, Bishop, Reno

USFS Ranger Stations: Lone Pine, Bishop, Mammoth Lakes, Lee Vining, Bridgeport, Markleeville, Carson City, Sparks, Truckee, Susanville

The auto club (CSAA) maps of the region are also excellent. Most relevant are: *Bay and Mountain Section, Northeastern California, Eastern Sierra,* and *Desert Area.* Several visitor centers also offer maps and reliable local information:

Interagency Visitor Center, 1 mile south of Lone Pine

Mammoth Lakes Ranger Station and Visitor Center, Mammoth Lakes

Mono Lake Committee Information Center and Bookstore, center of Lee Vining

Mono Basin National Forest Scenic Area Visitor Center, north of Lee Vining

Lake Tahoe Visitor Center, South Lake Tahoe

The larger cities have Chambers of Commerce that offer information on accommodations and points of interest. Several volunteer groups offer field trips year-round and welcome nonmembers: the Northern Nevada and the California native plant societies, Audubon chapters, Sierra Club chapters, and some museums. Most have no permanent phone numbers, but you can try to locate them through Forest Service or BLM offices, visitor centers, or local newspapers. The telephone number for the Lahontan Audubon Society in Reno is (775) 324-2473. Points of interest are listed from south to north, near or on Highway 395 unless otherwise noted. In the valleys, spring and fall days are cooler and more pleasant than midsummer.

SOUTHERN SIERRA

The valleys and lower elevations of the southern Sierra are Mojave Desert, in contrast to the Great Basin Desert or High Desert that includes most other Eastern Sierra valleys. Here Creosote Bush Scrub dominates the valley lands astride Hwy 395. In a wet year, Creosote Bush and Burro-bush shade masses

Redrock Canyon. (Photo by Gerhard Schumacher)

of annual wildflowers that will grow, flower, and die before the desert soil has dried from the spring rains.

REDROCK CANYON STATE PARK, HWY 14   Colorful cliffs consisting of up-lifted, tilted, and eroded layers of whitish clay, red sandstone, brown lava, and pinkish tuff. Joshua Trees, Sidewinders, kangaroo rats, and other plants, reptiles, and mammals characteristic of the Mojave Desert. Pleistocene fossils. Many raptors.

WALKER PASS, HWY 178, JOSHUA TREE WOODLAND AND CREOSOTE BUSH SCRUB   An extensive stand of Joshua Trees (a multi-trunked variety) and the plants associated with them, 5 miles west of Hwy 395. To see some circular *creosote rings,* which may be older than the oldest Bristlecone Pine, park at the historical marker at the intersection of Hwys 178/395 and look northwest up the slope. These circular clones began with one bush that sent out a ring of branches, some of which rooted and grew new bushes. Joshua Trees are also scattered along Hwy 395 toward Little Lake.

PINNACLES NATIONAL NATURAL LANDMARK   Weathered tufa that developed in a lake (now dry), by a chemical process similar to the one that produced Mono Lake's famed tufa towers. East of Ridgecrest, just south of the road to Trona (Hwy 178).

LITTLE LAKE   Large Desert Marshland. Waterfowl in winter.

FOSSIL FALLS   Recent lava, sculpted and polished by a long-vanished river that overflowed from Owens Lake during wetter times. Other volcanic debris: pale gray pumice and dark red cinder from the nearby cinder cone, Red Hill. From Hwy 395, about 2 miles north of Little Lake, follow Cinder Road east for 0.5 mile, turn south to the trailhead.

OWENS VALLEY

Between Little Lake and Owens Lake, Creosote Bush Scrub of the Mojave Desert blends with the Alkali Sink Scrub of the Owens Valley floor. As you drive north from Little Lake, you will notice that Creosote Bush becomes less abundant, while Greasewood and Saltbush become more abundant. North of Owens Lake, Creosote Bush Scrub disappears from the valley floor, replaced by Alkali Sink Scrub. (A *sink* is a low-lying, poorly drained area.) Hawks, quail, ducks, blackbirds, owls, jays, and the conspicuous Common Raven and Black-billed Magpie are among the birds that are permanent residents.

DIAZ LAKE   Three miles south of Lone Pine. A sag pond, a low point in a sunken block down-dropped between two parallel faults about a quarter-mile apart, within the Sierran Frontal Fault Zone. Resident birds and migrants in season.

INTERAGENCY VISITOR CENTER   South of Lone Pine at the junction of Hwys 395 and 136. Exceptional collection of books, maps, exhibits, and current information on the Eastern Sierra, plus a dramatic view of Mount Whitney.

SIDEROAD TO HORSESHOE MEADOW AND THE COTTONWOOD LAKES Provides almost aerial views of trough-shaped Owens Valley, which has been down-dropped between the uplifted Sierra Nevada and the White-Inyo Mountains. The highest road in the southern Sierra (over 10,000 feet) takes you to Subalpine Forest and Meadows and within a short walk of the Foxtail Pine on the ridge south of Horseshoe Meadow. An almost level 3-mile hike leads to the Cottonwood Lakes, where each spring Golden Trout eggs are spawned artificially and packed out to the Mount Whitney Hatchery. From

Alabama Hills. (Photo by Tom Ross)

Whitney Portal Road, 3 miles west of Lone Pine, turn south onto Horseshoe Meadow Road and continue for 20 miles. Passable summer and fall only.

GRAVE OF 1872 EARTHQUAKE VICTIMS   One mile north of Lone Pine. The low ridge at the gravesite is a *fault scarp,* the abrupt break where the land to the east has dropped down. This fault scarp, marked with clumps of green growth fed by springs, runs parallel to and a few hundred feet west of Hwy 395 for about 2 miles.

ALABAMA HILLS   An outstanding example of jointed granite, weathered while it was covered with alluvium. You may recognize the Alabama's weird, rounded shapes as the location for many movies and commercials. Similar forms occur in the Buttermilk country near Bishop, below Conway Summit, along Hwy 120 west of Benton, and on the east shore of Lake Tahoe. From Hwy 395, 6 miles north of Lone Pine, turn west onto a graded road that swings west and then south.

THE GREAT EASTERN ESCARPMENT OF THE SIERRA NEVADA   The deeply incised, precipitous granite wall rising west of Owens Valley was produced by the erosion of a steep fault scarp. The Sierra Crest rises highest between

Lone Pine and Big Pine, with many 14,000-foot peaks between Mount Langley and the Palisades.

EASTERN CALIFORNIA MUSEUM   Collections and exhibits covering natural history, native peoples, recent history. The museum is located on Grant St. in Independence, 3 blocks west of Hwy 395.

ONION VALLEY   At 9200 feet, a small, luxuriant Montane Meadow. Stands of Foxtail and Limber pine on the trail to Kearsarge Pass. From the center of Independence, follow to its end the road west that climbs the canyon of Independence Creek.

OAK CREEK TRAILHEAD   Sweeping views of Owens Valley and the Inyos, with black oaks, wildflowers among the brush. High Desert Riparian Woodland along lower Oak Creek. About 2 miles north of Independence, turn west to the Mount Whitney Fish Hatchery (pond with waterfowl and herons), then take the right fork to the Oak Creek trailhead.

BIG PINE VOLCANIC FIELD   Red cinder cones and black basaltic lava flows at the base of the mountain ranges on both sides of the valley. Although these volcanic features look as fresh as if they had erupted only a few years ago, they are up to 300,000 years old. A prominent fault scarp facing east, as much as 80 feet high, extends from the Poverty Hills north to Crater Mountain. Much of the movement occurred long ago, although there is evidence that the mountain side rose about 10 feet during the 1872 earthquake. North of Independence 8 miles, turn west on Black Rock Springs Rd., then north on the paved road (the old highway) that parallels Hwy 395. Just north of Aberdeen, pick up Tinemaha Rd., a dirt road that continues north around the west side of the Poverty Hills and rejoins Hwy 395 north of Fish Springs Hatchery.

TULE ELK VIEWPOINT   East of Hwy 395, 17 miles north of Independence, a short dirt road leads to a viewpoint from which there is a fair chance of seeing elk in the fields below from July to October. Between Independence and Big Pine, watch for elk near Hwy 395.

PALISADE GLACIER   The largest Sierra glacier. So closely does Hwy 395 hug the base of the Sierra that the crest and its glaciers are hidden. To see them, from Big Pine take Hwy 168 east. If you look back at the Sierra from the far side of Owens Valley in summer, you can recognize Palisade and the other glaciers as the broad white patches just below the dark, ragged peaks— patches that persist through the summer. From the far side of the valley you can also best appreciate the Sierra's massive character and the steepness of its eastern flank.

Palisade Glacier below North Palisade. (Photo by David Hamren)

ANCIENT BRISTLECONE PINE   A short walk among the oldest living things on earth. High-mountain insects, birds, and small rodents; Mountain Bluebirds are common. From Big Pine, take Hwy 168 to Westgard Pass and turn north on the road to Schulman Grove, 10,000 feet. Only 23 miles, but allow at least two hours. Road impassable in winter and spring.

KLONDIKE LAKE   This shallow lake has Tundra Swans in winter, nesting and migrant birds in season. North of Big Pine 3 miles, take the dirt road east about 1 mile. Information board at turn.

KEOUGH HOT SPRINGS   One of many hot springs along the Sierran Frontal Fault Zone. North of Big Pine 7 miles.

OWENS VALLEY PAIUTE-SHOSHONE INDIAN CULTURAL CENTER Located on West Line St. in Bishop. Outstanding exhibits of Paiute-Shoshone customs and native plant foods.

BISHOP CREEK CANYON Here are U-shaped canyons, moraines, cirques, knife-edged ridges—much evidence of past glaciers. In early October, aspen slopes are a mass of quivering yellow-gold. In the 6000-foot climb from the valley to the road end, you will find most of the trees described in chapter 5. Many insect species are found in the varied habitats. From Bishop, take Hwy 168 west to the three forks of Bishop Creek. A steep, winding road leads to North Lake. Just before reaching the lake, at the top of the grade, there is an outstanding view back across the Middle Fork canyon, where the relations among Sierra granites and the older rocks it intruded are easily seen and wonderfully exposed. A now vertical layer of older, reddish metamorphosed sediments separates 200-million-year-old light gray granite on the east from 90-million-year-old dark gray granite on the west. Near the North Lake road end, scattered meadows within the lodgepole forest are bright with wildflowers. Paved roads lead to the canyons of the South and Middle forks of Bishop Creek.

VOLCANIC TABLELAND North of Bishop, formed by enormous eruptions of glowing hot volcanic ash. (See the entry for the Long Valley caldera viewpoint, below.) From Bishop, drive north on Hwy 6 a little over 1 mile until the highway takes a right angle to the east; proceed straight ahead for about 2 miles. Where the road forks three ways, take the left fork, Chalk Bluff Rd., which runs along the base of the pinkish tan cliffs of Bishop Tuff. The prominent white exposure at the base of the cliffs is not chalk at all, but pumice erupted from the Long Valley caldera before the mighty explosions of Bishop Tuff. Pinyon-Juniper Woodland with scattered Jeffrey Pines. White-throated Swifts nest in Owens Gorge.

FISH SLOUGH Lush Desert Marshland fed by springs, with tall cattails and tules, and dragonflies. Wading birds and raptors can be seen here year-round. Protected pupfish habitat. Active faults have broken and tilted large masses of Bishop Tuff, both east and west of the road. From Bishop, drive north on Hwy 6 a little over 1 mile until the highway takes a right angle to the east; proceed straight ahead for about 2 miles. Where the road forks, take the right fork, Fish Slough Rd.

OLD BENTON Wintering Tundra Swans, hawks, ducks, and geese in the wet meadows. From Hwy 6 north of Bishop, drive west on Hwy 120.

NORTH OF BISHOP   Alkali Meadows along Hwy 395. Shadscale Scrub above Round Valley on Old Hwy 395.

SHERWIN GRADE   North of Bishop, Hwy 395 roadcuts expose the sparkling pink color beneath the tan weathered surface of the Bishop Tuff. Pinyon pines signal a bit more rainfall and the cooler temperatures of the High Desert. Sweeping views to the south emphasize the trough shape of Owens Valley, the deepest valley on the American continent, lying 2 vertical miles below the highest peaks on either side.

LONG VALLEY, MAMMOTH LAKES

Between Sherwin and Conway summits, two or more sagebrush-covered, curving ridges of gravel and boulders issue from the mouth of almost every canyon, looking as young as if the glaciers had deposited them yesterday. These *glacial moraines* mark the last high stand of the glaciers. Most canyons also have groves of aspen, which turn brilliant gold in fall. Black-tailed Jack Rabbits are common; Least Chipmunks occur in open sage and bitterbrush flats.

ROCK CREEK CANYON   From Toms Place, the road climbs from Pinyon-Juniper Woodland and riparian woodlands to Montane and Subalpine forest and meadows. At 10,300 feet, the road end at Little Lakes Valley is the highest in the Sierra and one of the few places where you can reach Subalpine Meadows and stands of Whitebark Pine by car. A typical broad, flat-bottomed U-shaped glacial valley. Short, easy walks lead to treeline. Indian Paintbrush, lupines, and other wildflowers in profusion throughout the summer; high-mountain birds and rodents. Miles of golden aspen in fall.

LONG VALLEY CALDERA VIEWPOINT   Marked by a parking area, north of the sideroad to Crowley Lake. An outstanding site to view the 20-mile-long ancient caldera, the source of catastrophic eruptions 760,000 years ago. Gigantic glowing avalanches of gas-charged froth compacted and cooled to form the pinkish tan volcanic rock forming the Volcanic Tableland. Between Round Valley and the June Lake junction, many roadcuts expose the sparkling pink color beneath the rock's weathered surface. Known as Bishop Tuff, this rock blankets about 450 square miles.

WHITMORE HOT SPRINGS, BENTON CROSSING, AND BENTON   A sideroad takes you into a desert world totally unlike the mountain world to the west and into several desert plant communities—Sagebrush Scrub, Alkali Meadow, Desert Marshlands, and Pinyon-Juniper Woodland. Shorebirds

Little Lakes Valley. (Photo by Ed Cooper)

nest and Spadefoot Toads breed in the Alkali Lakes and in the marshes along the Owens River. Migrating waterfowl rest and feed in spring and fall. Birds of the sagebrush country include Common Nighthawks, Horned Larks, Black-billed Magpies, shrikes, and thrashers.

MCGEE AND CONVICT CANYONS   West of Hwy 395. Spectacular exposures of the old Paleozoic metamorphic rocks into which molten granite intruded. Dark gray, white, rust, and blue-gray layers of old sediments, squeezed and folded until many now stand vertical. Fossil graptolites (tiny marine organisms) in Convict Canyon date some rocks as Ordovician, 500 million years old. High Desert Riparian Woodland follows the creeks down into the sagebrush.

HOT CREEK GEOLOGIC SITE   About 3 miles east of Hwy 395. Hot springs and gas vents along a 2-mile segment of Hot Creek. Monkeyflowers and aquatic buttercups abound.

MAMMOTH LAKES BASIN   This area features lakes, streams, forest, and meadows. From the Sagebrush Scrub of Long Valley, Hwy 203 and Lake Mary Rd. lead to Jeffrey Pine and White Fir at the lower elevations, Moun-

Mammoth Mountain. (Photo by T. J. Johnston)

tain Hemlock and Whitebark Pine at the higher. The Montane Forest and Montane Riparian Woodland attract many nesting birds in summer. From the high-altitude road ends (9000 feet), short, gentle trails lead to small meadows as well as to dry, sagebrush-covered hillsides alive with wildflowers. Montane Chaparral on the lower slopes of Mammoth Mountain, Mountain Hemlock on the shaded west shore of Horseshoe Lake.

MINARET SUMMIT   Viewpoint and nature trail. Along Hwy 203 west of Mammoth Lakes, Montane Forest with big old Red Fir grades into Subalpine Forest with stunted Whitebark Pine and, above treeline, sagebrush spangled with flowers. Along the jeep trail trending north from Minaret Summit, you might see Mountain Bluebirds, Clark's Nutcrackers, White-tailed Jack Rabbits, Golden Eagles, Cooper's Hawks, and other birds of prey. The pumice covering the ridge is carpeted with multi-colored dwarf flowers. Although Ritter Range peaks to the west rise 2000 – 4000 feet higher, Minaret Summit is on the Sierra Crest, which divides waters flowing east from those flowing west.

Although Sotcher Lake and Reds and Agnew meadows are west of Minaret Summit and therefore not strictly Eastern Sierra, they are so close it would be a shame to miss them. Because the slope below the summit is alive with springs and small streams, wildflowers are everywhere. An exceptional number of bird species nest in the forest. Squirrels, chipmunks, and deer are common, and bears sometimes visit the campgrounds. Near Sotcher an active Beaver dam has raised the lake level and drowned some Lodgepole Pines. The dead trees attract many birds, including White-headed and Pileated woodpeckers.

CASA DIABLO   East of the Hwys 395/203 junction. Geothermal plant producing electricity from natural hot water. Steam plumes from vents are conspicuous in winter.

JEFFREY PINE FOREST, INYO CRATERS, AND AN OBSIDIAN DOME   Between Mammoth Lakes and June Lake, a 100-square-mile Jeffrey Pine forest straddles Hwy 395. Watered by storm clouds that funnel over low saddles on either side of Mammoth Mountain, these pines thrive in the pumice soil. Site of a Pandora Moth infestation in 1979. To see some volcanic features, about 5 miles north of the Hwys 203/395 junction, take the road west to the Inyo Craters and a nature trail. Just north of Deadman Summit, Glass Flow Rd. branches west to an obsidian (volcanic glass) dome.

MONO BASIN, JUNE LAKE

JUNE LAKE LOOP ROAD   Begins and ends at Hwy 395. Sagebrush Scrub, riparian woodlands, Montane Forest and Meadow. Classic U-shaped glacial canyon with huge, long moraines reaching far into the sagebrush.

MONO CRATERS   The flat-topped, steep-sided Mono Craters east of Hwy 395 are a chain of obsidian domes and stubby obsidian flows rising 2400 feet above Pumice Valley. The eruptions that built them spanned about 35,000 years, although most were active within the last 10,000 years. They are part of a north-south alignment of domes and craters (including the Inyo Craters and the obsidian dome mentioned above) that extends from Negit Island in Mono Lake 30 miles southward to Mammoth Mountain. Much of the pumice blanketing the area came from explosions associated with the craters and the rise of these domes.

PANUM CRATER, MONO LAKE TUFA STATE RESERVE, AND BLACK LAKE Take Hwy 120 east. About 3 miles east of Hwy 395, a graded road branches left to Panum, the northernmost of the Mono Craters. A short walk takes you up the pale gray pumice rim that encircles its black, jagged obsidian plug. Two miles farther on Hwy 120, a dirt road leads north to South Tufa parking area at the Tufa State Reserve near the shore of Mono Lake. Exhibits explain the lake's unique features; a trail takes you to the shore and tufa towers. Up to

Mono Craters. (Photo by Robert C. Frampton)

50,000 California Gulls nest on the islands. The lake's abundant brine shrimp and Alkali Flies feed up to 750,000 migrating grebes in October and tens of thousands of phalaropes in late July and August. Continuing east beyond Sagehen Summit, Hwy 120 drops down to the meadows of Adobe Valley, fed by River Springs. Pronghorn have been reintroduced to this area. Desert Marshlands and an extensive Alkali Meadow system reach from River Springs to Black Lake, where there are nesting, migrating, and wintering birds in season.

TIOGA PASS, SADDLEBAG LAKE   Take Hwy 120 west (open in summer) to Yosemite National Park. Exhibits at the Mono Lake Ranger Station include the plant communities of Mono Basin. Riparian woodlands along Lee Vining Creek shelter nesting birds. If you are very lucky, you may see a Bighorn Sheep on the brush slopes, although their pale gray color provides superb camouflage. A paved road branches north to Saddlebag Lake. A motorboat taxi crosses the lake to a mile-wide basin of small lakes and Alpine Rock and Meadow Communities; short-stemmed flowers bloom from August into September. Grooved, glacier-polished granite. At Tioga Pass there is a large Subalpine Meadow with Belding's Ground Squirrels, marmots; Pikas inhabit the adjacent slopes.

MONO LAKE COMMITTEE INFORMATION CENTER AND BOOKSTORE   In Lee Vining. The center offers canoe trips, nature walks, and news on Mono Lake's recovery.

MONO BASIN VISITOR CENTER   The USFS visitor center is located north of Lee Vining. It offers outstanding exhibits, films, and programs on natural history and native peoples.

Bighorn Sheep. (Photo by David Dunaway)

MONO LAKE COUNTY PARK   Shaded picnic area, and a boardwalk cross-ing wet meadows among the tufa towers. Desert Marshlands and Alkali Meadows border the shores of the lake, which harbor many insect and bird species. Great Horned Owls nest in tufa. North of Lee Vining, take Cemetery Rd. east.

BODIE LOOP RD.   Varied insects and plants in the sagebrush and small meadows. North of Conway Summit, take the road to Bodie; return to Hwy 395 via Cottonwood Canyon and Hwy 167.

BRIDGEPORT TO MINDEN/GARDNERVILLE

In season, raptors and wetland birds are abundant in the huge meadows west of Bridgeport. Most are irrigated, some are natural Alkali Meadows. High Desert Riparian Woodland borders the Walker River. Between Topaz Lake and Gardnerville there is extensive pinyon forest.

LEAVITT MEADOWS   Arthropods, along Hwy 108 toward Sonora Pass.

TOPAZ LAKE   Resident and migrating birds in season.

Bridgeport Valley below Sawtooth Ridge. (Photo by Beverly F. Steveson)

CARSON VALLEY

From Gardnerville north, the Sierra Nevada is not as easily recognized as it is to the south, where it stands as a gigantic, massive block west of Hwy 395. Here it is splintered: although the Sierra Crest lies 15 miles west on the far side of Lake Tahoe, the Carson Range bordering Carson Valley is also considered part of the Sierra because the major rock type in both ranges is granodiorite of about the same age. Along the fault zone at the base of the mountains, aspen and willow mark springs—hot (such as Walley's and Hobo Hot Springs) as well as cold. The valley is well watered by the East and West forks of the Carson River. In spring many kinds of birds nest here; in winter populations of hawks and eagles peak, and deer come down from the mountains to feed on Antelope Bitterbrush and other shrubs.

GENOA   At a gravel pit near Genoa, an extraordinary fault plane—a steep, smooth granite face with vertical striations—is exposed. Friction from move-

Fault scarp near Genoa. For scale, note the man left of center. (Photo by Roy Bailey)

ment along the fault plane has polished and grooved the surfaces (slicken-sides). Rarely does one see these highly polished surfaces; most are buried or eroded away. For about 10 miles here at the base of the Carson Range, the Genoa fault scarp stands about 30 – 44 feet high. Evidence of the fault is abundant: hot springs, valley floors tilted toward the scarp, many recent small scarps. About 2 miles north of Minden, take Genoa Lane west to Genoa, then Foothill Lane south for 1.3 miles to the pit. From Genoa north, Jacks Valley Rd. (Hwy 206) is a great backroad for exploring and birding in the Carson Valley.

CARSON CITY, NEVADA STATE MUSEUM   Fine collections, exhibits, and books on all aspects of Eastern Sierra natural history, native peoples, and history.

WASHOE VALLEY

OLD HWY 395 AND FRANKTOWN RD.   Two backroads west of Hwy 395, north of Carson City, about 2 miles north of Lakeview Summit. Sagebrush bordered by pine forest. You'll find Davis Creek County Park 0.5 miles before you rejoin Hwy 395.

WASHOE LAKE STATE PARK   Although in some years the lake may be dry, Little Washoe Lake is likely to have shorebirds and waterfowl year-round.

STEAMBOAT SPRINGS   The springs rise at the northern end of the Genoa Fault, near the junction of Hwys 431 and 395. The high temperatures of Steamboat Springs (up to 442°F at depths over 3000 feet) are now harnessed to produce electricity. The springs probably derive their heat from a deep body of magma—perhaps the same body that erupted pumice and four rhyolite domes nearby between 1 million and 3 million years ago. The site of frequent low-risk earthquake swarms.

RENO, TRUCKEE MEADOWS

Truckee Meadows extends north to the Truckee River. The Christmas Bird Count here often finds about 100 species.

HUFFAKER OVERLOOK   Here one gets a wonderful view of the geologic "big picture"—the Carson Range uplifted along the Sierran (Carson Range) Frontal Fault Zone and the Truckee Meadows valley down-dropped below. The Sierra continues to rise in this region, about 1 cm per year, and small earthquakes are frequent. The Carson Range has a partially exposed core of granite that intruded into older rocks about 90 million years ago. Mount Rose is darker than other Carson Range peaks because of the volcanic rocks that cover its summit. Huffaker Hills is composed of volcanic domes and flows. A nature trail takes you to outstanding exhibits on birds, plants, clouds, geology, and human history. South of Reno, exit east from S. Virginia St. onto Longley Lane, turn right onto Huffaker Lane, then right onto Offenhauser. The path takes off at the north end of Huffaker Park.

RENO   Wetlands and parks between Reno and Carson City attract many birds. For local information, call (775) 324-BIRD, the Reno Bird Alert, sponsored by the Lahontan Audubon Society. Virginia Lake is a natural sag pond with a fault scarp up to 20 feet high on its west side. Many birds, especially in winter; resident Canada Geese, occasionally pelicans. From Virginia St., drive west on Plumb Lane, turn left on Lakeside Dr. At Oxbow Park, High Desert Riparian Woodland borders the ponds and small islands at this oxbow bend of the Truckee River. Butterflies, amphibians, wildflowers, and more than 100 bird species. From Hwy 395 turn west onto Second St., then left on Dickerson Rd., and continue to the road end. Moana Municipal Pool has natural warm water. Over an area of about 5 square miles, the Moana Lane area has warm-water wells that measure 160–205°F. In winter you can see the steam rising from ditches and other vents. Rock Park is on the Truckee River, about 3 miles east of central Reno. Streamside trees and shrubs shelter many kinds of birds, bats and other small mammals, amphibians, and reptiles year-round. From Hwy 80 in Sparks, take Rock Blvd. south.

LAKE TAHOE AND VICINITY

Lake Tahoe was formed by the same forces that formed most other Eastern Sierra valleys. The Tahoe Basin dropped down along faults on either side, while the Sierra block rose on the west and the Carson Range on the east. The bottom of the lake is lower than the Carson Valley floor, which is another down-dropped block. About 2 million years ago lava from Mount Pluto dammed the basin on the north, creating the lake. Although glaciers scoured the canyons and deposited moraines on the Sierra slope west of Tahoe, the resulting landforms are difficult to recognize because they are densely forested. But to the south, between Conway and Sherwin summits, these same landforms—U-shaped canyons and lateral moraines—are easily recognized because few trees obscure their shapes.

MOUNT ROSE HWY, RTES 431 AND 28   About 11 miles south of central Reno, Hwy 431 leads west. The large buildings south of the road are the geothermal plant producing power from Steamboat Springs. Among the many hot springs that rise along the Sierran Frontal Fault Zone, Steamboat and Casa Diablo near Mammoth Lakes have been harnessed to produce electricity. About 7 miles from Hwy 395 is Galena Creek Park. Riparian woodland, trails, one-mile nature trail, warblers in summer.

As Hwy 431 climbs the Carson Range, in typical Montane Forest succession, the first trees are Jeffrey Pine; then, higher, White Fir, Lodgepole Pine, Red Fir; and close to the summit, Mountain Hemlock and Western White and Whitebark pines of the Subalpine Forest. Near the summit is one of the first high-altitude weather observatories in the United States, built in 1905. Both the flower-filled Tahoe Meadows and a high-altitude sagebrush slope lie just over the summit. Beyond the summit 4.5 miles, where there is a fine view of the Tahoe Basin, three rock types are clearly evident on the skyline: dark metamorphics far to the south, pale granite directly opposite, and a mass of black volcanic rock covering all the ridges to the north (the hill behind you is also volcanic, andesite). The tops of Mount Tallac and Jacks Peak are old metamorphic rock. Drive south on Hwy 28 and along the east shore, where Sugar and Ponderosa pine predominate. Less than 1 mile north of the Hwys 50/28 junction, you will find the Spooner Lake entrance to Lake Tahoe Nevada State Park. The park consists of more than 10,000 acres of forest, lake, and meadow with many species of mountain animals and wildflowers. Canada Geese nest at Spooner Lake, which is a sag pond, formed by faulting. Miles of trails; short nature trail.

To make a loop trip back to Reno, take Hwy 50 east over Spooner Summit or Hwy 207 over Daggett Pass and Kingsbury Grade. At Spooner Summit a

Southern end of Lake Tahoe. Pyramid Peak and Mount Tallac left of center on Sierra Crest skyline. (Photo by Jim Hildinger)

short walk along the Tahoe Rim Trail will acquaint you with Sagebrush Scrub and Montane Chaparral. East of the summit, the south-facing slope traversed by Hwy 50 has a large stand of Montane Chaparral with scattered Jeffrey Pine.

HWY 89, WEST SIDE OF LAKE TAHOE   West side trees are taller and denser than those on the east side and include more species, such as Red Fir and Incense Cedar, in addition to the Sugar and Ponderosa pines that blanket the east shore of the lake.

PAGE MEADOWS   Spring wildflowers. From Hwy 89, 2 miles south of Tahoe City, turn west on Pineland Dr., then right on Rd. 15N60 or 16N48.

EAGLE ROCK   About 4 miles south of Tahoe City, Eagle Rock, like Cave Rock on the east shore, is the eroded neck of an old volcano.

SUGAR PINE POINT STATE PARK   Just north of Meeks Bay. Nature Center.

EMERALD BAY AND D. L. BLISS STATE PARKS   Bordering Lake Tahoe and linked by trails. Waterfowl, raptors, songbirds, squirrels, and chipmunks are common. Good examples of jointed granite. The half-mile Balancing Rock self-guided nature trail emphasizes the connections among all things—rock, lichen, soil, bacteria, shrub, insect, bird, dead tree, living tree. The Eagle Falls trail follows Montane Riparian Woodland. Glaciers shaped Emerald Bay and deposited moraines on both sides of the bay. Although its glacial polish has eroded away, the granite is rounded and smooth. Entrances to both parks on Hwy 89, between Meeks Bay and Camp Richardson. Displays at Inspiration

Emerald Bay. (Photo by Jim Hildinger)

Point Vista explain much about the lake and its origin. Fannette Island is closed to people for several months in spring, during birds' nesting season.

LAKE TAHOE VISITOR CENTER, USFS   Maps, books, and information on roads; hiking trails and bike paths; excellent exhibits about local mountain weather and climate. Boardwalks and bridges enable you to meander along Taylor Creek's riparian woodland and through a wet, lush meadow and marsh, bordered by pines and aspen. Abundant birds of many species, small mammals, and deer. Self-guided trails have outstanding interpretive displays, including a stream profile chamber emphasizing the life cycle and needs of fish. From South Lake Tahoe, take Hwy 89 north 3 miles.

MEYERS SOUTH TO HOPE VALLEY AND CARSON PASS, HWYS 89/88   Montane Forest grading into Subalpine Forest. Trees range from squatty Sierra Juniper to Mountain Hemlock on the north slopes, and from Ponderosa Pine at lower elevations to Jeffrey Pine at higher. From Luther Pass the road drops down to the wonderful Mountain Meadows of Hope Valley, headwaters of the Carson River's West Fork. Abundant moisture nourishes lavish summer wildflowers and expansive groves of aspen. At the junction, go west on Hwy 88. More huge meadows bordered by Quaking Aspen and Subalpine Forest with a sagebrush understory. Granite at Carson Pass; most of the granite in this region is covered by reddish, dark brown, or blackish layered volcanics. Nearing the crest, east of Red Lake, Red Lake Rd. leads to quiet small meadows and streams.

CARSON PASS   From the pass an easy trail leads to Frog Lake, in typical Subalpine Forest. Many species of montane and subalpine wildflowers.

HOPE VALLEY TO GROVER HOT SPRINGS STATE PARK   From Hope Valley (see entry for Meyers, Hope Valley, and Carson Pass, above), continue on Hwy 89 as it follows the west fork down-canyon, then turns south to Markleeville. Jeffrey Pine forest with sagebrush understory. Grover Hot Springs State Park has two swimming pools. Springs drain into a large wet meadow, surrounded by Montane Forest with Montane Chaparral on the southeast-facing slope; self-guided trail. Ground squirrels, gophers, mice, and shrews have dug many tunnels in the soft soil. Their predators, hawks and Coyotes, often hunt the meadow. Hot Springs Creek has cut a channel through the terminal moraine at the east edge of the meadow. Sagebrush, bitterbrush, cottonwoods, Incense Cedar, willows, Jeffrey Pine. Many species of birds. From Markleeville, drive west on Hot Springs Rd. 4 miles.

MONITOR PASS   As Hwy 89 leads from Markleeville through a pinyon forest, you may suddenly realize that the land looks drier, the vegetation more sparse, the trees smaller, the colors gray-green instead of deep green. Monitor is only a few miles east of Carson Pass, but what a difference: The wide, open slopes of Sagebrush Scrub on Monitor Pass signal that you are no longer in the Sierra Nevada but rather in the Great Basin Desert biogeographic region, where sagebrush extends far to the east. Miles and miles of it at lower elevations and occasionally, as here, at higher elevations, where a surprising number of flowers bloom among the shrubs from mid-June to August. Sagebrush, which can tolerate a wide variety of harsh conditions, also forms an understory in Pinyon-Juniper Woodland. Near the pass, soaring raptors are common, and huge slopes of aspen gleam golden in fall. From the Sierra Crest it is a several-hour drive west to the Central Valley, but from Monitor Pass east, it takes only 10 minutes to drop 3000 feet to Antelope Valley.

NORTHERN SIERRA

North of Lake Tahoe, the Sierra's crest and passes dwindle to lower elevations, allowing moisture-bearing winds to come over the lower crest and shower more water on the east slope. Yuba (6701 feet) and Fredonyer (5748 feet) passes, for example, don't seem like passes at all because they are not much lower than the crest itself. Near the crest, east and west slope forests differ little from each other.

HALLELUJAH JUNCTION   Junction of Hwys 395/70, 23 miles north of Reno. Sagebrush, Antelope Bitterbrush, Utah Juniper. Many insect species in summer. In the vicinity, you are likely to see Mule Deer in spring and fall, Pronghorn during most of the year. The Pronghorn forage on some nonnative plants, such as Crested Wheatgrass and Sweetclover, which were planted to stabilize the soil and to provide forage after the area burned in 1973. This is Mule Deer winter range. Three underpasses, built to avoid accidents between deer and cars, funnel migrating deer between their winter range to the east and summer range to the west. Auto pull-offs are located at each of the underpasses, between 3.5 and 5.5 miles south of the junction.

DONNER PASS HIGHWAY   Old US 40. West of Reno on I-80, take the Soda Springs exit east to Donner Pass Hwy, which rejoins I-80 east of Donner Lake. Donner Memorial State Park, at the east end of Donner Lake, has a nature trail in Montane Forest; good selection of books in the visitor center.

DONNER SUMMIT WALK   Just before Donner Spitze Hütte, turn right onto the paved road and park. About 25 yards from the main road, follow the rocky dirt road between two iron posts for a short way. In midsummer low clumps of wildflowers enliven the gray granite. Trees range from a wind-battered old Sierra Juniper, lodgepole, and willows to graceful Mountain Hemlock on the north-facing slope above. Back on the highway, as you descend you will see solitary Jeffrey Pine growing out of cracks in the granite.

BOREAL RIDGE SKISPORT MUSEUM   Features exhibits on snow and on the history of winter sports. At Donner Summit about 12 miles west of Truckee, on the south side of I-80.

DONNER CAMP PICNIC AREA   A large Montane Meadow bordered by Sagebrush Scrub and Montane Forest with a few monster Jeffrey and Lodgepole pine. Many birds and wildflowers; the purple-blue Camas Lily, which blooms in late spring and early summer, is especially notable. From I-80, drive north on Hwy 89 about 2 miles.

SIERRA VALLEY   A huge marsh and meadow area, alive with birds and wildflowers, that drains north into the Feather River. Several plant communities occur within short distances. A typical Sierra Montane Forest borders the valley's western edge, while sagebrush and juniper of the High Desert border its eastern edge. Desert Marshland, Alkali Meadows, and wet Mon-

Donner Lake. (Photo © Tom Lippert)

tane Meadows adjoin each other, with Sagebrush Scrub not far away. In early June, look for a blue expanse of Camas Lilies mixed with gleaming yellow buttercups. Dyson Lane, crossing the marsh, offers rich opportunities for birding: ducks, White-faced Ibis, Prairie Falcons, American Bitterns, Black Terns, Sandhill Cranes, and Soras, to name only a few. During winter, Sierra Valley provides resting and feeding habitat for many species of waterfowl, other water birds, Bald and Golden eagles, and several species of hawks. Deer also winter here. Follow Hwy 89 north to Sierraville, at the southern end of Sierra Valley, where the highway turns west. About 4 miles west of Sierraville, turn north on A-23 for about 10 miles, then east on Dyson Lane to Hwy 49, north on 49 to Hwy 70, and east to Hwy 395 at Hallelujah Junction. (Tip: Find the most detailed map of this area that you can, since roads may not have signs.)

YUBA PASS   At 6701 feet, on Hwy 49. In this northern portion of the Sierra, there is little difference between the higher forests of the east and west slopes. The forest east of Yuba Pass resembles a west slope forest in many ways—the trees are tall and close together, and species such as Sugar Pine (which do not occur on the east slope of the southern Sierra) spill over from the west slope.

BIZZ JOHNSON RAIL TRAIL   In Susanville, a trail for hiking, riding, and bicycling. The trail traverses the Susan River Canyon, following the old Fernley and Lassen Branch Line of the Southern Pacific Railroad for 25 miles from Susanville to Mason Station. From the Sagebrush Scrub around Susanville, the trail leads to a Montane Forest of pines and firs. In the Montane Riparian Woodland bordering the river, you will find the flora and fauna common to this habitat. Birds include kingfishers, quail, orioles, warblers, and hummingbirds; from the trail you can look into the *tops* of trees. Susanville trailheads lie at the south end of Lassen St. and at the end of Miller Rd., below a small cliff of volcanic rocks.

HONEY LAKE WILDLIFE AREA   A portion of the former large, shallow natural Honey Lake. Although much water has been diverted, this lake still has enough water to provide habitat for shorebirds, waterfowl, and raptors. It is an important waterfowl wintering area. In fall, up to 25,000 migrating waterfowl rest here; in spring, up to 50,000. Newly hatched birds April to May, Bald Eagles in winter, resident raptors. Nesting birds include Canada Geese, Sandhill Cranes, American Avocets, Black-necked Stilts, and Long-billed Curlews. Tundra Swans and White-faced Ibis are among the 200 bird species known at Honey Lake. Montane Forest with Montane Chaparral on the southeast-facing slope. To reach the Dakin Unit from Susanville, take Hwy 395 to Standish, turn south on Lambert Lane, east on Mapes, and south on Dakin. Hunting allowed in season.

EAGLE LAKE   A 13-mile-long natural lake, a remnant of Pleistocene Lake Lahontan. Famed for its Ospreys, breeding colonies of Western and Eared grebes, and the Eagle Lake Trout. Although Eagle Lake lies a few miles beyond Hwy 36, the northern boundary of our area, we include it because of its unusual features. Osprey nesting sites are protected from human disturbance from May 15 to September 15; however, during that time you can see these raptors flying and hunting. If the lake remains ice-free, the Ospreys will stay all winter; if the lake freezes, they winter at Honey Lake. Bald Eagles, White Pelicans, cormorants, and gulls in early summer. Five native fishes besides the Eagle Lake Trout can tolerate these alkaline waters: Lahontan Creek and Lahontan Lake tui chubs, Tahoe Sucker, Lahontan Redside, and Lahontan Speckled Dace. Rainfall varies from 30 inches at the southern end of the basin to 12 inches at the northern end. It is no surprise, then, that Montane Forest dominates the south shore, sagebrush and juniper trees the north shore.

Take Hwy 36 west from Susanville, turn north on A-1, Eagle Lake Rd., which traverses large expanses of Montane Chaparral.

BEYOND THE MAIN ROADS

Our aim here is to introduce you to the main roads and the lay of the land. After that, the fun really begins, and it's all up to you. Miles of backroads lead through the sagebrush up into the canyons, where waterfalls spill out of lakes and snowfields feed cascading streams. Dozens of trails beckon you to explore the forests and high ridges, to discover your own special places that are unmarked and unmapped—places where the meadows are lush or the views immense. Really getting to know this dramatic land at the edge of the Great Basin might just take a good part of your lifetime.

RECOMMENDED READING

Trail guides and natural history books are flooding off the presses at such a rate that any recommended list will soon be out of date. (We provide a list anyway.) You can learn about new books by sending for catalogs from publishers that specialize in outdoor and nature books, such as the following:

Artemisia Press, PO Box 8058, Mammoth Lakes, CA 93546

Mono Lake Committee, PO Box 29, Lee Vining, CA 93541

Sierra Club Store, 730 Polk St., San Francisco, CA 94109

University of California Press (California Natural History Guides), 2120 Berkeley Way, Berkeley, CA 94720

University of Nevada Press (Fleischmann Series in Great Basin Natural History), MS 166, University of Nevada, Reno, NV 89557-0076

Wilderness Press, 2440 Bancroft Way, Berkeley, CA 94704

At the end of each chapter, we recommend books for further reading. We also recommend the following general books.

Clark, Jeanne L. 1992. *California Wildlife Viewing Guide.* Helena, Montana: Falcon Press.

———. 1993. *Nevada Wildlife Viewing Guide.* Helena, Montana: Falcon Press.

Farquhar, Francis P. 1965. *History of the Sierra Nevada.* Berkeley: University of California Press.

Gaines, David. *Mono Lake Guidebook.* Rev. ed. 1989 by Lauren Davis. Lee Vining, Calif.: Kutsavi Books.

Grayson, Donald K. 1993. *The Desert's Past: A Natural History of the Great Basin.* Washington, D.C.: Smithsonian Institution Press.

Hall, Clarence A., Jr., ed. 1991. *Natural History of the White-Inyo Range, Eastern California*. Berkeley: University of California Press.

Houghton, Samuel G. 1994. *A Trace of Desert Waters*. Reno: University of Nevada Press.

Irwin, Sue. 1991. *California's Eastern Sierra*. Los Olivos, Calif.: Cachuma Press.

Mono Basin Ecosystem Study Committee. 1987. *The Mono Basin Ecosystem: Effects of Changing Lake Level*. Washington, D.C.: National Academy Press.

Schoenherr, Allan A. 1993. *A Natural History of California*. Berkeley: University of California Press.

Smith, Genny, ed. 1993. *Mammoth Lakes Sierra*. 6th ed. Mammoth Lakes, Calif.: Genny Smith Books.

Smith, Genny, and Jeff Putman, eds. 1995. *Deepest Valley: Guide to Owens Valley*. 2d ed. Mammoth Lakes, Calif.: Genny Smith Books.

Status of the Sierra Nevada: Sierra Nevada Ecosystem Project, Final Report to Congress. 1996. Wildland Resources Center Report 36. Davis: UC Davis, Centers for Water and Wildland Resources.

Storer, Tracy I., and R. L. Usinger. 1963. *Sierra Nevada Natural History*. Berkeley: University of California Press.

Trimble, Stephen. 1993. *The Sagebrush Ocean: A Natural History of the Great Basin*. Reno: University of Nevada Press.

Whitney, Stephen. 1979. *A Sierra Club Naturalist's Guide to the Sierra Nevada*. San Francisco: Sierra Club.

FOR CHILDREN

Eyewitness Books. New York: Alfred A. Knopf. Titles include *Bird, Butterfly and Moth, Eagles and Birds of Prey, Insect, Mammal, Rocks and Minerals, Tree, Volcano and Earthquake, Weather*. Close-up photos and lively captions emphasize important concepts.

CHAPTER THREE

GEOLOGIC STORY

MARY HILL

No mountain range in the Americas has had a greater impact on human history than the Sierra Nevada. The reasons for this impact are almost wholly geologic: It was the discovery of gold in the Sierra foothills that altered the course of the region's history. If that momentous discovery, and the gold rush that followed it, had never occurred, the West would have been settled much more slowly. The settlers would have been farmers and ranchers seeking a better life for their families, and not wild young men bent on finding their fortunes.

Stretching along more than half the length of California, and rising more than 14,000 feet (4200 m) at its highest point, the Sierra is a barrier to storm-laden Pacific winds, forcing them to drop much of their moisture on the western slope of the great range and leaving the eastern side in a "rain shadow." Consequently, eastern California and the land beyond is desert or near-desert. Had the Sierra Nevada not been pushed so high by mighty earth forces, today's Eastern Sierra might look quite different, greener and wetter.

The gold rush sent California's name around the world. It linked the U.S. states and territories together quickly, as thousands of Argonauts packed their belongings in covered wagons and headed for the gold fields, braving hostile Indian tribes, waterless deserts, and their own ignorance about the western lands. Some came by sea; they were joined in the Sierra not only by the covered wagon parties but also by gold seekers arriving by other routes from faraway places. This first worldwide gold rush showed clearly how events in one place could reverberate throughout the world.

When the forty-niners arrived, they found that the easy pickings in the gold fields (if such back-breaking work can be called easy) had been scooped up by those who had come before—the miners of 1848. Already by 1850, many placer mines were petering out, and miners began to press for help from the new science of geology. Where, they asked, could they find more rich gold deposits?

A rough geologic map published in the 1850s served to outline Sierra geology but did not point the exact way to rich new lodes—in fact, it was meant as a warning to "get-rich-quick-ers." So with high scientific ideals and the votes of men who no doubt thought they were voting for signposts to more gold, the California legislature in 1860 established the Geological Survey of California, whose mission was to furnish maps and diagrams of the state's geology, together with a "scientific description of the rocks, fossils, soils, and minerals, and of its botanical and zoological productions." It was to be the very model of a modern scientific survey. J. D. Whitney, a geologist renowned for his 1854 volume on the mineral resources of the United States and his report on the copper deposits of Michigan, was appointed State Geologist.

Not until 1863 did the Survey get around to working in the Sierra Nevada, having surveyed the state as a whole first. It was too little and too late. Although the work done by the Survey was monumental by any standard, the legislature, seeing no new gold deposits forthcoming from its work, provided fewer and fewer funds. Whitney's academic approach, his lack of understanding of politics, and his abrasive personality—combined with the governor's and legislature's misunderstanding of the role of science in the search for gold—doomed the Geological Survey. Its major work on geology was published in 1865; less than a decade later, in 1874, the Survey was disbanded before its other reports were published. Whitney secured the funds to publish the rest himself.

Although there was to be no state survey with the word *geology* in it for more than eighty years, the State Mining Bureau (later the Division of Mines), set up by the legislature in 1880, did a certain amount of geological work. Until the 1960s, however, its mandate was the practical charting of gold and mineral resources. In spite of the state's lack of interest in its own geology, there has been worldwide scientific interest in the geology of the Sierra. In the century and a half since the gold rush, countless geologists have worked on various parts of its geology, publishing thousands of books and scientific papers. This may seem like a huge amount of research on one mountain range. And so it is—but remember that geology is studied by piecing together bits of a wide variety of information. Gaining insight into the Sierra is rather like putting together a jigsaw puzzle, each part making the whole a little clearer—all the while trying to fit the picture of the Sierra Nevada into the larger picture of the Earth.

The Sierra Nevada is an immense mountain range, 400 miles (640 km) long and 50–80 miles (64–130 km) wide. It stretches from Tehachapi Pass northward nearly to Mount Lassen, an active volcano, where younger lava flows near Lake Almanor cover the bedrock granite and define its northern boundary. "Una gran sierra nevada," wrote Pedro Font in his journal as he looked east from a hill near San Francisco Bay in 1776. A "great snowy range" indeed. The Sierra rises gradually from near sea level, in California's Great Central Valley, to a lofty snow-covered crest that drops abruptly to the arid valleys to the east—Owens Valley, Mono Basin, Carson Valley, Truckee Meadows. Near Mount Whitney the topographic relief measures close to 2 miles (3.2 km) in a horizontal distance of only 6 miles (9.7 km).

The highest peaks rise in the southern part of the range in the Whitney region. Here, between Cirque Peak and Kearsarge Pass, the Sierra Nevada culminates not in one spirelike peak but in a massive granite wall more than 12,000 feet (3660 m) high and 21 miles (34 km) long. The Whitney region includes six peaks and several "needles" over 14,000 feet (4270 m), plus others well over 13,000 feet (3965 m). Mount Whitney itself, at 14,494 feet (4418 m), towers more than 10,000 feet (3048 m) above Owens Valley. But far more impressive than the number of very high peaks is the massive nature of the Sierra. For most of its length, as the pioneers and explorers learned to their dismay, not a single low pass exists between Donner Summit and Walker Pass. Most of the passes are higher than many mountain peaks: Tioga Pass at 9945 feet (3034 m), Sonora at 9624 feet (2936 m), Ebbetts at 8730 feet (2664 m), and Carson Pass at 8573 feet (2616 m). For 170 miles (240 km), from Gomez Meadow north of Olancha Peak to Ebbetts Pass, the entire crest stands well above 9000 feet (2750 m) without a single low gap. And for almost 90 miles (160 km), from Trail Pass to Duck Pass, the crest stands well over 11,000 feet (3350 m). Elevation gradually decreases to the north. North of Reno the Sierra Crest drops to altitudes between 7000 and 8000 feet (2100 and 2400 m), and its passes are considerably lower: Beckwourth Pass at 5212 feet (1593 m) and Fredonyer Pass at 5748 feet (1755 m).

A QUICK VIEW OF THE SIERRAN PAST

The story of the Sierra Nevada is about 570 million years long—about 12 percent of the age of the Earth (about 4.6 billion years). As shown in figure 3.1, the story is quite complex. The first 4 billion years of Earth's history are missing in the record of Sierran rocks. The oldest pages we can decipher are imprinted on rocks that were laid down about 570 million years ago. Earlier pages in the related White-Inyo Mountains tell a story that probably included the Sierra, but in the Sierra Nevada itself, these earlier pages are missing. Later pages, the story that younger rocks reveal, are clearer and easier to read, as one might expect.

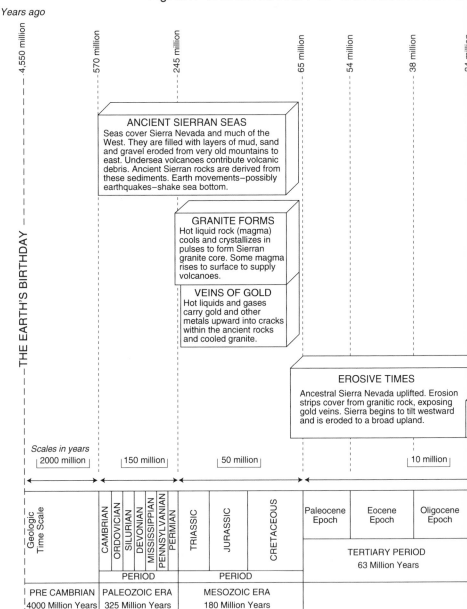

Fig. 3.1. THE LONG PAST OF THE SIERRA NEVADA

Years ago

4,550 million — THE EARTH'S BIRTHDAY

570 million | 245 million | 65 million | 54 million | 38 million

ANCIENT SIERRAN SEAS
Seas cover Sierra Nevada and much of the West. They are filled with layers of mud, sand and gravel eroded from very old mountains to east. Undersea volcanoes contribute volcanic debris. Ancient Sierran rocks are derived from these sediments. Earth movements—possibly earthquakes—shake sea bottom.

GRANITE FORMS
Hot liquid rock (magma) cools and crystallizes in pulses to form Sierran granite core. Some magma rises to surface to supply volcanoes.

VEINS OF GOLD
Hot liquids and gases carry gold and other metals upward into cracks within the ancient rocks and cooled granite.

EROSIVE TIMES
Ancestral Sierra Nevada uplifted. Erosion strips cover from granitic rock, exposing gold veins. Sierra begins to tilt westward and is eroded to a broad upland.

Scales in years

| 2000 million | | 150 million | | 50 million | | 10 million |

Geologic Time Scale	CAMBRIAN	ORDOVICIAN	SILURIAN	DEVONIAN	MISSISSIPPIAN	PENNSYLVANIAN	PERMIAN	TRIASSIC	JURASSIC	CRETACEOUS	Paleocene Epoch	Eocene Epoch	Oligocene Epoch

PERIOD — PERIOD

TERTIARY PERIOD
63 Million Years

PRE CAMBRIAN	PALEOZOIC ERA	MESOZOIC ERA
4000 Million Years	325 Million Years	180 Million Years

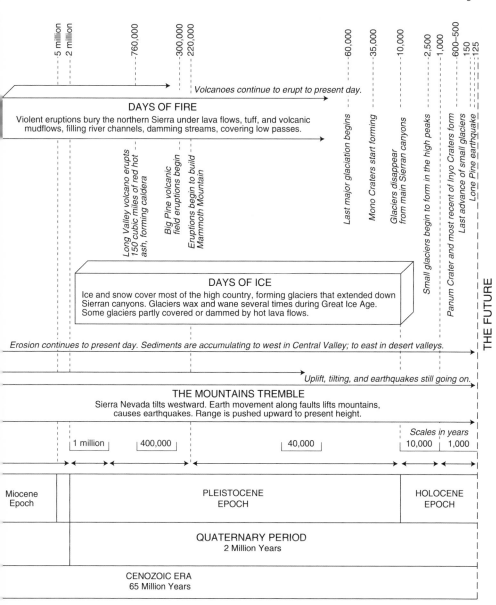

Drawn by Adrienne E. Morgan. Mary Hill 1997.

The rocks tell us that during much of the first 400 million years, shallow seas covered most of the area we know as the West. Quantities of sand, lime, and mud poured into these seas, just as rivers today pour sand and mud into the Pacific Ocean. These layers were consolidated into sandstone, limestone, and shale. Eventually the land rose above the sea for a time (or sea level dropped), the layers were tilted, the land sank again, and more sediments accumulated. This happened more than once. Widespread volcanic eruptions added layer after layer of fragmental volcanic material upon the older layers. Finally, molten granitic rock pushed upward in the core of the range, bringing with it the gold veins that so changed human history.

As the mountains stood exposed to the ravages of wind and weather, gradually the ancient rocks laid down in the seas so long ago (now *metamorphosed,* changed by the heat and pressure of mountain building) were worn off the mountain tops, exposing the granite core. Only remnants of the ancient rocks remain today—patches on the crests of the mountains, bands in the foothills. Some of the gold was worn from the mountains, too, and carried down the rivers of yesterday.

Most of the framework of the Sierra Nevada was now complete, although it did not greatly resemble the Sierra of today. Its crest was around 3000 feet (1000 m). On the west, a shallow sea edged with lagoons lapped at its foothills. The climate of the foothills was subtropical; deciduous forests mantled the mountain slopes.

About 20 million years ago, following this period of intense erosion, another volcanic episode began, covering much of the land with lava flows and volcanic debris. Volcanoes erupted along the Sierra Crest north of Yosemite, blowing volcanic ash far and wide. These ashy explosions were followed a few million years later by volcanic mud flows and, still later, by more red-hot ash and other volcanic eruptions. As the eruptions increased in frequency, the mountains began to rise along faults, eventually reaching heights far greater than before and forming the range we know as the Sierra Nevada. But we might not have recognized it, had we been there, for today's mountains have been sculpted by the ice sheets that began forming near their crests about 2 million years ago and continued to reshape the peaks until about 15,000–20,000 years ago.

Neither the latest volcanic episode nor the most recent spell of faulting and earthquakes has ended. Volcanoes have erupted within the last 600 years—a few seconds of time, geologically speaking—and the Sierra is still rising.

Just when people first inhabited the Sierra Nevada is not precisely known, though it was certainly long before the Argonauts of 1849. Remnants of ancient settlements are scattered up and down the foothills. Certainly people were here 5000 years ago, as a skull of that age was found in the gold country on the western slope of the Sierra. In the long span of Earth's story, human beings arrived only yesterday; their tale is the last sentence of the most recent page of Earth history.

HOW IT ALL HAPPENED

It is impossible to explain how the Sierra Nevada came to be, as we understand it now, without referring to *plate tectonics,* a unifying principle that helps explain the geologic history of our world. As more and more evidence is discovered, the principle becomes clearer and its details refined. Only since the late twentieth century have earth scientists had an overall idea of how the Earth is constructed and how it came to be as it is. They had been piling up facts about Earth for some three centuries, gathering knowledge of what had happened, but without much idea as to why or how. Then in the 1960s some imaginative geologists, who were willing to speculate on the forces propelling Earth's geologic story, reviewed new studies of the sea floor and arrived at a process they called plate tectonics (*tectonics* is from the Greek word for "builder"). This was a revival of an earlier theory of "continental drift" put forth early in the twentieth century but dismissed by many for lack of evidence.

Allowing that perhaps the continents have indeed moved, plate tectonics envisions the Earth's crust as being made up of several large, rigid plates — some carrying both land and sea — and perhaps a dozen smaller plates. Data from earthquake epicenters, the positions of volcanoes, and other phenomena outline the locations of these plates. Each of the plates is slipping around on Earth's *mantle,* its viscous inner layer. We can envision the Earth as a vast jigsaw puzzle, with the pieces in constant slow motion. Because the plates are rigid, as they move — sliding past or colliding with one another — they damage and change each other, principally along their borders.

PLATES THAT MOVE BY SLIDING   When plates slide past one another, as they do along a fault zone such as the San Andreas, small earthquakes are common, with now and then a large one, all of them arising from centers not far down in the Earth's crust. Mountains may rise along the zone of slippage.

PLATES THAT MOVE AWAY FROM ONE ANOTHER   If the plates are moving away from one another, hot, molten, iron-rich rock wells up from the inner Earth to fill the space between. (When cool, this becomes the dark, fine-grained volcanic rock called *basalt.*) This is the situation in the Atlantic Ocean, whose sea floor is "spreading" as molten material wells up along the plate margins in the center of the ocean.

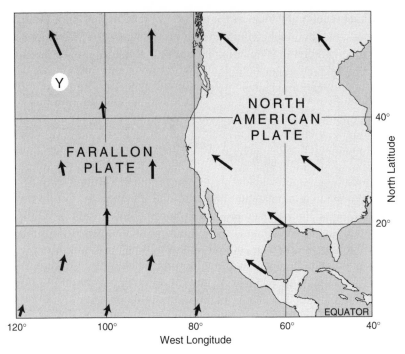

Fig. 3.2.   Relative motions of the Farallon and North American plates about 100 million years ago. The Farallon plate slid under the North American plate for many millions of years (see also fig. 3.4). Arrows show directions of movement; length of arrows shows amount of movement during 10 million years. Yellowstone (Y), shown here west of the continent, is a hot spot that has apparently remained in a constant position while the plates carrying the continents moved over it. Note that 100 million years ago the North American continent was 40° east and somewhat south of its position today. Of course, at that time the continent's coastline was not shaped as it is now; we show it this way only for comparison. (Based on Engebretson, Cox, and Gordon [1985], Special Paper 206, Geological Society of America; from Paul Bateman in *Deepest Valley* [1995]; drawing by Annie Kook)

PLATES THAT CRASH INTO ONE ANOTHER   If the plates are colliding, one plate may slide under the other (fig. 3.2). In such cases, where cold rocks that may have been on land or under the sea for millions of years are *subducted*— pushed and pulled down under the plate opposite, into the deeper, hotter portions of the Earth—they usually melt. This subduction zone becomes, essentially, a gigantic recycling center. Such a recycling process produced the granite of the Sierra Nevada.

ANCIENT SIERRAN SEAS

During the first 300 million years of Sierran history, the area that today we call the Sierra Nevada was at the edge of the North American continent. The

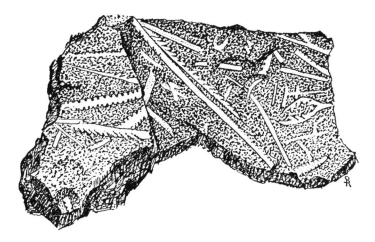

Fig. 3.3.   Fossil graptolites from Mount Morrison (actual size). Graptolites formed branching colonies. Large populations of them floated in the sea about 500 million years ago, and they are common fossils in Ordovician and Devonian rocks. Formerly, scientists believed that graptolites had been extinct for 370 million years, but in 1989 marine biologists found living graptolites in the South Pacific. (Drawing by Susan A. Rinehart in *Mammoth Lakes Sierra* [1993])

continent and the adjacent sea were traveling together on a single plate, the North American plate. The process was not entirely uneventful, since the rocks along the plate margin were deformed. It is difficult to tell exactly what happened, because later events were so vigorous they erased or overprinted much of the evidence of those ancient times. But we do know a little about the ancient seas.

The old Paleozoic rocks, those formed more than 200 million years ago, tell the story of sediments laid down in ancient seas: mud, sand, and gravel eroded from older mountains to the east. Over the years they were changed to sandstone, limestone, and conglomerate, and then metamorphosed by heat and pressure into rocks such as quartzite (from sandstone) and marble (from limestone). The oldest of these Paleozoic rocks yet found in the Sierra are near Big Pine in Owens Valley, where fossil-bearing rocks are 570 million years old. Fifty miles (97 km) north, the striking panorama that stretches across the mountains behind Convict Lake includes rocks nearly 500 million years old. Among the spectacular layers of vari-colored rock are some that contain the remains of *graptolites,* small organisms that floated freely in the seas of half a billion years ago (fig. 3.3).

SUBDUCTION AND VOLCANOES

Around 210 million years ago, at the end of the Paleozoic Era, something new was happening. The motion of the North American plate changed, so that the plate encountered another, cooler plate—the Farallon, containing more oceanic crust and less continental crust. The Farallon plate was crashing into and under the North American plate. Mountains formed along the plate margin, and a chain of volcanoes developed atop the overriding plate. A great thickness of the volcanic rocks of this age lies in the high parts of the wildly beautiful Ritter Range, west of Mammoth Lakes. If you hike from Agnew Meadows west toward Shadow and Thousand Island lakes, you will pass through a series of volcanic and sedimentary rocks laid down in the sea as much as 100 million years ago. Once they formed a thick blanket of volcanic debris, which may represent the collapse of an ancient volcanic caldera near Jackass Lakes. Now metamorphosed and eroded, they exhibit bands of sparkling minerals.

Volcanoes within the ocean basins, such as the Hawaiian volcanoes, have as their sole source molten rock (*magma*) from Earth's mantle. These volcanoes produce dark, iron-rich basaltic lavas. Most volcanoes on the continent are caused by the subduction of one plate beneath another. Such volcanoes produce lighter-colored lava, because they have as their source not only the magma from Earth's deeper reaches, but also new magma formed from the partial melting of the continent's rocks. The continental rocks, being largely sedimentary, give a different chemistry to the fluid magma. Where it erupts onto the surface, this mixed magma cools as lighter-colored, silica-rich, rhyolite lava. Where it remains below the surface, it slowly cools and solidifies as granite. If the heat, pressure, and chemically active fluids generated by subduction do not completely melt the old rocks adjacent to the granite, they often change and recrystallize them into metamorphic rocks. During this time of subduction, the Sierra Nevada may have resembled today's Cascade Range that extends from northern California to British Columbia. A still active subduction zone underlies this range, which is capped by active volcanoes.

The two stacked plates, the North American over the Farallon, thickened the Earth's crust, and the deep root of granite magma caused the mountains to rise. As they rose they were also being eroded. The old rocks of the North American plate, atop the granite core, were rapidly worn away, and eventually the cooled granite was exposed. The resulting sedimentary debris was carried downstream, westward into the ocean, and later became part of the California Coast Ranges.

THE GRANITE HEART

Almost everyone who walks the Sierra's deep canyons or high country is overwhelmed by the expanse and majesty of the granite. A vast body of granite underlies the Sierra Nevada. Only a small amount of this granite is exposed at the surface; it extends downward an unknown distance. Granite bodies form when molten rock pushes upward from great depth, melting and assimilating old rocks and shouldering others aside, but never reaches the surface. The bodies eventually cool and crystallize. During late stages of cooling, fluids from the magma, carrying silica and other substances such as gold and silver, migrate along fractures and cracks into the older rocks and are deposited as quartz veins. The fabled gold deposits of the Mother Lode formed in this way.

Although it may seem that the Sierra is one gigantic mass of granite—a mass called the Sierra Nevada *batholith*—actually it is made up of hundreds of individual granite bodies called *plutons* (named for Pluto, Roman god of the underworld). More than 100 plutons lie within the Yosemite National Park area alone. They range in size from less than 1 sq mile (2.5 sq km) to more than 500 sq miles (1300 sq km). The plutons *intruded* (pushed their way upward and outward into older rocks) at different times.

Plutons may differ slightly from each other in mineral and chemical composition, but they all belong to the family of rocks called granite. In this chapter we use the term *granite* as embracing the entire granitic rock family. Family members include diorite, monzonite, granodiorite, and one member also named granite; any one of these rocks used in construction or as a memorial stone would be sold as "granite." All of them contain slightly different proportions of the minerals quartz and feldspar, with a sprinkling of the dark minerals biotite (a mica) and hornblende—giving the rocks a salt-and-pepper look. They range in color from very light, with many light-colored minerals (the "salt") and almost no dark minerals (the "pepper"), through gray to a very dark rock with almost no light-colored minerals. The grains in most granites are about the same size. An exception is the Cathedral Peak Granite of Tuolumne Meadows, which consists of blocky white feldspar crystals up to 1 inch (2.5 cm) across set in a matrix of much smaller grains. The various granites have counterparts, with the same mineral and chemical composition, among the rocks that erupt from volcanoes. But because the granites cooled very slowly, giving crystals much more time to grow, granites look quite different from their relatives—the fine-grained, sometimes glassy, volcanic rocks.

It took more than a hundred million years for the batholith to form. The oldest granites were emplaced about 210 million years ago; they form the Wheeler Crest escarpment northwest of Bishop and a span of the escarpment that extends south from Lee Vining Canyon to June Lake. The next oldest plutons, emplaced about 160 million years ago, occur here and there in the Eastern Sierra, the White-Inyo Range, and in the western foothills of the Sierra. The great majority of the plutons were emplaced much later, dur-

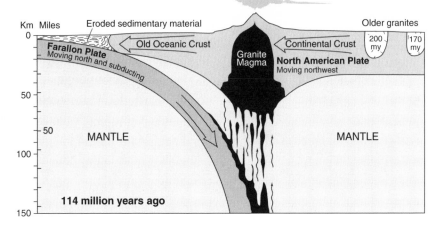

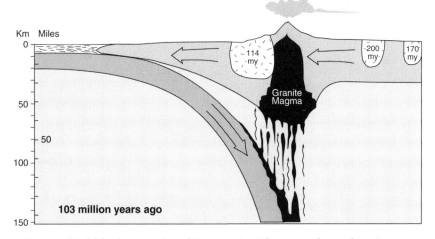

Fig. 3.4. Model for the generation of Sierra granites. The most widespread granites intruded between about 85 and 115 million years ago. Hot basaltic magma rose from the subduction zone to mix with and melt lower crustal materials. The most fluid portion continued upward into the upper crust, some of it breaking through the surface in volcanic eruptions, most of it cooling and solidifying underground as granite. (From Paul Bateman in *Deepest Valley* [1995]; drawing by Annie Kook)

ing a 40-million-year span about 120–80 million years ago; the most abundant intrusions occurred 115 million years ago, 102 million years ago, 98 million years ago, and 90 million years ago (fig. 3.4). Of these, the oldest are on the western side of the Sierra, while the plutons are progressively younger as one goes east. Comparison of the isotopic ages of plutons shows that even though the granites were emplaced episodically, the place where magma was

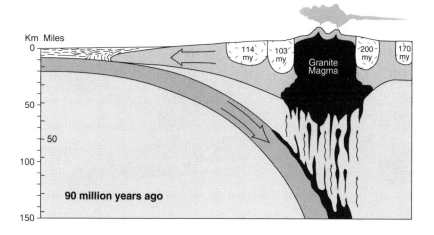

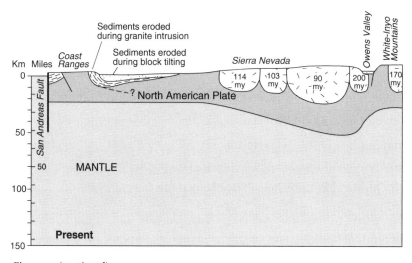

Fig. 3.4. (*continued*)

generated moved steadily eastward—or so it seems—across the Sierra at a
constant rate of 2.7 mm per year during those 40 million years. Possibly there
was a hot spot at that time (as there is now below Yellowstone National Park),
and the plate carrying what is now the Sierra Nevada rode westward over it.
This would explain why the older granites, which were emplaced first, are toward the west and those emplaced later are progressively farther east.

It is not possible to guess which plutons are the oldest merely by looking
at them. Geologist Paul Bateman, who has spent many years working in the
high country west of Bishop, suggests that in the clear light of early morning you try standing on the outskirts of town and look west. At that time of

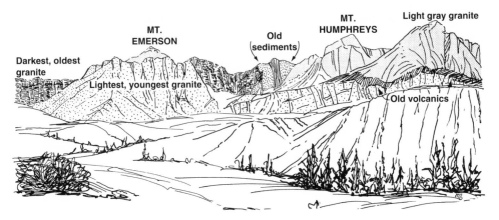

Fig. 3.5.   Bodies of Sierra granite west of Bishop. In the light of the morning sun, different plutons can be distinguished by their shades of gray. (From Paul Bateman in *Deepest Valley* [1995])

day, when shadows do not obscure the colors, you should be able to distinguish the light gray granite of Mount Tom, the still lighter, younger granite of Mount Emerson, and older, darker granite farther south (fig. 3.5).

JOINTING AND SHEETING

Jointing, a characteristic of granite, has had a great influence on Sierra scenery. *Joints* are more or less even, parallel planes along which rock cracks—vertically and horizontally. In the Sierra, joints run in two principal directions, which are nearly at right angles to each other. One set trends more or less northeast; the other, which is more closely spaced, trends northwest. A third, less conspicuous set of joints runs east-west. Aerial photographs show clearly the criss-cross pattern made by joints. Other things being equal, where rocks are fine-grained, joints are closer together; where rocks are coarse-grained, joints are farther apart. We know that joints are not the product of cooling, because they cut across granitic bodies of different ages. We also know that the joints developed before the period of extensive weathering that followed the emplacement of the granite and before the granite was exposed, because the joints themselves are commonly deeply weathered.

Joints have provided avenues down which streams run, etching the joints even more deeply. In the High Sierra, some master joints look like plowed furrows. Near the state border south of Lake Tahoe, along State Route 89 north of Markleeville, the mountains display ribs of granite, left after erosion along the adjacent joints wore down the intervening rock. Joints have helped create some of the most spectacular scenery of the Sierra, including its dramatic eastern front.

Although the Alabama Hills near Lone Pine have been called the oldest mountains in the United States, this is far from true. Their granite is the same age as that of the mountains towering over them. True, the Alabama Hills look old, but they achieved that look by being buried for millions of years. While they were underground, water penetrated along the numerous joints, cutting the granite into blocks and corroding the blocks along the joints by chemical action. During the wet and warmer times of the Great Ice Age, between ice advances, this type of chemical weathering was particularly effective. Then, when the glaciers melted and erosion stripped away soil and rock from the Sierra's eastern slope and washed them into Owens Valley, erosion also uncovered the Alabama Hills. The jointed blocks had by that time become rounded, like giant loaves of bread. Many Western films and commercials have been made in this setting; in fact, a scenic road in the Alabama Hills leads to "Movie Flat."

A special form of jointing, called *sheeting* or *sheet jointing,* is the cracking of rock along curved surfaces more or less parallel to the surface of the rock. Sheeting is a local phenomenon. It can be observed on a single rock outcrop or on a single mountain, but there is no regional sheeting pattern. Sheeting is probably the result of unloading—that is, relief from the pressure of the heavy glaciers as the ice melted.

WEARING DOWN THE MOUNTAINS

It took millions of years for the granite to cool. While it was cooling, wind and weather were eroding the older rocks that covered it—the ancient metamorphic rocks, tilted and twisted, and younger volcanic deposits probably similar to those in the Cascade Range today—and wearing down the mountains. Much of the cover was stripped away, its fragments carried downstream to the adjacent valleys. In the Central Valley these fragments formed deposits many thousands of feet thick. Only here and there do patches remain in the mountains. Streams also cut into the granite core, and gold-bearing veins gave up some of their gold, which was then washed into the streams that drained the mountains. By 50 million years ago, the range had been worn fairly low, to perhaps no more than 3000 feet (1000 m) in elevation. A quiet shallow sea edged the mountains on the west. The climate in the foothills at this time was subtropical, and frosts were a rarity.

VOLCANOES RISE

Once again plate movement changed. The neighboring oceanic plate, which was being subducted beneath the North American plate, was almost wholly consumed. A new neighbor, the Pacific plate, moving in an entirely different manner, took its place. The Pacific plate was sliding along faults (notably the

San Andreas) in a northwesterly direction. This change in motion altered the stresses on the Sierra Nevada. Instead of being compressed, the land was now able to expand, allowing the mountains to rise along faults. They rose slowly at first, but sped up, and they may still be speeding up. The present rate, about 1.5 inches (4 cm) per century, may seem small, but it is faster than erosion that is wearing down some of the mountains, so that locally there is a net increase in elevation.

The new plate motion allowed volcanoes to break through, especially along the mountain crest, spreading volcanic debris over the northern part of the range. Today thick beds of volcanic ash and mud flows lie on the western slope of the Sierra, remnants from eruptions that began 20 million years ago. Volcanic deposits probably also covered the eastern slope, but much has been eroded away, although San Joaquin Mountain has a remnant stack nearly 2000 feet (610 m) thick. Conspicuous deposits occur in the Carson Range, on the summit of Mount Rose, and at volcanic centers northeast and south of Lake Tahoe. The volcanoes that erupted these thick beds of volcanic debris have doused their fires, but Sierran volcanoes are far from dead, and the mountains themselves are still rising.

RECENT VOLCANOES

Although volcanoes north of the Sierra have erupted in the twentieth century—Mount Lassen and Mount St. Helens—no Sierran eruptions have been seen in historic times. Several volcanic centers are located in the Eastern Sierra, but these are by no means all of the recent volcanoes of the eastern slope. To the south are the Coso and Big Pine volcanic fields, and a line of eruptive centers extends intermittently northwestward past Reno. The Sierra Nevada ends where its rocks are covered by lavas of the Cascade Range, including still active Lassen Peak.

About 300,000 years ago, between Sawmill Creek and Big Pine, eruptions began from some of the forty vents that constitute the Big Pine volcanic field. The Fish Springs cinder cone, a tidy volcanic cone that has been cut by faulting, is about 314,000 years old. A large basalt flow at Sawmill Creek, 110,000 years old, looks as if it might have cooled only yesterday.

The region between Owens Valley and Bridgeport Valley contains the most conspicuous and photogenic of all the Sierra's volcanic features. The most thoroughly studied of these features are Long Valley Caldera and its neighbors near Mammoth Lakes, and the Mono-Inyo volcanic chain south of Mono Lake. The story of these later volcanic features begins with the eruption, 3.6–2.2 million years ago, of extensive dark-colored lava throughout much of the Long Valley area. Remnants of these lava beds include the Adobe Hills, north of Long Valley, and scattered flows and cinder cones in the High

Sierra. The eruptions that formed them preceded the mighty eruptions that created Long Valley Caldera. Other lavas, intermediate in color and chemical composition between dark basalt and light rhyolite, erupted about 3 million years ago north and northwest of Long Valley at Bald Mountain and along San Joaquin Mountain Ridge. San Joaquin Mountain and Two Teats are erosional remnants of large volcanic domes.

Dark-colored lavas, which are hotter and contain more iron and magnesium than light-colored lavas, tend to flow more readily and release gas nonexplosively. Light-colored lavas are cooler, silica-rich, and therefore less fluid and slower moving than darker lavas. Light-colored lavas are also more explosive because they retain gases and water vapor more effectively, until the pressure is released suddenly when they break through the Earth's surface. A volcano that produces light-colored, viscous lava tends to build up gas pressure until it explodes. Consequently, Hawaiian volcanoes that expel dark-colored lavas are less dangerous to observe than explosive volcanoes like Mount St. Helens, which erupts light-colored lava.

At Long Valley very light-colored lava, *rhyolite,* began erupting near Glass Mountain about 2 million years ago. A series of domes and flows, surrounded by fans of volcanic block flows and ash beds as much as 3000 feet (1000 m) thick, testifies today to eruptions lasting about 1 million years. About three-quarters of a million years ago, the roof over a magma chamber in Long Valley ruptured, and about 150 cu miles (600 cu km) of incandescent ash and molten rock blasted out—about 2500 times the amount blown out by Mount St. Helens in its eruption of May 18, 1980 (fig. 3.6). The eruption partially emptied the magma chamber, caused its roof to collapse and the chamber to sink in on itself, and formed a *caldera,* now an oval depression 10 by 20 miles (17 by 32 km). The incandescent particles—ash and pumice—that were thrown out covered 580 sq miles (1500 sq km) of the surrounding countryside and piled up more than 600 feet (200 m) thick. The ash deposits, called Bishop Tuff (*tuff* is consolidated volcanic ash), within Long Valley Caldera are more than 4500 feet (1400 km) thick.

In a volcanic eruption, the volcanic vent may look as if it is on fire, but it is not really burning. The red, fiery color is even more frightening than fire: Rock heated to incandescence reaches a much higher temperature than burning wood or coal. The temperature of the explosive ash flows from Long Valley Caldera has been calculated at about 1500°F (800°C). The tiny, glowing glassy fragments welded together as they cooled, settling to perhaps half their original thickness as they compacted.

Clouds of fine ash from the Long Valley eruption drifted with the wind. Volcanic ash fragments from that eruption have been identified as far east as Nebraska, in southwestern California, and in cores drawn from the bottom

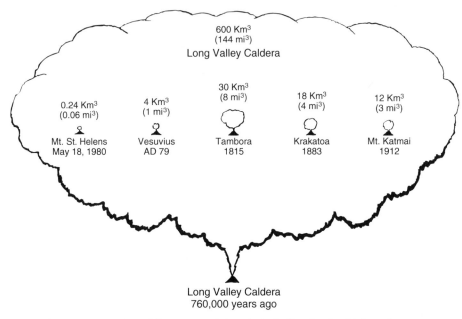

600 Km³
(144 mi³)
Long Valley Caldera

| | | 30 Km³
(8 mi³) | 18 Km³
(4 mi³) | 12 Km³
(3 mi³) |

0.24 Km³
(0.06 mi³)

4 Km³
(1 mi³)

Mt. St. Helens
May 18, 1980

Vesuvius
AD 79

Tambora
1815

Krakatoa
1883

Mt. Katmai
1912

Long Valley Caldera
760,000 years ago

Fig. 3.6. Comparison of the amount of ejecta (principally volcanic ash) from the eruption of several volcanoes. Mount St. Helens produced a quarter of a cubic kilometer of magma in its May 18, 1980, eruption. In Alaska, Mount Katmai produced 12 cubic kilometers in its 1912 eruption, while the historic eruption of Mount Vesuvius in A.D. 79, when hot ash buried Pompeii, produced only 4 cubic kilometers. This diagram illustrating the volume of ejecta from the Long Valley Caldera encompasses all of these eruptions. About 760,000 years ago 600 cubic kilometers of ejecta—2500 times the amount from Mount St. Helens—blasted out of the caldera, providing enough material to pave a 12-lane highway to the moon. (Drawing by Doris Marsh)

of the Pacific Ocean. However, not all the ash was expelled in one huge burp. Incandescent ash flows followed one after another, sometimes so quickly and traveling so fast that later ones overtook earlier ones. Today these ash flows are beautifully exposed in the Owens River Gorge and on the Volcanic Tableland, particularly in roadcuts on Sherwin Grade north of Bishop, as well as in cuts south of June Lake.

Eruptions continued until finally, about 100,000 years after the collapse, a dome rose up in the center of the collapsed caldera. Lava erupted at that time was glassy, leaving deposits of *obsidian* (volcanic glass), prized by Native Americans for making arrowheads. While the dome was rising, the caldera filled with rain and surface water, forming a large crater lake. Glaciers from Sierra canyons flowed into it, breaking off icebergs, some of which carried rocks from the high country to the flanks of the dome. Pleistocene (Ice Age) Long Valley Lake persisted until its water spilled over the southwest caldera rim, where the rushing overflow cut the Owens River Gorge into the Volcanic Tableland. As the waters continued to cut the gorge and enlarge the exit from the caldera, eventually the lake drained. (Crowley Lake is not a remnant of this ancient lake but a man-made reservoir completed in 1941.) Some thousand centuries of quiescence followed, while the caldera stood with a dome in its center, surrounded by a moat similar to those of medieval castles. Then about half a million years ago light-colored rhyolite began to erupt in the moat and continued to do so on and off for 400,000 years (fig. 3.7).

While these later eruptions in Long Valley and those in the Big Pine volcanic field were going on, another volcanic center north of the caldera opened fire. It is not clear whether Long Valley Caldera and the new northern volcanoes derived their molten rock from the same source or not. The new volcanoes, a chain of vents about 28 miles (45 km) long, extend from Mammoth Mountain to Mono Lake. Basaltic lava began to erupt at least 300,000 years ago, from vents near the west moat of the caldera. The moat itself was filled to a depth of 800 feet (250 m), the lava surging around the dome and sending long tongues north and south within the moat. About 220,000 years ago, Mammoth Mountain began to build. At least twelve domes and associated flows of light-colored lava constructed the imposing volcano that today straddles the caldera rim. It ceased erupting about 50,000 years ago, but a few vents still emit steam.

Other basaltic vents in the chain erupted farther to the north near June Lake between 40,000 and 20,000 years ago and at Black Point, Mono Lake, about 13,000 years ago. In the interim, a conspicuous group of light gray rhyolite domes erupted along the 10.5-mile (17-km) Mono Craters chain lying south of Mono Lake. The older ones, at the southern end, began to form about 35,000 years ago; the youngest one, Panum Crater, erupted between A.D. 1345 and 1445.

At the southern end of the chain, the Inyo Craters—a line of domes and craters that expelled light-colored rhyolitic lava—began to rise. Two of them, Wilson Butte (fig. 3.8) and North Deadman Creek Dome, erupted lava so similar to the Mono Craters lavas that it is probable that they came from the same source. Also, they were formed at about the same time, geologically speaking, as the Mono Craters: 1350 years ago (Wilson Butte) and 6000 years ago (North Deadman Creek Dome). Other domes in the Inyo chain—Obsidian Dome, Glass Creek Dome, South Deadman Creek Dome, and two that are unnamed—all erupted in a short space of time, within a few years or perhaps even a few months. They were active in the late fourteenth century, about the same time Panum erupted, perhaps as late as A.D. 1472. Dating of trees leveled by eruptions provides dates of A.D. 1369, 1433, and 1469. It is possible that the two younger Inyo domes expelled magma that was a mixture of molten rock from Long Valley Caldera and the Mono

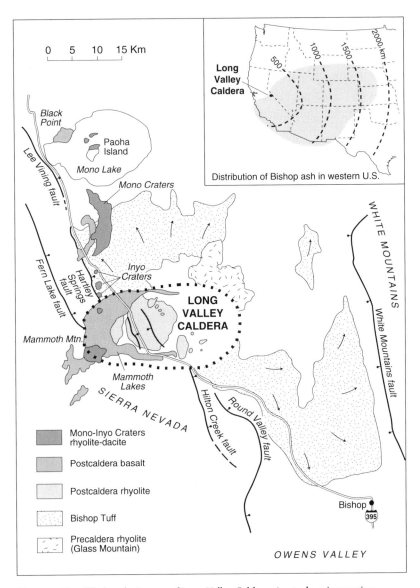

Fig. 3.7. Simplified geologic map of Long Valley Caldera. A cataclysmic eruption 760,000 years ago ejected flows of hot glowing ash, which cooled to form the Bishop Tuff, and formed Long Valley Caldera. Wind-blown ash from that ancient eruption covered most of the West (inset map). (Roy Bailey, USGS; modified by Annie Kook)

Craters magma chambers. About the time the Inyo domes were being formed (A.D. 1340–1460), three craters blasted out of the top and side of Deer Mountain, a rhyolite dome south of Deadman Creek. Rainwater collects in two of the small craters, now known as the Inyo Crater Lakes.

In Mono Lake, cinder cones burst forth, forming Negit Island, while a

Fig. 3.8. Wilson Butte, a rhyolite dome north of June Lake that formed about 1350 years ago. (Photo by Edwin C. Rockwell)

rhyolite eruption northeast of Negit flowed into the water to form small islands. Paoha, the larger, lighter-colored island, is a rhyolite dome covered by lake sediments and pushed up from the lake bottom. It is thought to have been emplaced sometime between 1720 and 1850, shortly before Europeans first explored the area.

Not far away, southwest of the caldera, at what is now Devils Postpile National Monument, molten basalt flowed downstream from a vent near Upper Soda Springs Campground, forming a pool of lava, possibly ponded behind a glacial moraine. Precisely when this happened is uncertain, but probably less than 100,000 years ago. The lava was very fluid and cooled slowly, congealing first along the bottom where it encountered cool granite and at the top where it was exposed to the air. Since it cooled from the top and bottom inward, the interior remained molten for some time. As the lava cooled, it shrank, causing stresses in the newly formed rock mass that were relieved by cracking. Under ideal conditions, a cooling, uniform mass would crack into hexagons, giving rise to long, near-vertical, parallel columns. But conditions were not ideal, and many columns are not bounded by hexagons. Conditions were nearest to ideal near the bottom of the flow, where the columns that are now the Devils Postpile formed.

Even so, only about half the postpile columns have six sides. The remainder have four, five, or seven; a very few have only three sides. Neatly geometrical columns are rare. The world's most spectacular group of postpiles is at Giant's Causeway, Northern Ireland, where huge columns are partly covered by the sea; only about half of them are six-sided. The largest columns in the Devils Postpile are a little over 3.5 feet (1 m) in diameter, but they average about 2 feet (0.6 m). Some columns are 60 feet (18 m) tall. After the lava cooled, forming the postpile, glaciers again moved into the area, exposing the postpile and polishing its top. Natural forces continue to attack the postpile. Three columns that had been leaning outward since 1909 fell during the 1980 earthquakes, and other columns cracked.

Seismological evidence and earth movements suggest that at least one hot magma body still lies beneath Long Valley Caldera. The evidence for this consists of the ages of the most recent eruptions; geophysical data that can be interpreted to indicate the persistence of a magma chamber beneath the caldera; a bulge of the ground surface since 1980 of at least 24 inches (60 cm); and the *spasmodic bursts* of 1980–83, a peculiar pattern on a seismograph that is believed to indicate the movement of hot fluids through fractures. In the Long Valley area in 1982, persistent *earthquake swarms,* thousands of tiny earthquakes, coupled with continued upward bulging of the caldera floor, caused geologists to think that an eruption was possible. The U.S. Geological Survey (USGS) issued a "potential volcanic hazard" notice. Besides uplift and earthquakes, a dormant fumarole at Casa Diablo Hot Springs began to spout again, and new hot springs broke out on the southern edge of the dome. By 1984, when no eruption had taken place and earthquake activity had diminished, the hazard notice was rescinded. Then in 1989 a six-month-long earthquake swarm and bulging beneath Mammoth Mountain caused volcanologists to reassess the volcanic potential, but the activity quieted down by the end of the year and no hazard notice was issued. The geophysicists who studied those events concluded that magma was being squeezed into fissures beneath Mammoth Mountain but that it congealed before reaching the surface.

In 1990 trees near Horseshoe Lake at the base of Mammoth Mountain mysteriously began to turn brown. After ruling out drought, disease, and insects as causes, geologists solved the mystery when they discovered significant amounts of carbon dioxide (CO_2) seeping through the soil, killing the trees. By 1996, trees in scattered locations on the mountain were dead. The area of dead trees totaled more than 150 acres. Geochemical analyses indicate that the CO_2 gas is derived from magma. Apparently large amounts of gas were trapped beneath the surface until 1989, when magma rising along a fault fractured the underlying rock, allowing the gas to escape to the surface.

These recent events remind us that the Eastern Sierra is still earthquake and volcano country; there is no reason to think either activity has ceased. The fumaroles and many hot springs in the area—Whitmore, Casa Diablo, Hot Creek—testify that the volcanic furnaces are probably only banked. Fortunately an extensive network of devices continually monitors changes that could indicate the beginnings of renewed activity. Constant monitoring will give ample warning, but the possibility of eruption is not to be disregarded.

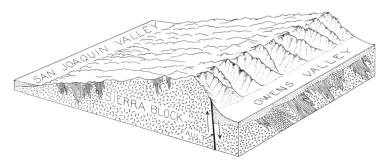

Fig. 3.9. Diagram showing the Sierra's steep, faulted east face and the uplift and tilt of the Sierra block along faults that mark the eastern boundary of the range. The arrows show the relative movement of the mountain block and the valley block during faulting. Debris eroded from the upfaulted mountains has been deposited on the downfaulted floor of Owens Valley to depths of thousands of feet. (Drawn by the late François E. Matthes)

THE MOUNTAINS RISE (AGAIN)

It is often true that "the mountains grow, unnoticed," as John Muir wrote, but it is also true that they grow by perceptible jolts. The Eastern Sierra provides ample evidence of the jolts. In 1872 one of California's three "monster earthquakes" lifted the mountains up several feet; destroyed every stone, adobe, and brick building in the little town of Lone Pine at the foot of the mountains in Owens Valley; and killed one-tenth of the town's population. (Its strength is estimated at magnitude 7.8, revised downward from 8.3.) It was a lesson in earthquake engineering: No wooden buildings were destroyed, except for "an insubstantial shed."

Earthquakes in the Eastern Sierra may not be as frequent as in the coastal area, but they do come, and they can be quite strong. State Geologist Josiah Whitney recognized this in 1872, when he visited Lone Pine to see the earthquake devastation. He saw the new scarp formed during the earthquake and concluded that faulting had lifted the mountains to their present height. Although he did not connect faults and earthquakes as intimately as we now do (that connection was made by scientists who studied the 1906 San Francisco earthquake), he recognized the importance of faulting in creating the mountains. Earth movements along faults, over millions of years, have jerked and tilted the Sierra westward, creating a steep eastern face (as much as 25 degrees) and a more gradual western slope (2 degrees). The central Sierra is essentially a fault-block range (fig. 3.9), unlike the folded mountains of the East Coast. Its ruggedness is the result of its geologic history. Earthquakes and faults have created the splendid mountain scenery.

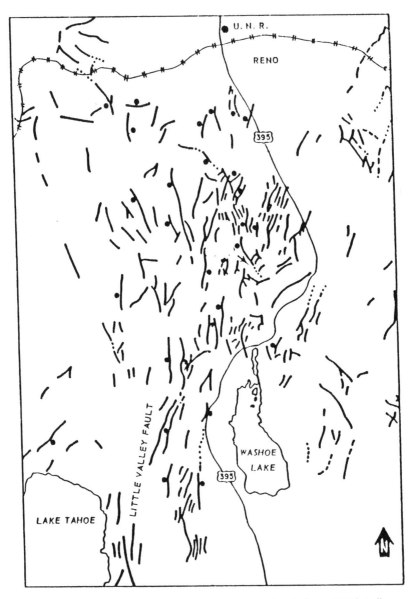

Fig. 3.10.   A complex swarm of faults extending from Washoe Valley and Little Valley to Reno, Nevada. Scale 1:250,000. (From *California Geology* [May 1975], by permission of D. Burton Slemmons)

The major fault lines along the Eastern Sierra are known as the Sierran Frontal Faults. It was along one of these fault lines that the mountains rose, relatively speaking, in the earthquake of 1872. Some individual faults have names, but many have no names at all. Figures 3.10 and 3.11, showing many of the faults south of Reno, give some idea of the multitude of faults in one

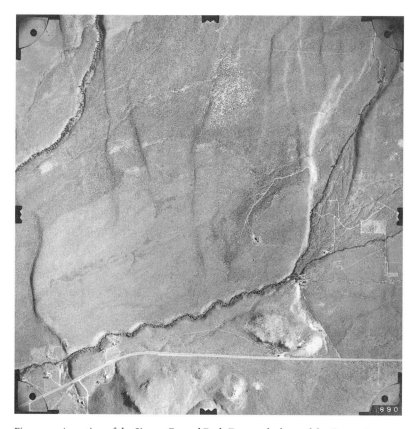

Fig. 3.11.   A portion of the Sierran Frontal Fault Zone at the base of the Carson Range, showing a swarm of fault scarps cutting recent glacial outwash west and north of Steamboat Hills. Faulting along the frontal fault zone continues today, and the Carson Range continues to rise. The north-trending fault scarps, ranging from several inches to several feet high, reflect faulting in the basement rock. This low-sun-angle photo accentuates these low scarps. The road at the bottom is Hwy 431, the Mount Rose Highway; just to its north is Whites Creek. The top of the photo is north. Scale 1:12,000. (From E. C. Bingler [1975], *Guidebook to the Quaternary Geology along the Western Flank of the Truckee Meadows, Washoe County, Nevada,* with permission of Nevada Bureau of Mines and Geology; photo by D. Burton Slemmons)

area. Many faults are concealed and can only be identified by seismic data, including seismographic records of earthquakes. (In the 1970s the federal government asked the state of California to estimate the cost of mapping all of the nation's geologic faults. The geologist who was asked to make the estimate said, "Between six thousand dollars for two graduate students to mark all mapped faults as a summer's job and three trillion dollars to map every fault in close detail.") What researchers would like to know about faults is where they all are and when they were last active. For example, the Fish Springs cinder cone in the Big Pine volcanic field has been cut by faults so

Fig. 3.12.   Fault scarp up to 80 feet high and small cinder cone south of Big Pine. Recent scarps such as the one shown here may be only tens of feet high, but where there has been repeated movement, they may be thousands of feet high. (Photo by Roland von Huene)

The Sierra Crest (opposite photo) rises 10,000 feet above Owens Valley, the result of innumerable up and down movements along the Sierran Frontal Fault Zone. The prominent white line at the base of the Alabama Hills is not a fault scarp; it is the Los Angeles aqueduct. Note that it curves around a small alluvial fan near Lone Pine (at left edge of photo). The straight white line across this fan, west of the aqueduct, is an east-facing scarp up to 20 feet high.

that it looks as if someone were planning to serve it for dessert (fig. 3.12). Obviously, in this case, some of the faults are younger than the cinder cone, which has been dated at about 314,000 years old.

We cannot always tell positively that mountains rise. It may be that adjacent valleys sink, or that both happen. Faulting not only has uplifted the Sierra (fig. 3.13), but also has raised the White-Inyo Mountains to the east and dropped Owens Valley down between, to be half filled with boulders, gravel, and sand from both mountain ranges. The top of the White-Inyo Mountains is about 3 miles (5 km) higher than the same surface beneath Owens Valley, which lies as much as 4000 feet (1200 m) below sea level. Other examples of valleys down-dropped between uplifted mountain ranges are the Carson and Washoe valleys, Lake Tahoe, and Truckee Meadows.

Fig. 3.13.   View westward of the Sierra's east face, a gigantic eroded scarp. The down-dropped Alabama Hills are in the middle ground, the Owens Valley floor in the foreground. (Photo by John S. Shelton)

THE TENDER SNOWFLOWERS

Faulting has not been the only creator of mountain scenery. While volcanoes were adding to the landscape, water and ice were sculpting the details. Josiah Whitney, fresh from studying the changes in the Sierra Nevada during the 1872 earthquake, concluded that the main sculpting of the mountains was due to faulting. It was faults, he contended, that had cut Yosemite Valley, faults that had created the steep eastern slope, and faults that had gashed the mountain canyons. He presented his views in a popular guidebook to Yosemite Valley. While Whitney was working out his theory, another observer was coming to an entirely different conclusion. He was John Muir, a scientist of less training than Whitney, but an acute observer, a talented writer, and a thoughtful reader. It was not sudden, cataclysmic faulting, he claimed, that had carved Yosemite Valley and scraped headwalls into the mountain crest. It was "the tender snowflowers noiselessly falling through unnumbered centuries, the offspring of the sun and sea."

Nonsense, Whitney roared. "A more absurd theory was never advanced than that by which it was sought to ascribe to glaciers the sawing out of these vertical walls and the rounding of the domes." In Whitney's view, Muir was nothing but "a mere sheepherder, an ignoramus" who knew nothing about geology. Nonetheless, during a hiking trip with Professor Joseph LeConte of the University of California, Muir convinced LeConte that glaciers had been important in creating Sierran scenery. But Whitney was not convinced, even after Muir demonstrated that the living glacier on Mount Lyell moved.

"There are no glaciers at all in the Sierra Nevada," Whitney wrote. He would be surprised to know—but perhaps not convinced—that a count made in the 1970s gave a total of 497 glaciers and 788 ice patches, although most of them are little more than handkerchief-size compared to Ice Age glaciers. The largest is Palisade Glacier west of Big Pine, which is less than 1 sq mile (2.5 sq km) in size; the smallest is an unnamed glacier of only 110 sq feet (10 sq m). Neither Whitney nor Muir knew at the time that the Sierran glaciers they saw were not relics of ancient, large ice bodies, but small glaciers formed during a Little Ice Age that began about 700 years ago when the climate cooled again after the large glaciers had melted away. Sierran glaciers reached their greatest recent advance probably around 1850; they have been melting back ever since.

Other geologists who have worked in the Sierra Nevada since Whitney and Muir's time have concluded that Whitney was correct in assigning a major role in the rise of the mountains to faulting, and that Muir was correct in assigning a major role to glaciers as sculptors of the landscape. Despite the controversy between the two men, their memorials stand together near Lone Pine. Mount Whitney, the highest U.S. peak south of Alaska (14,494 ft or 4418 m), is flanked by Mount Muir (14,015 ft or 4272 m) and Mount Russell (14,086 ft or 4293 m), named for Israel C. Russell, who studied the glacial history of Mono Lake.

The glaciers that had so drastically changed the landscape were not the tiny ones of Whitney's day but older, much larger ones, that took their timing from the huge ice sheets that pressed their way over much of the world during the last 2 million years. Although these giant ice sheets never reached California, mountain glaciers formed at the heads of many Sierra canyons. At the height of glaciation, the Sierra must have looked much as the Juneau Ice Cap in Alaska does today, with many high peaks and ridges protruding above the glistening ice. By the time these glaciers had melted, they had vastly changed the landscape.

Muir based his deductions about the Sierra's glacial history less on the living glaciers so derided by Whitney than on the evidence of glaciers past. What he saw on the smooth granite were scratches, gouges, and grooves. He saw gleaming rock surfaces, polished by the passing ice. On a larger scale, he saw V-shaped valleys transformed by glaciers into steep-sided U-shapes, with waterfalls plunging thousands of feet from the tops of the valleys to their bottoms. (Yosemite and Hetch-Hetchy valleys were his best examples.) He saw spoon-shaped basins *(cirques)* marking the heads of now vanished glaciers. All this he attributed to the grinding, erosive action of ice bodies hundreds of feet thick slowly moving downslope. The rocks frozen into the bottom and sides of the moving ice acted as large files, while the frozen sand smoothed and polished the bedrock it passed over.

What Muir did not recognize or did not emphasize was that the newly carved glacial valleys followed stream pathways that already existed. However, glaciers enlarged them, scooped them out, created cirques at their heads, and changed the cross-section of the valleys from a V to a U. François Matthes, who studied the Sierra Nevada (particularly the Yosemite region) after Muir and Whitney, pointed out that glaciation in the High Sierra had left a "biscuit board" topography: Where glaciers had cut into canyons' headwalls and then melted, surrounding mountain peaks looked like the bits of dough left over from cutting out round biscuits. Where the heads of two glaciers had once nearly met, a knife-edged ridge was left; where three glaciers had once headed, a sharp peak, or *matterhorn,* remained (named for the Matterhorn in the Swiss Alps, which was carved by glaciers in this way).

Besides noting what the glaciers had carved away, Muir also saw what they had deposited: rocks out of place, carried miles from their source by these great conveyor belts of ice. He also saw *moraines,* low ridges of boulders and gravel heaped up on the sides of valleys where they had melted out of the glaciers (fig. 3.14), and curved moraines that marked the ends of glaciers, where the glacial snout had pushed debris ahead of it like a snowplow. Eastern Sierra moraines are shorter than those on the west slope, attesting to the glaciers' shorter, stubbier shape. They are also more conspicuous because they have weathered little in the arid east slope climate and many of them are bare of trees or shrubs, while the western slope is densely forested. Conspicuous and easily recognizable, nevertheless they are sometimes overlooked. Near Mono Lake, for example, now vanished glaciers left rubble more than 800 feet (250 m) high. In any other setting, these would be impressive hills, but set here at the base of 13,000-foot (4000-m) peaks and just opposite the 9000-foot (2800-m) Mono Craters, they are nearly lost in the spectacular landscape. You can see moraines at many locations around Lake Tahoe (Fallen Leaf Lake lies behind a morainal dam), along Highway 395 on Mount Rose, west of Bridgeport, along Lee Vining Creek, and at many other canyons such as Sherwin, Laurel, Convict, and Pine Creek.

The moraines were not all formed at the same time. True, they were all (except those of the tiny recent glaciers) formed during the Great Ice Age, but studies have shown that the ice advanced and retreated several times, leaving moraines of different generations. Easiest to see, because they are

Fig. 3.14. Green Creek glacial moraines, south of Bridgeport. (Photo by John S. Shelton)

large and stand out sharply, are the moraines left by the glaciers of 60,000 – 20,000 years ago. They form low, voluptuous ridges, particularly striking when the sun is near the horizon — in the orange light of sunset or the cool light of early dawn. Walker Lake, west of Mono Lake, is enclosed by fresh-looking moraines left by glaciers of four different generations.

The Eastern Sierra was glaciated as far south as Cottonwood Pass, 10 miles (16 km) south of Mount Whitney. Glaciers extended down the canyons to as low as 5000 feet (1500 m). The glaciers melted away, to be sure, but while they were alive and well, they poured meltwater onto the adjacent lands. During the latter part of the Great Ice Age, two huge lakes covered much of what is now the western desert. Lake Bonneville filled most of Utah, while Lake Lahontan covered much of Nevada, with huge arms extending into Oregon and California. Besides these two vast lakes, smaller ones stood in many Great Basin valleys. Most of their blue waters have dried, leaving yellow sand and white salts.

LONG-LIVED MONO LAKE

Some lakes have not yet dried. Of the chain of lakes that led from Lake Mono to Death Valley, only present-day Mono Lake still has water, although it has been shrinking from natural as well as man-made causes. Ice Age Lake Mono, called Lake Russell for early geologist Israel C. Russell, at its peak about 13,000 years ago was 28 miles (45 km) long by 18 miles (29 km) wide. It covered more than 338 sq miles (875 sq km) and was as much as 900 feet (275 m) deep. It spilled east through Adobe Valley (its waters continuing south as the Owens River, which fed into ancestral Lake Owens), flowed farther south into Lake Searles, and eventually reached Lake Manly in Death Valley. (The Ice Age lakes are called Lake Owens and Lake Searles to distinguish them from the modern lakes, Owens Lake and Searles Lake.)

Like many Eastern Sierra valleys, Mono Valley, which contains Mono Lake, sank along faults as the mountains rose during the past 10 million years. Altogether, the difference in elevation between the granite bedrock of Mono Basin and the Sierra Crest is 11,000 feet (3400 m), yet the difference in elevation between today's valley floor and the mountain crest is only 6000 feet (1800 m). The reason for this is that Mono Basin has been filling up steadily with more than 4000 feet of silt, sand, and gravel washed in from the adjacent mountainsides.

Mono Lake is one of the few remnants of Ice Age lakes in the West that is still a year-round lake. Today, it is about 14 miles (23 km) long by 10 miles (16 km) wide; its greatest measured depth in the twentieth century was 169 feet (51.5 m). Mono Lake is one of the few places in the world where we can see the relationship between the height of an Ice Age lake and some of the glacial stages. Ancestral Lake Russell, which was hundreds of feet higher than today's lake, left its mark at various lake levels as terraces cut into the hills—giant "bathtub rings" that are clearly visible early and late in the day, when the low angle of the sun casts long shadows. In places, these terraces are cut into glacial moraines, showing very clearly the relative age of each.

Since the Ice Age the level of Mono Lake has fluctuated in response to the climate. Its highest level is estimated at just under 6500 feet (1981 m) above sea level. As the lake level fell in recent years, a tufa-coated tree stump was exposed that had grown on the shore about A.D. 1000, indicating that Mono Lake then stood at an elevation of about 6400 feet (1950 m). This was the time of the "climatic optimum," a warm, dry period when northern seas were free of ice. Mono Lake swelled again in size during one of the Little Ice Age advances, when new, small glaciers grew in the Sierra and the lake level reached 6455 feet (1968 m).

The water of Mono Lake is very alkaline, and since 1941 has rapidly become more so, as freshwater streams that would normally replenish it were diverted into the Los Angeles Aqueduct. Between 1941 and 1982 the lake's level dropped 45 feet (14 m). After numerous lawsuits filed by the Mono Lake Committee and its allies—notably the National Audubon Society and California Trout—in 1994 the State Water Resources Control Board modified

Fig. 3.15. Mono Lake tufa towers. (Photo by Beverly F. Steveson)

the license of the Los Angeles Department of Water and Power (DWP) to divert water from the lake. The board decreed that diversions be limited to allow Mono Lake to rise over 17 feet and to maintain the lake at 6392 feet above sea level on average. In addition, the board established minimum stream flows in all four diverted tributaries and ordered DWP to restore damaged stream and waterfowl habitats. See p. 213 for details of the lake's brine shrimp and Alkali Flies and pp. 292, 321, and 461 for the lake's critical importance to nesting gulls and to the grebes and phalaropes that rest there on their long migrations.

At Mono, Honey, and a few other Great Basin lakes, curious white and brown structures called *tufa towers* dot the shore (fig. 3.15). These knobby towers are deposits of calcium carbonate, as distinguished from *tuff,* which is consolidated volcanic ash. They formed underwater, when the lake level was higher, clustered around freshwater springs that bubble up from the lake bottom. A chemical reaction between the calcium in the fresh water and carbonates in the lake water causes calcium carbonate to precipitate. Tufa deposits are still forming underwater. Tufa forms are reminiscent of ruined spires, honeycombs, or coral. Some are 100 feet (30 m) high, giving an idea of how high the lake was in recent times past.

Mono Lake—or rather Lake Russell and Mono Lake—has persisted for more than 730,000 years. Part of geologist Ken La Joie's evidence for this age is that a hole drilled in 1908 struck the Bishop Tuff at 1400 feet (425 m). The drill did not penetrate any layers of salts, either above or below the tuff, indicating that the lake may not have dried up for at least 730,000 years.

THE SIERRAN FUTURE

The Sierra Nevada has been lifted by forces of Earth, sculpted by water and ice, and left to us as one of Nature's great gifts. Today the Sierra Nevada is being changed more by humans than by nature, and far more rapidly. Bulldozing, logging, diversion of water, thoughtless scarring of the land, and dumping of trash go on and on—day after day, week after week, year after year. Nature may make changes on a grand scale, but those vigorous events are spaced by hundreds, thousands, even millions of years. The hills are far from everlasting. Can we tread lightly enough?

RECOMMENDED READING

Bailey, Roy A. 1989. *Geologic Map of the Long Valley Caldera, Mono-Inyo Craters Volcanic Chain, and Vicinity, Eastern California.* Map I-1933. Washington, D.C.: U.S. Geological Survey.

Hill, Mary. 1975. *Geology of the Sierra Nevada.* Berkeley: University of California Press.

Huber, N. King. 1987. *The Geologic Story of Yosemite National Park.* U.S. Geological Survey Bulletin 1595. Washington, D.C.: U.S. Geological Survey.

Huber, N. King, and Wymond W. Eckhardt. 1985. *Devils Postpile Story.* Three Rivers, Calif.: Sequoia Natural History Association.

Purkey, Becky W., and Larry J. Garside. 1995. *Geologic and Natural History Tours in the Reno Area.* Nevada Bureau of Mines and Geology Special Publication 19. Reno: University of Nevada.

Tierney, Timothy. 1995. *Geology of the Mono Basin.* Lee Vining, Calif.: Kutsavi Press.

Tilling, Robert I., R. A. Bailey, and C. D. Rinehart. In press. *Earthquakes and Young Volcanoes along the Eastern Sierra Nevada.* Mammoth Lakes: Genny Smith Books.

CHAPTER FOUR

WEATHER AND CLIMATE

DOUGLAS POWELL AND HAROLD KLIEFORTH

Compared with other mountain ranges in the forty-eight continental states, the Sierra Nevada is the highest, yet its climate is relatively mild. Although some winters bring heavy snowfall, the storms alternate with many days of sunshine. The Sierra Nevada has a very unpredictable climate—several years of below "normal" precipitation may be followed by a year of abundant rain and deep snow. These factors have played a large part in determining which plants and animals are native to the region; some must be able to withstand months of heat, sun, and little rain, as well as months of freezing cold and snow. Some knowledge of Eastern Sierra weather and climate is basic to an understanding and appreciation of the region's unusual flora and fauna, the geological and biological processes that are active, and the past and present appearance of the terrain and biota.

Weather and climate are part of a continuous whole. *Weather* is the short-term condition of the atmosphere—an instant, an hour, a day, a week, maybe longer; *climate* is a generalization of weather conditions over time—a month, a season, decades, hundreds or thousands of years. The terms grade imperceptibly into each other. We use both in this chapter, with an emphasis upon climate.

MEASURING THE WEATHER

The first systematic daily weather observations and measurements in the Eastern Sierra and the western Great Basin were made in the mid-1800s by the medical officers and scientists of the various U.S. Army surveys. When semi-permanent forts were established, the Army Signal Corps began keeping weather records. Later, with the building of railroads in the latter half of the nineteenth century, weather stations were established in towns served by the railroad. Much later, the advent of aircraft and the need for meteorological reports led to the creation of a network of weather stations. Since most of these sites—at forts, railroad towns, and airports—were in valleys, for many years there were few quantitative records of mountain weather.

In 1901 Dr. James Church, a professor of classics at the University of Nevada, began his studies of weather and snowfall in the Carson Range east of Lake Tahoe. A few years later he established a weather station on the 10,780-foot (3285-m) summit of Mount Rose. Assisted by colleagues and students, he maintained the station for several years. They made twice-monthly visits, often in severe winter conditions, to take measurements and change the charts of recording instruments.

When disputes arose among various groups over regulation of Lake Tahoe's water level and over allocation of water from the Truckee River for power generation, flood control, domestic use, and agricultural irrigation, Dr. Church and his colleague, S. P. Fergusson, developed a technique for measuring the depth and water content of the mountain snowpack. They used a portable hollow metal tube 1.5 inches (3.8 cm) in diameter with a cutting edge, attachable units each 30 inches (76 cm) long, and a spring-scale to measure the amount of water held in the early spring snowpack. This information was used to estimate the resulting runoff, providing the basis for a rational plan to regulate the level of Lake Tahoe and the flow of water in the Truckee River. Snowy areas throughout the world still use this method today, and James Church is known as the father of snow surveying.

Following the pioneering work of Dr. Church, in 1965 atmospheric scientists of the Desert Research Institute established a remote-recording weather station on the summit of Slide Mountain, at 9650 feet (2941 m) near Mount Rose, and a network of precipitation measuring sites in a 1000-square-mile (2600-sq-km) area in the Lake Tahoe–Truckee River watershed. The data from this network provide valuable information on the spatial and temporal variation of rain and snowfall in the region. At the Tahoe Meadows station at 8540 feet (2600 m), the average annual precipitation is about 53 inches (135 cm). Extremes range from 22.24 inches (56.5 cm) in 1975–76 to 96.91 inches (246.2 cm) in 1982–83, when total snowfall was 695 inches (1765 cm) and the greatest snow depth was 19.7 feet (6 m). During each of two months, January 1969 and February 1986, the total monthly precipitation was 37 inches (94 cm) of water. Figure 4.1 shows the seasonal liquid precipitation at Tahoe Meadows from July 1968 through June 1998. The maximum monthly and annual precipitation, snowfall, and snow depth amounts at Tahoe Meadows are all records for Nevada. At a nearby site, Incline Lake (8200 ft or 2500 m), 76 inches (193 cm) of snow fell in a 24-hour period on February 17, 1986. On the same date at the Cliff Ranch (5250 ft or 1600 m) on Franktown Road in Washoe Valley, 7.36 inches (18.7 cm) of rain was measured. Today there are

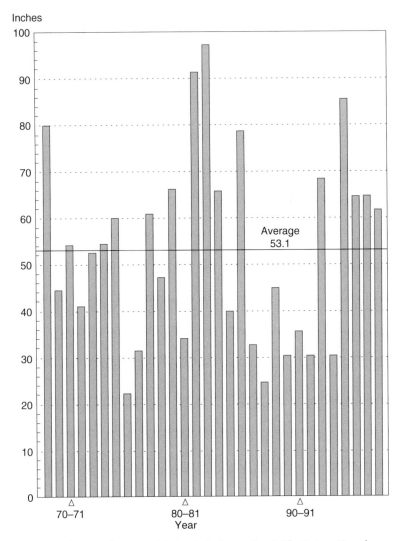

Inches

Average
53.1

70–71   80–81   90–91
Year

Fig. 4.1.   Tahoe Meadows, precipitation in inches, 1968–98. The Eastern Sierra has huge fluctuations in yearly amounts of precipitation. For example, at the Tahoe Meadows weather station (8540 feet) on the Mount Rose Highway on the west side of the Carson Range, precipitation has ranged from 22 inches (1975–76) to 97 inches (1982–83). In 1982–83 total snowfall was 695 inches and the greatest snow depth was almost 20 feet. (Precipitation years are calculated from July 1 to the following June 30.) The maximum annual precipitation, snowfall, and snow depth at Tahoe Meadows are records for Nevada. (Diagram by Annie Kook)

additional weather stations in the Eastern Sierra, including several remote-recording sites in the high mountains.

WEATHER'S IMPACT ON HISTORICAL EVENTS

The huge Sierra Nevada, with its high altitude and rugged topography, and the adjacent semi-arid Great Basin were formidable barriers to exploration of the West. It was not until 1827 that any of the early explorers—Canadian, Mexican, or American—managed to cross the Sierra. That honor went to the American fur trapper Jedediah Smith and two companions, who started in the Central Valley and crossed the Sierra and the Great Basin from west to east, heading for the fur trappers' rendezvous in Utah; they barely survived the ordeal. In November 1844 the first wagon train, the Murphy-Stephens Party, crossed the Sierra; that same year, John C. Frémont and Charles Preuss became the first American explorers to view Lake Tahoe. Deep winter snow, added to all the other hazards, often led to tragedy, as in the case of the notorious Donner Party in 1846–47. Their failure to cross the Sierra that winter resulted from their tardy arrival to the eastern slope and from the early arrival of winter snowfall.

After the discovery of gold in California in 1848, stagecoach routes were established across the Sierra, but for several months during winter there was little or no communication between communities on either side of the Sierra. To remedy this, in January 1856 "Snow-shoe" Thompson fashioned a pair of heavy, 10-foot-long (3-m) wooden skis and, with a single wooden pole and a heavy pack, crossed the mountains from Placerville to Genoa carrying mail to the people on the east side (fig. 4.2). He continued these trips with little compensation for twenty years until his death in 1876. In the pioneering annals of the Sierra, Snow-shoe Thompson's strength, endurance, and selfless dedication are unmatched.

At about this time, as Berry and Wentworth recount in *Lost Sierra* (1991), skiing became a popular form of recreation in isolated mining camps, such as La Porte in Plumas County. Miners and loggers organized races and experimented with the use of various waxlike substances to coat the undersides of their skis and make them slide faster. Much later someone thought of building a rope tow to haul skiers uphill, and by the 1920s small winter sports centers operated in Truckee and the Tahoe Basin. After World War II, when the ski troopers of the famous Tenth Mountain Division returned home, the sport of skiing grew in popularity. In February 1960 Squaw Valley hosted the Winter Olympics. There had been a dearth of snow in the Sierra all winter, when suddenly, just a few days before the competitions, a fortuitous meteorological event occurred: A big storm brought several feet of snow, and just

Fig. 4.2.   Snow-shoe Thompson, legendary mail carrier. Thompson left Norway at age ten, but crafted from memory his "Norwegian snow skates," or "snow shoes," as they were then called. (From Wm. M. Thayer [1887], *Marvels of the New West*)

as the opening ceremonies began, the clouds parted and the sun shone on the majestic scene.

Weather made more difficult one of the great engineering feats of the nineteenth century, construction of the Central Pacific Railroad. Its route from Sacramento to Reno followed the transverse ridge between the North Fork of the American River and the South Fork of the Yuba River to Donner Pass, then dropped down the east side via Coldstream Ravine, Donner Lake, and Truckee Canyon. More than 10,000 Chinese with picks, shovels, and wheelbarrows worked year-round, laying track and driving tunnels, between January 1863 and April 1868, including the severe winter of 1866–67. To cope with the heavy snow and avalanche-prone slopes at the higher elevations, hundreds of carpenters built snowsheds over the tracks— 40 miles of them, according to David Myrick in his book *Railroads of Nevada and Eastern California*. Today, with rotary snowplows usually able to keep the tracks clear, only about 6 miles of snowsheds remain in use. However, heavy snow can still overwhelm all the machinery we possess. On January 14, 1952, near Donner Summit, deep snowdrifts blocked the streamliner *City of San Francisco* for three days. Rescuers on skis and on snowcats finally reached the passengers, who had had little food and water and no heat at all.

Even with all the snow removal equipment now available, snow is still a serious hazard on the major trans-Sierra route, Interstate 80. It is ironic that the increase in traffic—and especially the increase in the number, size, and speed of huge trailer trucks—has led to more frequent road closures during winter snowstorms. Jack-knifing and skidding out of control are perpetual hazards in snow country, because it takes only light snow or ice for vehicles of any kind to get into trouble.

GEOGRAPHIC FACTORS DETERMINING EASTERN SIERRA CLIMATE

The location of the Sierra Nevada—its latitude, orientation, and proximity to the Pacific Ocean—largely determines its climate. It lies in mid-latitudes, between 35° and 40°N, placing it squarely in the path of the prevailing westerly winds (the terminology refers to the direction from which the wind is blowing), especially at upper levels of the atmosphere. Because the entire range is within 200 miles (320 km) of the moderating waters of the Pacific Ocean, the Sierra Nevada enjoys more moderate temperatures, drier summers, and warmer winters than locations at similar latitudes mid-continent, such as the Rocky Mountains. The west coasts of continents between latitudes 30° and 45° have *Mediterranean climates*—wet winters, dry summers. Other high mountain ranges in similar climatic zones are the Andes of central Chile and the Atlas Mountains of northwest Africa. Because Mediterranean climates occupy only about 4 percent of the Earth's land area, California's winter-wet, summer-dry precipitation pattern is distinctive. The eastern slope of the Sierra is largely within this Mediterranean climate zone. However, the valleys and mountain ranges farther east are in a transitional zone of more evenly distributed precipitation throughout the year, with less exposure to the moderating maritime influence of the Pacific.

The orientation of the Sierra, more or less at right angles to the prevailing moisture-bearing winds, greatly influences airflow and precipitation (fig. 4.3). Because storm winds come predominantly from the west or southwest and because the Sierra Crest is aligned essentially north-south, creating a mountain barrier 400 miles (650 km) long and 8000–14,000 feet (2440–4270 m) high, air masses are forced to rise up the windward (west) slope of the range.

Because the Sierra Nevada extends over 5 degrees of latitude, temperature gradually increases from north to south at similar elevations. In contrast, *precipitation* (the combined total amount of rainfall and the water equivalent of snowfall) generally *decreases* from north to south because the frequency of storm systems declines toward the south.

ELEVATION FACTORS

As altitude varies, temperature and precipitation also vary from east to west. Throughout its entire length, the Eastern Sierra slope has spectacular changes in elevation. The rise from mountain foot to mountain crest is particularly abrupt between Mono and Owens lakes—as much as 8000–10,000 feet (2440–3048 m) in less than 10 miles (16 km), a rise equaled at few places in the world. Significant trends in mountain weather and climate reflect increasing altitude: Temperature decreases, precipitation increases, and winds become stronger. Although these trends are typical of all high mountains, in the Eastern Sierra they are especially marked because of the magnitude of elevation change within short horizontal distances. The combination of abrupt elevation change and diverse topography leads to great climatic variation from the base to the summit of the range, with myriad individual microclimates.

A mountain range that rises thousands of feet above its adjacent lowlands, as does the Sierra Nevada, greatly influences airflow and precipitation, especially if its crest is oriented more or less at right angles to the prevailing moisture-bearing winds. Airflow from the west and southwest is forced to rise over this 400-mile-long (650-km) mountain barrier, resulting in *orographic uplift.* The airflow first ascends the long, gradual, windward slope of the Sierra, then descends the much steeper, leeward east slope. As the ascending air cools, conditions are favorable for precipitation; as the descending air warms, conditions are *un*favorable for precipitation. Consequently, leeward slopes are drier than windward slopes, especially if the elevation changes are many thousands of feet, as in the Eastern Sierra. This phenomenon of drier, often much drier, leeward slopes and valleys, called the *rain shadow effect,* has two basic causes. First, a portion of the moisture moving upslope is dropped on that slope as rain and snow, leaving less available for the downslope. One often hears such colorful phrases as "clouds wrung completely dry" or "precious water robbed" by the rugged barrier of the Sierra Nevada. Contrary to such notions, however, much of the moisture passes over the crest and falls on the upslopes of mountains farther east, such as the Wasatch Range in Utah, 500 miles (805 km) away. Second, and more important, as descending air warms, clouds may evaporate and precipitation may decline rapidly. Observers in Owens or Carson Valley in winter can frequently observe clouds thinning or disappearing as they move down from the crest. In fact, as the air flowing down the eastern slope warms, it robs the region of some of its moisture.

The eastern slope of the Sierra and its adjacent valleys are indeed in the rain shadow of the Sierra Crest. For example, the difference in average annual precipitation with change in elevation on windward and leeward slopes is clearly evident along Interstate 80 as it crosses the Sierra from west to east. In Sacramento at sea level, annual precipitation is 18 inches (46 cm); at Auburn (1200 ft or 365 m) it rises to 34 inches (86 cm); at mile-high Blue Canyon, to 68 inches (173 cm); and between 7000 and 9000 feet (2135–2745 m) near Donner Summit it reaches 70–75 inches (178–191 cm). On the leeward eastern slope, averages drop to 28 inches (71 cm) at Truckee (6000 ft or 1830 m), 10 miles (16 km) from the summit; and to 7 inches (18 cm) at Reno (4500 ft or 1370 m), 30 miles (50 km) farther downslope. In the southern Sierra, the average precipitation drops from 30–40 inches (75–100 cm) along the 12,000-foot (3660-m) crest to 5 inches (13 cm) on the 4000-foot (1220-m) floor of Owens Valley 10 miles (16 km) east. Lesser rain shadow effects also occur when airflow descends several thousand feet down steep slopes, as in the Lake Tahoe Basin and in the Grizzly and Diamond mountains of northeastern California.

Elevation also affects air pressure. Air pressure decreases with increasing altitude because there is less atmosphere above. At 10,000 feet (3048 m) the pressure is about one-third less than at sea level; oxygen content is also one-third less. Less oxygen at higher elevations has little effect on weather events, but can significantly affect human health and comfort. Above 10,000 feet (3048 m) many people will suffer some form of *mountain sickness*—shortness of breath, fatigue, headache, nausea. Oxygen deficiency is the culprit. Individuals differ greatly in their response to less oxygen; some will feel symptoms at lower altitude, while nearly everyone is affected at 14,000 feet (4270 m). In the Sierra mountain sickness is seldom serious; most bodies acclimatize in a few days. People at high altitude also need to adjust their cooking to compensate for the fact that the boiling temperature of water decreases as air pressure decreases.

THE SEASONS

Figure 4.3 shows the principal directions of the flow of air masses that affect the Eastern Sierra's weather. Throughout the year airflows from the Pacific Ocean hold sway over the region, though polar and subtropical flows from continental regions may occur at any season. Continental polar or Arctic flow occurs infrequently in winter and spring; subtropical flow from Gulf regions takes place mainly in summer and early fall.

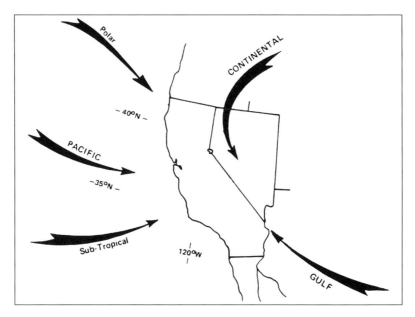

Fig. 4.3. Principal airflow patterns that determine Eastern Sierra weather. The arrows indicate the directions of air movement at levels between 10,000 and 20,000 feet (3–6 km) above sea level. Storms that approach California from the southwest bring warm, moist air from the Hawaiian region of the Pacific. Storms from the northwest bring cold, moist air from the Gulf of Alaska. (Drawing by Harold Klieforth)

WINTER

Winter usually comes to the Eastern Sierra in a series of steps in which air temperatures drop several degrees and the snow level lowers with successive storms. The transition from autumn to winter often is heralded by a windstorm during the first half of November, which strips most of the dried leaves from deciduous trees in the valleys. Snowfalls may occur in October, but it is generally mid-November before mountain snow begins to pack and beckon the skiers. At most weather stations 75 percent or more of the annual precipitation falls between November and April; it is concentrated between December and March, and January is usually the wettest month. Amounts of rain and snow vary greatly from year to year. This seasonal distribution and variable precipitation is typical of Mediterranean climates in the Northern Hemisphere.

Winter, of course, is the coldest period, but the length and severity of cold varies greatly from year to year. Sometimes continental polar air from Canada pours in, bringing a week or more of temperatures much colder than normal. Such periods occurred in January 1937, January 1949, December 1972, February 1989, and December 1990. However, even in the wettest and coldest winters, periods of clear, warm, sunny days often last a week or longer.

SPRING

Spring brings longer days and warmer temperatures, with melting snow at lower elevations, but firm snow and sunny weather for cross-country skiing in the high country. Depending upon the amount and timing of precipitation, there may be lavish displays of wildflowers in the valleys, with the bloom moving upslope as the season progresses. The most profuse spring wildflower displays occur if ample rains come in late fall and early winter. Leaves and blossoms generally appear on the valley floors in April, later at increasing elevations.

Spring is seldom a season of continuous balmy weather, particularly in the northern part of the region. Just as the long-awaited signs of spring appear—warmer air, melting snow, rising streams, blossoms, and new leaves—a blustery, winterlike storm may bring back a chill. Episodes of clouds and chilly winds may persist into mid-June, in spite of warm, sunny interludes. The following sequence frequently occurs: Warm weather from mid-March into early April brings out vulnerable blossoms on fruit trees such as peach and apricot; then later in April temperatures drop well below freezing and no fruit matures. Commercial orchards are limited to a few well-protected areas. Ripe fruit is seldom picked from home-grown peach or apricot trees more than one year in five. From Owens Valley north, killing frosts in the valleys may occur anytime in May and during the first part of June. A common saying in the Eastern Sierra is that spring does not last long enough or begin soon enough; others say that there are really only three seasons: winter, summer, and fall.

SUMMER

The transition from spring to summer may be sudden. Summer in the valleys usually begins about the third week of June. Warm and dry westerly and northwesterly airflow from the Pacific dominates throughout the summer, the primary vacation time in the mountains. Campers and hikers cannot help noticing how air temperature decreases with altitude, a key feature of Eastern Sierra climate. When the lower valleys are in the 90s or low 100s°F (32–38°C), it is often delightfully warm, but not hot, at mid-altitudes and chilly on the summits. July and August temperatures tend to be remarkably even from day to day, with little variation in highs and lows. However, daily ranges in temperature are large, often on the order of 40°F (22°C) or more,

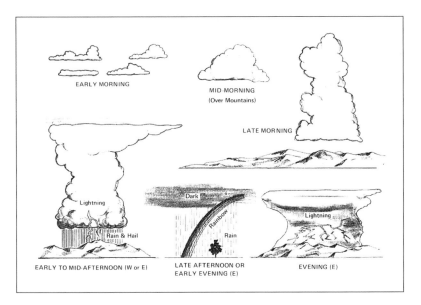

Fig. 4.4. Development of summer afternoon thunderstorms. In summer the growth of towering cumulus clouds by midday often portends an afternoon of thunder, lightning, rain, and hail, beginning with patchy, turreted clouds in the early morning. As the sun heats the mountain slopes and the warming air rises, cumulus clouds may develop over the peaks, becoming towering cumulus by late morning as the clouds continue to rise. The clouds may develop flat cirrus tops shaped like anvils, and if droplets in the cloud tops turn to ice crystals, streamers of ice clouds will stretch in the direction of the winds at those altitudes. By mid-afternoon, lightning may flash, while local showers of rain and hail pelt the peaks and ridges. If downdrafts of colder air, displaced by the rising warm air, predominate, the thunderstorm may lose its source of energy and cease by late afternoon. Or, as the sky brightens in the west, over the mountains, clouds may thicken over the leeward valleys to the east, and by early evening a brief thundershower may dampen portions of the valley floors. By early evening the sky may clear and stars appear; as the storm moves eastward over Nevada, lightning may continue to flash in the mountains to the east. (Drawing by Harold Klieforth)

when clear skies permit radiational cooling at night. Frost at dawn in the high country is not uncommon.

As the summer sun heats the mountain slopes, updrafts and thermals form, leading—if there is sufficient moisture in the air—to vertical cumulus cloud development over the peaks (fig. 4.4). When towering cumulus clouds *glaciate* (when the droplets in cloud tops turn to ice crystals), rain, hail, lightning, and thunder may hit the Sierra, reaching their maximum intensity in late afternoon. In the leeward valleys brief thundershowers may occur in the early evening. Clearing occurs near sunset, and night skies are mostly clear. From year to year this daily pattern varies greatly; some summers are almost free of afternoon clouds and thunderstorms.

In the Eastern Sierra one distinctive element of summer climate at all elevations is the abundant sunshine. In the lowland valleys 80–90 percent of the days throughout July and August may be sunny. Percentages will be less at higher elevation, but are often above 70 percent. Few mountain areas in the world have so much sunshine. It was the ever present sun, shining on the light-colored granite, that led John Muir to name the Sierra *The Range of Light.* Although usually regarded as desirable, sunshine at high elevation exposes people to injurious ultraviolet rays. Visitors to the high country need to protect their skin and eyes.

Summer mountain weather is not always pleasant and sunny. The principal exception to the prevalent Pacific pattern is airflow from the south—either from the subtropical Pacific off the coast of Baja California or from the gulfs of California and Mexico (fig. 4.3). This flow, when dominant, brings moist, unstable air, cloud buildup, and thunderstorms. The main hazards of such weather are lightning strikes, local flash floods, and lightning-caused wildfires. Southern airflow is frequent in some years—such as 1955, 1956, 1973, 1983, and especially 1984—and absent or nearly so in other years, such as 1993, 1994, and 1995. Tropical storm sequences can engulf the Sierra in clouds for several days and nights or even a week; heavy rains accompanied by thunderstorms and cool, wet weather cause streams to rise and increase backpackers' risk of *hypothermia* (subnormal body temperature that can be fatal).

The saying that in the Sierra "it never rains at night in the summer" is simply not true. Though not as common in the Eastern Sierra as in many other mountain ranges, night rain does fall; it can be heavy and cold, and snow can fall in the high country even in July and August. Campers should be prepared for these possibilities. However, one adage about weather *is* valid: "Do not use the adverbs *never* or *always. Never* use either one, it is *always* a mistake."

AUTUMN

A gradual transition from summer to autumn usually occurs in September. Autumn generally extends from mid-September to early November, sometimes longer at the lower elevations. The weather may be consistently sunny and calm for weeks. Late-season tropical thunderstorms may still occur, and early-season Pacific storms may bring light snow to the high country, but these events are generally short interruptions in the usual succession of calm, bright sunny days and clear starry nights. Days are mild, even at high elevation, and nights are cool to crisp. In September frost is common in the high country; in October freezing temperatures may occur everywhere.

Autumn is many people's favorite season in the Eastern Sierra. The weather is consistently delightful, and there are fewer people along mountain trails; fall colors are an additional bonus. California is not blessed with vivid fall foliage, but the Eastern Sierra offers one of the better displays in the state. At higher elevations, aspen and willow turn golden usually beginning in late September. The change in color moves downslope, culminating in the lower valleys with cottonwoods and other deciduous trees, and often lasts well into November. Another splash of color comes from the yellow flowers of rabbitbrush, prominent along valley roadsides.

From August through October hurricanes and tropical storms that form off the southwest coast of Mexico can send tropical moisture into the Sierra. This can cause heavy rain, even snow, at high elevations. One such storm at the end of September 1982 caused record high flows on Bishop Creek, damaging hydroelectric power installations. These formidable storms are most likely to occur in September or early October. They are infrequent but are somewhat predictable several days in advance.

In late fall, after the first major Pacific snowstorms in mid- or late November, a period of mild, clear weather may persist well into December at all elevations in the Eastern Sierra. This is the true "Indian summer," a time during which the Pilgrims celebrated the first Thanksgiving in late November in New England.

A PATTERN TO THE SEASONS

Most years there is a predictable pattern to the Sierra seasons: Pacific storms interspersed with clear periods in winter; short blustery, changeable periods in spring; dry, sunny summers with occasional thunderstorms; and long pleasant autumns, before the cycle begins again. But everyone should be aware of the important exceptions to this pattern and be prepared for the uncommon, but not unprecedented, events.

WEATHER ELEMENTS

To understand more fully the complex subject of Eastern Sierra weather and climate, it is helpful to look in more detail at the chief elements of weather—temperature, precipitation, and wind.

TEMPERATURE

Throughout the entire region at all elevations, the month of July has the highest average temperature and January the lowest, though the hottest and coldest single days may occur on either side of those two months. Keep in mind that at similar elevations temperatures increase from north to south.

For example, from north to south, Susanville, Reno, and Bishop are all in lowland valleys within elevations of 4100–4500 feet (1250–1370 m). Average monthly temperatures throughout the year are about 2°F (1°C) higher in Reno than in Susanville, and about 7°F (3.5°C) higher in Bishop than in Reno. Vegetation zones occur at higher elevations from north to south; treeline is about 1000–1500 feet (305–455 m) higher in the Mount Whitney area than in the Tahoe Basin. The *snow line* (above which precipitation falls as snow rather than rain) is, on average, progressively higher from north to south, often as much as 3000 feet (915 m) higher.

This temperature change occurs largely because the sun's path is progressively higher above the horizon as latitude changes from 40°N to 35°. The higher the sun above the horizon, the greater the total amount of solar radiation reaching Earth, and the warmer the temperature. Other factors—exposure to wind, location on ridge tops or in canyon bottoms, degree of cloudiness—can alter this basic relationship, but the length of the Sierra, extending over 5 degrees of latitude, is enough to cause a rising temperature gradient from north to south.

Temperature also changes with altitude; it decreases as elevation increases. The average *lapse rate* (generally, the decrease of temperature with altitude) of approximately 3.6°F per 1000 feet (6.5°C per 1000 m) accounts for the very large temperature differences among valley, foothill, mid-mountain, and crest areas. Moreover, at midday in summer, as the sun heats the slopes, the warming air rises. With this strong thermal activity, the lapse rate on the slopes approaches the *adiabatic value* (the rate at which air cools as it rises) of about 5.4°F per 1000 feet (10°C per 1000 m). For example, when it is 90°F in Bishop at 4000 feet (32°C at 1220 m), it would likely be 68–58°F at 10,000 feet (20–14°C at 3048 m), and between 54°F and 36°F at 14,000 feet (12–2°C at 4270 m). These approximations are subject to numerous local influences, but they are useful rules of thumb. The large gradient in average air temperatures contributes to a long-lasting snowmelt period and persistent soil moisture, so that wildflowers in the high country may last long into the summer.

Why does temperature change with altitude? Two physical processes are responsible for this phenomenon. Short-wave radiation from the sun passes through Earth's atmosphere with little direct heating effect on the air. Upon striking Earth's surface, short-wave energy is absorbed and converted to long-wave radiation, which moves upward. Thus the atmosphere is heated from below, primarily at or near sea level. Thus, the farther from the source of heat, the colder the air temperature. Moreover, the atmosphere thins with increasing altitude, and its ability to absorb and retain heat is reduced.

Accordingly, one might expect to find the highest summer temperatures at the lowest and southernmost locations. This is indeed the case, with the highest recorded maximums (about 115°F or 46°C) in a belt from southern Owens Valley to Mojave (3800–2800 ft or 1160–855 m). However, the coldest temperatures in winter do not occur at the highest elevations. Cold air is heavier than warm air and drains downslope; it accumulates in valleys, especially in a broad-floored valley that narrows and is constricted at its lower end, bottling up and collecting cold air drainage. Boca (5600 ft or 1705 m), near the junction of the Little Truckee and Truckee rivers, meets these criteria and has recorded the coldest readings ever in California: −45°F (−43°C) on January 20, 1937, and −43°F (−42°C) on February 7, 1989. Such extreme lows can occur when frigid continental air, originating far to the north in Canada or Alaska, is trapped in a constricted valley such as Boca. On clear nights radiation upward from the valley floor enhances the cooling process.

At Lake Tahoe, by far the most popular mid-mountain area in the Eastern Sierra, temperatures in all seasons are mild for its 6230-foot (1900-m) elevation. Major causes are the high percentage of winter sunshine, relatively mild Pacific air entering the area during much of the year, and the moderating influence of the lake itself. Skiers in the Tahoe Basin and elsewhere in the Eastern Sierra know that daytime temperatures are significantly higher than those at comparable locations farther inland, as in the Wasatch Range and the Rocky Mountains. Average maximum and minimum temperatures at Tahoe City, which has a long weather record, are 38° and 19°F (3° and −7°C) in January, 78° and 44°F (26° and 7°C) in July. The south shore of the lake is similar; Glenbrook on the Nevada east shore is slightly warmer, probably because of downslope winds and a higher percentage of sunshine. Downstream 15 miles (24 km) along the Truckee River and 400 feet (120 m) lower than the lake, the town of Truckee is colder in winter and warmer in summer. In autumn, spring, and summer, Truckee attracts some national notoriety by having occasionally the lowest temperature of the hundreds of stations regularly reporting on a National Weather Service network—again the result of cold air drainage and radiation cooling. Just a few miles downriver from Truckee lies Boca, the site of record lows.

PRECIPITATION

Most precipitation in the Eastern Sierra results from two airflows. The more important is due to the passage of cyclones with associated fronts from the Pacific during fall, winter, and spring. The other comes from the flow of moist tropical air from the southeast and southwest in summer and early fall. *Cyclones* are centers of low pressure generally moving from west to east into which air moves in a counterclockwise spiral. Cyclones are associated with low pressure, rising air, cloud formation, and precipitation. *Anticyclones* (clockwise whirls of high pressure) are associated with sinking air, few clouds, and fair weather. *Fronts* are boundaries between converging air masses from different regions; those affecting the Eastern Sierra usually form over the Pacific between contrasting cold marine polar air from the Gulf of Alaska and warmer, moist subtropical air from lower latitudes. Both *cold fronts* (cold air moving underneath and lifting warmer air) and *warm fronts* (warm air sliding up over colder air) pass over the region.

In summer the *Pacific Anticyclone* (a very large area of high pressure, perhaps 1000 miles or 1600 km across) lies just west of the California coast. It brings an onshore flow of cool marine air, stratus clouds, and fog to the coast, and clear, dry air to the Sierra Nevada. During fall, winter, and most of spring this *Pacific High* usually moves south, opening the door for cyclonic storms to enter the Eastern Sierra region. The frequency and intensity of storms vary greatly from year to year. If the Pacific High remains well south of its summer position, the year is wet; if not, the year is dry.

Three general rules are useful in understanding annual averages and spatial distribution of precipitation in California. All of them apply to the Eastern Sierra.

1. Amounts of precipitation generally decrease from north to south. This results from a declining frequency of cyclonic storms. Not only does the total number of storms decrease southward, but in the north storms begin earlier in the fall and cease later in the spring.

2. Amounts increase with higher elevation (the orographic effect).

3. Amounts decrease on the leeward side of major mountain barriers (the rain shadow effect).

The orientation of any particular location will modify these generalities. For example, a slope facing southwest into the prevailing storm winds will be wetter than other areas in the same general location.

Using these rules, you could expect the northern Sierra would be wetter than the southern at comparable elevations, and that the greatest precipitation would fall at or near the crest. Such is the case. From Lake Tahoe north, annual totals along the crest are close to 70 inches (178 cm). Southward, they gradually decrease to as low as 10–20 inches (25–51 cm) near Tehachapi Pass. On the valley floors, at comparable elevations, from north to south, Susanville averages 14 inches (35.6 cm), Reno 7 (17.8 cm), Bishop 5.5 (14 cm).

Near Lake Tahoe there are two rain shadows. From the Sierra Crest west of the lake, in a 3000-foot (915-m) descent to Tahoe City (6230 ft or 1900 m) on the west shore, average precipitation declines from about 70 inches to 32 inches (178 cm to 81.3 cm) and drops even more to 18 inches (45.7 cm) at Glenbrook on the east shore. The Carson Range, which borders the eastern shore of Lake Tahoe, causes the second rain shadow. As the air rises up the western slope of the Carson Range, precipitation rises to 53 inches (135 cm) at Tahoe Meadows (8540 ft or 2600 m) and declines to 10 inches (25 cm) at Carson City, Nevada (4700 ft or 1435 m), at the eastern base of the range.

In high mountain regions the snowpack begins to accumulate in the fall of one calendar year and builds to a maximum in late winter or spring of the next calendar year. Consequently, it is desirable to calculate the precipitation year from July 1 to June 30 in order to include the continuous accumulation of snow in one year's record. The very big snow year of 1983 began on July 1, 1982, and ended on June 30, 1983. In California's winter-wet, summer-dry climatic regime, the same timing is used to encompass the one continuous rainy season, from November to April, in one weather year.

For complex reasons, difficult to explain and predict, California has huge fluctuations in yearly amounts of precipitation. The variation is most extreme in valley locations with low average totals. At Bishop (4150 ft or 1265 m) the wettest year (1969) brought 17.28 inches (43.9 cm), while the driest year (1960) brought only 1.68 inches (4.3 cm)—a range of 300–30 percent of average. The range is less at higher elevations with higher annual averages, but still varies from about 200 percent to 40 percent of normal. At Tahoe Meadows (8540 ft or 2600 m) precipitation varies from 22 inches to 97 inches (55.9 cm to 246.4 cm), averaging 53 inches (134.6 cm). A series of dry years, such as 1987–92, cumulates in a severe water deficiency, with much stress on plants, animals, and humans. Much effort has gone into trying to predict wet and dry years: Are there cycles of so many years? Be wary of enthusiastic claims about cycles, regardless of source; they are difficult to prove with existing data.

In a twenty-five-year study of precipitation in the Lake Tahoe–Truckee River watershed, researchers found that, on average, near 5280 feet (1610 m) elevation about 25 percent of the annual precipitation falls as snow. At 6500 feet (1980 m) about half falls as snow, while at 8540 feet (2600 m) snowfall accounts for 85–90 percent of the annual total. The percentages of snow at these elevations probably increase gradually to the north and decrease to the south. However, these percentages change with different storms. Depending upon the direction from which cyclonic storms come into the Eastern Sierra, snow may fall as low as 2800 feet (855 m) at the town of Mojave

Plate 1.   Carson Valley. (Photo by Larry Prosor)

The Eastern Sierra is a world apart

from lands to the west. In contrast to its 60-mile-wide western slope that sweeps leisurely upward from the Central Valley, the Sierra Nevada's eastern slope drops abruptly. Carson Range peaks drop 4000–5000 feet to Carson Valley's irrigated ranch lands. Mount Whitney in the southern Sierra drops a precipitous 2 miles to the floor of Owens Valley.

Here where the eastern slope of the Sierra merges with the western edge of the Great Basin,

Plate 2. Mount Williamson (14,375 feet), Owens Valley. (Photo by Ed Cooper)

desert valleys of long summers and snow-spangled mountains of long winters lie side by side—a land of immense views and overarching sky. Nowhere along Highway 395, which hugs the base of the Sierra, can you escape the mountain crest looming thousands of feet above. And nowhere in the mountains can you escape looking down on arid valleys to the east.

Plate 3. *Above:* Lake Tahoe. (Photo by Ed Cooper)
Left: Walley's Hot Springs, Carson Valley. (Photo by Larry Prosor)

Faulting is key to understanding

the Eastern Sierra, its topography, its weather, and its arid nature. Faulting has uplifted and tilted the Sierra Nevada and down-dropped the valleys. The Tahoe Basin dropped down along faults on either side, while the Sierra block rose on the west and the Carson Range block rose on the east. The lake bottom is lower than the adjacent Carson Valley, which is another down-dropped block. Truckee Meadows, Mono Basin, Long Valley, and the Owens, Bridgeport, and Washoe valleys are all down-faulted valleys. Hot springs, seeps, and fault scarps along the base of the Sierra mark fault zones.

Plate 4. *Top:* Olancha Peak (12,123 feet), southern Sierra. (Photo by Ed Cooper) *Above:* The Sierra Wave, a distinctive cloud pattern, above Mono Lake. (Photo by David Hamren) *Right:* Thunderclouds above Truckee Meadows. (Photo by Harold Klieforth)

The Eastern Sierra's arid nature and weather patterns

are shaped largely by the towering, massive Sierra. As the prevailing moisture-carrying westerly winds rise to cross the Sierra, the air cools and drops much of its moisture on the western slope. Because the region lies in the Sierra's rain shadow, far less rain and snow fall on the eastern slope and even less on the adjacent eastern valleys.

Plate 5.   *Top:* Sagebrush, Carson Valley. (Photo by Larry Prosor) *Above:* Sagebrush near Reno. (Photo by Harold Klieforth)

Looking eastward is another key

to understanding this complex region—eastward toward the vast arid region known as the Great Basin, a gray-green "sagebrush ocean" that stretches across Nevada and half of Utah.

Plate 6.   *Top:* Spring flowers, Owens (dry) Lake dust in the distance. (Photo by Ed Cooper) *Above:* Folded ancient rocks, Convict Canyon. (Photo by Susan Moyer) *Right:* Clark's Nutcracker. (Photo by Diana F. Tomback)

At first acquaintance

the Eastern Sierra strikes some people as bleak and forbidding, but upon longer acquaintance you—like so many others—may find it dramatic, peaceful, and teeming with life. Rocks, sky, space, colors, hidden oases, and, as you will see on the following pages, brilliant insects and unexpected flowers.

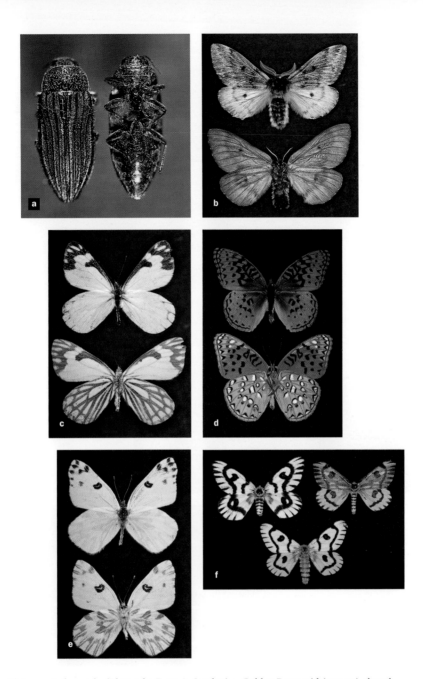

Plate 7.   Arthropods. (Photos by Evan A. Sugden) a. Golden Buprestid (19 mm), dorsal and ventral aspect. b. Pandora Moth (W 75 mm), dorsal and ventral aspect. c. Pine White (W 33 mm), dorsal and ventral aspect. d. Apache Fritillary (W 54 mm), dorsal and ventral aspect. e. Becker's White (W 34 mm), dorsal and ventral aspect. f. Buck Moths: *from top left*, Hera Moth (W 65 mm), Common Sheep Moth (W 65 mm); *bottom*, Nuttall's Sheep Moth (W 65 mm).

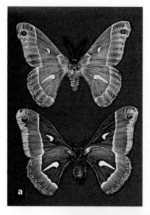

Plate 8. Arthropods. (Photos by Evan A. Sugden)
a. Glover's Silk Moth (W 110 mm), dorsal and ventral aspect. b. Silver-aproned Bee Fly (12 mm). c. Mimics and Look-alikes: *from left*, Yellow-faced Bumble Bee, queen (20 mm), worker (16 mm); Sierran Bumble Bee Moth (W 32 mm); Sacken's Robber Fly (18 mm). d. Mimics and Look-alikes: *from left*, Red-skirted Bumble Bee, queen (19 mm), worker (11 mm); Fire-tipped Hover Fly (13 mm); Fernald's Robber Fly (17 mm). e. Mimics and Look-alikes: *from top left*, Golden Paper Wasp (17 mm), Locust Clearwing (W 33 mm), Aerial Yellowjacket (16 mm), Black-mouthed Soldier Fly (13 mm), Graceful Carder Bee (13 mm), Zebra Hover Fly (14 mm).

Plate 9.   Alpine plants. (Photos by Ann Howald) a. Alpine Columbine, blue form, *Aquilegia pubescens*. b. Sierra Primrose, *Primula suffrutescens*. c. Alpine Laurel, *Kalmia polifolia*. d. Creeping Penstemon, *Penstemon davidsonii*. e. Alpine Gold, *Hulsea algida*. (Photo by J. R. Haller) f. Alpine or Dwarf Lupine, *Lupinus lepidus*. g. Sky Pilot, *Polemonium eximium*. (J. R. Haller) h. Rock Fringe, *Epilobium obcordatum*.

Plate 10.   Montane Forest and Mountain Meadows plants. (Photos by Ann Howald)
a. Alpine Buttercup, *Ranunculus eschscholtzii*. b. Camas Lily, *Camassia quamash*. (Photo
by J. R. Haller) c. Little Elephant's Heads, *Pedicularis attolens*. d. Alpine Gentian, *Gentiana newberryi*. e. Alpine Shooting Star, *Dodecatheon alpinum*. f. Snowplant, *Sarcodes sanguinea*. g. Sierra Penstemon, *Penstemon heterodoxus*. h. Pinedrops, *Pterospora andromedea*.

Plate 11.   Mountain Meadows plants. (Photos by Ann Howald) a. Corn Lily, *Veratrum californicum*. b. Common Monkeyflower, *Mimulus guttatus*. c. Meadow Lupine, *Lupinus polyphyllus*. d. Meadow Paintbrush, *Castilleja miniata*. e. Fireweed, *Epilobium angustifolium*. f. Arrowleaf Butterweed, *Senecio triangularis*. g. Sierra Gentian, *Gentianopsis holopetala*. h. Cow Parsnip, *Heracleum lanatum*.

Plate 12.   Shrubs of Montane Chaparral and Sagebrush Scrub. (Photos by Ann Howald)
a. Greenleaf Manzanita, *Arctostaphylos patula*. b. Tobacco Brush, *Ceanothus velutinus*.
(Photo by B. Orr) c. Bitter Cherry, *Prunus emarginata*. d. Antelope Bitterbrush, *Purshia tridentata*. e. Desert Peach, *Prunus andersonii*. f. Spiny Hop-sage, *Grayia spinosa*. g. Rubber Rabbitbrush, *Chrysothamnus nauseosus*. h. Cotton Thorn, *Tetradymia axillaris*.

Plate 13.   Sagebrush Scrub and Pinyon-Juniper Woodland wildflowers. (Photos by
Ann Howald) a. Prickly Poppy, *Argemone munita*. b. Scarlet Gilia, *Ipomopsis aggregata*.
c. Mules Ears, *Wyethia mollis*. d. Sulphur Buckwheat, *Eriogonum umbellatum*. (Photo
by J. R. Haller) e. Fragrant Evening Primrose, *Oenothera caespitosa*. f. Blazing Star,
Mentzelia laevicaulis. g. Rose Penstemon, *Penstemon floridus*.

Plate 14. Plants of Riparian Woodlands, Alkali Meadow, and Desert Marshlands. (Photos by Ann Howald) a. Copper Birch, *Betula occidentalis.* b. Quaking Aspen, *Populus tremuloides.* c. Basin or Showy Goldenrod, *Solidago spectabilis.* d. Interior Wild Rose, *Rosa woodsii.* e. Alkali Cinquefoil, *Potentilla gracilis.* f. Yerba Mansa, *Anemopsis californica.* g. Crisped Thelypodium, *Thelypodium crispum.* h. Annual Indian Paintbrush, *Castilleja minor.*

Plate 15.   Plants of Joshua Tree Woodland
and Creosote Bush Scrub. (Photos by
Ann Howald) a. Showy or Interior Gold-
enbush, *Ericameria linearifolia*. (Photo by
J. R. Haller) b. Desert or Apricot Mallow,
Sphaeralcea ambigua. c. Bladder Sage,
Salazaria mexicana. d. Bush Sunflower,
Encelia virginensis. e. Cheesebush, Burro-
brush, *Hymenoclea salsola*. f. Blue Sage,
Salvia dorrii. (J. R. Haller) g. Creosote
Bush, *Larrea tridentata*. h. Beavertail Cac-
tus, *Opuntia basilaris*.

or rain may fall as high as 10,000 feet (3048 m). Polar flow brings low snow lines; subtropical flow (often called Hawaiian or "Pineapple Express" storms) brings high snow lines; central Pacific flow is intermediate. Throughout any one season all three types of flow may occur.

There is a common belief that rain on snow causes significant snowmelt, which aggravates flooding during warm storms. This question has been much studied, especially in the Donner Summit area, where nearly every winter brings rain-on-snow episodes. Results indicate that rain may melt and wash away a shallow snowpack (2 ft or 60 cm, or less) but passes *through* a deeper snowpack—especially one containing 10 inches (25.4 cm) or more of water in frozen form—and runs off with very little snowmelt. The overwhelming source of water in warm storms in most cases is the falling rain itself, not snowmelt.

Annual snowfall totals near the crest of the Eastern Sierra, particularly from Mammoth Mountain north, may reach 25–35 feet (7.7–10.8 m). Wet years may bring more than 50 feet (15 m). Maximum accumulations generally occur immediately east of the crest. Precipitation from clouds may be heaviest there, and cloud momentum does not stop at the summit—but another factor enters in. Prevailing westerly winds pick up newly fallen, loose snow, carry it across the summit, and deposit it on the eastern slopes, particularly in *cirques,* the semicircular depressions carved by glaciers at the heads of canyons. This becomes a self-perpetuating process. Glaciers originate from snow accumulation at canyon heads and erode bowl-like cirques, which then serve as even more efficient catchment basins for falling and windblown snow. The cirques then become favorable locations for the present active glaciers (all of the fifty to sixty small live glaciers in the Eastern Sierra are in north-facing or northeast-facing cirques) and for possible rebirth of other glaciers if average snowfall increases and/or summer temperatures become colder.

The heavy snowfall, varied relief, and long snow season have led to the development of major ski areas, such as Heavenly Valley, Squaw Valley, Alpine Meadows, and Mount Rose in the Tahoe area and Mammoth Mountain at the head of the Owens River. In lean years however, these resorts generate man-made snow, especially if the beginning of the natural snow season is delayed.

The physical setting of Mammoth Mountain is important for an understanding of its relatively heavy snowfall. Its summit (11,053 ft or 3369 m) rises directly above two low passes, Minaret Summit (9175 ft or 2796 m) on the north and Mammoth Pass (9300 ft or 2835 m) on the south. Extending southwest from the base of the mountain is the deep, broad canyon of the San

Joaquin River's Middle Fork. Incoming southwest storm winds are channeled up the canyon, unimpeded by topographic barriers, and unload their moisture on Mammoth Mountain, making it an island of greater snowfall than higher areas to the north and south. The heavy snowfall and high altitude produce a long ski season, often lasting from November to early July. The low passes also provide corridors for wind and moisture to penetrate the extensive tableland that lies east of Mammoth Mountain. Travelers along Highway 395 can often see cloud streamers pouring through these gaps in the crest; snowfall from these clouds accounts for the large Jeffrey Pine forest east of the highway.

The direct relationship between precipitation and stream runoff is obvious, yet rivers react on different time scales than meteorological events. Eastern Sierra rivers, even after a few years of much below normal precipitation, continue to flow, though at a severely reduced rate. The hydrologic response to rain and snowfall lags, as water percolates slowly through the soil and rock strata before eventually reaching springs and creeks. In spring and early summer streamflow increases as a result of snowmelt and runoff; after a winter of deep mountain snows the increase may be very pronounced. The most damaging and dangerous conditions spring up quickly, after prolonged periods of warm winter rains or after intense and heavy summer thunderstorms. These variations in streamflow have profound effects on domestic and agricultural water supplies, and on recreational activities such as fishing, rafting, and hiking.

THE PREVAILING WESTERLIES AND OTHER WINDS

Wind is air in motion; it results from the unequal heating of Earth's surface, the rotation of the Earth, and the complex interaction of air masses. California lies near the southern edge of the westerly wind belt, a region between 30° and 60°N latitude. In both winter and summer the dominant winds are the prevailing westerlies (fig. 4.3). The rotation of the Earth sets the direction of these prevailing winds. As they blow across the Pacific Ocean, they pick up moisture and carry it inland.

In winter they accompany a continuing series of storms from the Pacific across California, over the Sierra Nevada, and farther east across the Great Basin. During many of these winter storms, temperatures are relatively moderate, because the winds are bearing warm, moist maritime air. During winter the storm belt may shift southward, bringing more Pacific storms to the state. Or if the storm-bearing winds are deflected northward by the Pacific High, clear, sunny days will occur over much of California and western Nevada.

In addition to the prevailing westerlies, winds blow in from other directions. At any season winds from the northwest may bring cold, marine polar air, and winds from the southwest may bring warm, moist subtropical air. Storms from the southwest generally produce the most rain and snow, because they bring air that originated over warmer water and has a high moisture content. In winter and spring winds from the northeast, although infrequent, may bring frigid continental air masses from Canada to the Eastern Sierra. These are the winds that cause subzero temperatures in the Rocky Mountains. By the time these air masses reach the Eastern Sierra, however, their temperatures may be somewhat moderated by the mountain ranges east of the Sierra. In summer and fall the principal exception to the westerlies is winds from the southwest and southeast, which bring subtropical moist, unstable air masses from the gulfs of Mexico and California and from the Pacific off the coast of Baja California (fig. 4.3).

From August through October hurricanes that form off the southwest coast of Mexico can influence Sierra weather. Moving northwestward over colder water, they lose hurricane wind velocity (75 mi or 120.6 km/hour) but retain high winds and much moisture. Occasionally these weakened, but still potent, systems move far enough north to get caught in a westwardly upper atmospheric flow and move northeastward into the Sierra, especially the southern portion. They can cause heavy rain, and even snow, at high elevations.

The strongest winds experienced in the Eastern Sierra are those that precede and accompany winter storms. As a Pacific cold front approaches from the west, the airflow pattern in the lee of the mountains is made visible in the forms of spectacular lenticular clouds and turbulent roll clouds as illustrated in fig. 4.5.

DAILY AND LOCAL WINDS

During most of the summer an afternoon breeze often arises in the Sierra about noon and arrives at the eastern valleys about two hours later. It is related to the afternoon sea breeze that makes San Francisco cool and foggy during much of July and August and to the gusty breeze that often penetrates inland as far as Sacramento or even the Sierra foothills. In the mountains the breeze diminishes by late afternoon; in the valleys, by early evening.

Mountains also generate their own breezes: A cycle of valley and mountain breezes occurs in canyons. In the morning, as the sun warms the ground and the air next to it, the warming air rises. Cooler air moves in to replace it and rises in turn. This upslope or *valley breeze* tapers off as the sun goes down. Later, as colder air from aloft sinks toward the now cooled ground, there is a downslope movement of air known as a *mountain breeze*. By midmorning this cold air starts to warm and rise, beginning the cycle over again.

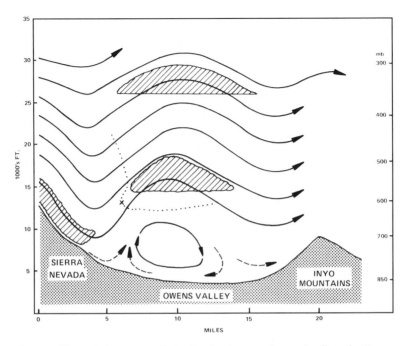

Fig. 4.5. Mountain lee waves, a distinctive cloud pattern, known locally as the Sierra Wave. The figure shows a typical pattern of airflow and cloud forms in a strong lee wave. Air flowing over the Sierra Crest plunges downward, then upward and downward again in a series of crests and troughs. The temperature and wind velocity of the air mass and the profile of the underlying downwind terrain determine the shape of the lee waves. Up- and downdrafts in a strong lee wave often rise and fall 2000 feet per minute, sometimes exceeding 4000 feet per minute. Sailplane pilots ride the updrafts to record heights. Harold Klieforth, co-author of this chapter, is still co-holder with Larry Edgar of the world altitude record for two-person sailplanes: 44,500 feet above sea level, while riding the Sierra Wave over Owens Valley. (Drawing by Harold Klieforth)

WEATHER WISE

The Sierra Nevada is one of the friendliest mountain ranges for people. Most summer and fall days are pleasantly warm and sunny; the nights, above freezing. Access is easy; dozens of roads lead to high-elevation trailheads. But just because the Sierra climate is usually mild, many climbers, backpackers, and hikers are notoriously unprepared for bad weather.

Conditions in the Sierra can be unexpectedly harsh and brutal. If you come prepared, the Sierra will reward you with safe journeys and days of happiness. But if you fail to respect one of the world's great mountain ranges — if you fail to prepare for its black moods, bitter weather, or high altitudes — it is unforgiving. You may pay with your life, as did four of the five climbers on Mount Ritter one Memorial Day weekend. Or the two boys, in separate

Fig. 4.6.   Locating a campsite. Due to the nightly movement of cold air draining from the upper slopes and settling in canyon bottoms, temperatures at dawn in a meadow or alongside a lake may be more than 20°F colder than at sites a few hundred feet upslope in the trees. (Drawing by Harold Klieforth)

parties, who died one Labor Day weekend less than 10 miles (16 km) from road ends, when an unexpected storm soaked their clothes and their sleeping bags.

With a basic understanding of temperature, precipitation, winds, and the seasons, campers can make smarter choices about what to wear and where to locate their campsites (fig. 4.6). Due to the nightly movement of cold air draining from the upper slopes and settling in canyon bottoms, temperatures at dawn in a meadow or alongside a lake may be more than 20°F (10°C) colder than those found a few hundred feet upslope in the trees. (Camping in meadows is undesirable anyway, because trampling easily damages fragile meadow plants.) The dry air throughout the Eastern Sierra causes unprotected skin to chap and crack easily, and dehydration is a hazard unless one has a good supply of water. In the high country the changeable, unpredictable weather that can bring rain and freezing temperatures at any time can also cause frostbite or hypothermia. Clothing that protects against rain and cold wind is a must.

DANGER SIGNS

How do you know when storms are approaching? First of all, in preparing for a trip, pay attention to radio and television weather reports and forecasts. Learn all you can from weather maps and satellite photographs of cloud patterns, and develop an understanding of the limits of weather forecasting. If a major storm is imminent, or if other weather-related phenomena threaten to make a planned trip hazardous or unpleasant, either postpone your expedition or go to a safer area.

Knowing the current weather and forecast conditions, you should then watch the clouds and their development, along with changes in prevailing winds. From autumn through spring the formation of lens-shaped wave clouds with increasing winds usually precedes a Pacific storm (fig. 4.5). In summer the growth of towering cumulus clouds by midday usually portends an afternoon of thunder, lightning, rain, and hail showers, especially on the peaks and ridges (fig. 4.4). If you feel strong winds and see many clouds from the southwest, expect heavy rain or snowfall. In all cases—whether on a long day hike or a backpacking trip of several days—consider the consequences of possible weather scenarios and be prepared with the food, clothing, and equipment necessary for survival.

SIGNS OF CLIMATE CHANGE

Flying over, driving through, or hiking the Eastern Sierra, one sees abundant evidence of very different climates in years past. The clouds, the runoff, and the vegetation all tell us about today's climate. The U-shaped valleys, cirques, and moraines conjure up a time when glaciers dominated the landscape. Former shorelines of Pyramid Lake, Walker Lake, and Mono Lake stand at much higher elevations than today's shorelines—evidence of a time when water covered much of the Great Basin. Besides the evidence of landforms, biologic evidence has much to tell us about past climates, for scientists are piecing together sequences of past climate variations. In the Black Rock Desert north of Pyramid Lake paleontologists are unearthing the skeletons of woolly mammoths; west of Pyramid palynologists have found a packrat midden 30,000 years old in which pollen and other vegetative remains indicate a climate with more moisture than at present. In Lake Tahoe dendrochronologists have determined that conifer stumps found below the present lake surface suggest a much drier period in the region as recently as 6000 years ago. In the White-Inyo Mountains east of Owens Valley ancient Bristlecone Pines provide a record of climatic variations over the past 9000 years. A one-hundred-year-long dry period about A.D. 1100 tells us we should expect such dry periods again.

FUTURE CLIMATE CHANGE: WHAT IF . . . ?

Global warming may bring significant changes locally in the Eastern Sierra and throughout the world, perhaps in our lifetime. In recent years there has been much concern that global warming is the likely consequence of the continuing increase in carbon dioxide caused by the combustion of oil and coal. Some scientists have predicted a higher snow level in the Sierra as the storm tracks shift northward, which would reduce summer runoff and lower water supplies. Some climatologists point to recent increases in average monthly air temperatures at many locations as evidence that global warming is already proceeding rapidly. But there is also persuasive evidence that these changes are occurring mainly in urban areas, where the increased warmth results from the expanse of concrete and asphalt, and from traffic—the so-called urban heat island. Meanwhile, climatic observations from rural areas show little increase in warmth.

A few scientists who have examined the effects of recent volcanic eruptions, such as El Chichón in Mexico and Mount Pinatubo in the Philippines, have discussed a possible contrary scenario of global cooling; both eruptions cooled the stratosphere and reduced solar heating of the Earth's surface. Studies of ice cores from the Greenland Ice Cap are showing that climate change often occurs in a relatively short time rather than gradually, as formerly believed. Since climate change is influenced by many complex interactions, what happens even a few years from now may not be predictable. What is certain is that we will surely continue to be surprised and amazed by future observations and discoveries of the atmosphere and by the geophysical events, chemical processes, and biological phenomena that affect its behavior.

GIFTS

The absence of low clouds, the altitude, and the distance from centers of urban air pollution give the Eastern Sierra exceptionally good visibility. On fair-weather days it is not unusual for people to be able to see ridges and peaks 70–100 miles (115–160 km) away. One of the great joys of mountain camping is to lie in one's sleeping bag and look at the stars. In the clear air, their abundance and brightness are amazing—another gift, among so many others, that the Eastern Sierra offers. The Sierra is indeed remarkable for its seemingly endless sunshine, even in winter and even in the high country, its pronounced seasonal changes, and the enormous variations in weather patterns from year to year.

RECOMMENDED READING

Berry, William B., and Chapman Wentworth. 1991. *Lost Sierra: Gold, Ghosts and Skis.* Soda Springs, Calif.: Western America SkiSport Museum.

Schaefer, Vincent J., and John A. Day. 1981. *A Field Guide to the Atmosphere.* New York: Houghton Mifflin.

Stewart, George R. 1941. *Storm.* Lincoln: University of Nebraska Press.

CHAPTER FIVE

PLANT COMMUNITIES

ANN HOWALD

Eastern Sierra plant life echoes the region's topographic and climatic extremes. Compact alpine cushion plants scarcely an inch tall inhabit rocky crags that tower over spirelike hemlock forests. Rich flower-filled meadows lie below dry slopes of prickly chaparral. To the east, ancient Bristlecone Pines that have persisted for millennia crown peaks of the White Mountains, while at the base of the mountains, in the Owens Valley, dusty little saltbush shrubs eke out a brief existence on the shore of an alkali playa. To compound the diversity, plants common to the Sierra Nevada, the Great Basin, and the lower Mojave Desert all intermingle in the Eastern Sierra.

The rain shadow cast by the two-mile-high Sierra Crest exerts a profound influence on plant life to the east. Storms moving inland from the Pacific drop most of their moisture on the western slope, creating a desert climate east of the crest. The vegetation is sparser and lower-growing than that of the western slope; and it is greatly influenced by an influx of desert species. Some plant communities that are widespread on the western slope, such as Montane Forest and Montane Meadow, are far less extensive on the eastern slope, and moisture-loving plants are less abundant. These contrasts are more pronounced south of Lake Tahoe. North of Tahoe the crest is lower, allowing more rain and snow to pass eastward; thus the vegetation of the west slope and that of the east slope are more alike. Annual precipitation is greater to the north, and also temperatures become cooler and evaporation rates lower, resulting in moister conditions that enhance plant growth.

Within the Eastern Sierra, climate, rainfall, and soils vary tremendously. In April, when winter snows lie 15 feet (4.6 m) deep in the high mountain forests, the first spring flowers are opening in desert valleys just a few miles to the east. Soils range from rocky and nutrient-poor to loamy and nutrient-rich, from acidic beneath conifer forests to alkaline in desert basins. The growing season for plants ranges from six weeks in alpine sites to six months or more in the lowlands. Yearly precipitation (rain and snow) commonly exceeds 30 inches (76 cm) in the wettest sites, yet rarely exceeds 3 inches (7.6 cm) in the driest. Low humidity and clear skies allow temperature fluctuations of 40°F (4°C) or more in a single day. The lowest temperatures for the region may reach −10°F (−23°C); the highest may exceed 110°F (43°C).

This complex and variable set of environmental conditions creates many different habitats for plants. As every home gardener knows, a few plants will grow almost anywhere, but most will grow only within certain limits of sunlight, moisture, temperature, and soil conditions. For example, Creosote Bush can tolerate the extreme heat and aridity of the low desert but will not survive in a damp meadow. Willows, even those found in the desert, are restricted to sites with moist soils. Hemlocks and firs can survive the coldest mountain winters, but cannot tolerate desert heat.

Competition with other plants and seed dispersal ability are two additional factors that influence where plants grow. Even species with very broad geographic ranges may be limited by competition. For example, Fireweed, a tall pink-flowered perennial herb, tolerates a wide range of arctic and alpine conditions. However, it is most abundant in recently burned or otherwise disturbed sites where few other plants grow, suggesting that it is a poor competitor. Its wide distribution throughout the northern latitudes is probably facilitated by its tiny, fluffy, wind-dispersed seeds.

HOW DO PLANTS COPE?

East of the Sierra, rain may not come for months on end. In both alpine and desert climates, plants have to cope with extremes—extreme cold and wind in the high mountains, extreme heat and cold in the desert. When even home gardeners, making use of fertilizers and drip irrigation systems, have a hard time growing plants, how do native plants manage to cope with below-freezing temperatures, high winds, months of no rain at all, or blazing white alkali soils? Birds can fly away when winter approaches. Deer and mountain sheep can migrate upslope and downslope with the seasons, to more favorable conditions. But plants must remain rooted, coping with whatever the winds and the clouds bring. Some of their coping mechanisms are among the wonders of the natural world.

For example, Alpine Everlasting escapes high winds and low temperatures by producing tiny leaves in a dense mat; only the flowering stalks stand above the ground surface. This growth form allows the plant to resist desiccation from wind and to benefit from the summer warmth reradiating from the rocky barrens where it grows. In autumn the aboveground parts die back.

The buds that will resprout in spring lie underground, avoiding the worst winter cold.

Wetland plants like Common Tule must cope with "wet feet"—roots that are always in water. This might not seem to be a problem, for all plants need water to survive, but roots also need to "breathe." Breathing is more difficult for roots that can't obtain air from the spaces between soil particles, as do most roots. Tules are able to breathe underwater because they have special chemistry and tissues that convey air to the roots, thus avoiding the root rot that may occur in overwatered potted plants.

In contrast, desert plants must cope with drought—long summers, or even whole years, with scarcely a drop of rain. Cacti are champion drought-tolerators. Their thick stems store water that is soaked up after every rain by an expansive network of fine roots. Instead of water-wasting leaves, they have green stems to conduct *photosynthesis,* the chemical process plants use to make their own "food." Cacti spines deter grazers that would otherwise find the plants a succulent feast.

WHAT IS A PLANT COMMUNITY?

A plant community is a group of species that grow together in sites that are environmentally similar. For example, sagebrush, rabbitbrush, and bitterbrush grow together in well-drained High Desert sites, making up the Sagebrush Scrub community. Another community, Eastern Sierra Riparian Woodlands, consists of trees, shrubs, and herbs that grow together along the banks of streams. Hundreds of plant species occur in the Eastern Sierra, but none is found in all of this region's communities; most are restricted to one or a few.

Throughout its range, each plant community has a certain recognizable "look." Often this is because the largest and most common plant species are the same wherever the community occurs. These most obvious plants are referred to as "dominants." For example, certain tall pines and firs are the dominant species of Montane Forest: Their height, density, and color create this community's characteristic aspect. Dominant species are not necessarily large. The dominant species of Mountain Meadows consist of low-growing sedges, rushes, grasses, and broad-leaved herbs.

Sometimes the border between two plant communities is abrupt—for example, where dense Montane Forest halts sharply at the edge of a Mountain Meadow. Often this results from a correspondingly abrupt change in soil type or soil moisture. More often a broader transition zone, or *ecotone,* forms the boundary between two communities. Within this transition zone environmental conditions are such that species from both communities can grow there. An ecotone may show a gradual transition in species composition as one passes through it from one community to another. In ecotones between communities that are similar in aspect—between Montane Forest and Subalpine Forest, for example, or between Shadscale Scrub and Sagebrush Scrub—it takes a practiced eye to notice the transition. With communities that are quite different in aspect, such as Montane Meadow and Sagebrush Scrub, a hillside ecotone that includes both dull, gray-green sagebrush shrubs and brightly colored meadow herbs can be recognized readily.

Although defining and naming plant communities help us understand a region's vegetation, any such classification is bound to be somewhat artificial, and many different interpretations are possible. The plant communities discussed here are defined broadly in order to emphasize readily observable differences. Remember, though, that each of these communities has many distinctive subtypes, which we do not describe. As you look at the vegetation of the Eastern Sierra, most of the time you will find it easy to recognize plant communities. However, don't be surprised if some sites have vegetation that can't be so easily classified.

Plant communities in this chapter are arranged more or less according to their elevation, beginning at the summits of the highest Sierran peaks and ending with the desert communities in the lowest valley bottoms. No single transect across the Eastern Sierra would pass through all of its plant communities. Pages 98–102 show representational transects through the Eastern Sierra at five different latitudes. To assist in identifying and understanding the communities, we introduce them grouped according to whether they are dominated by trees, shrubs, or herbs.

PLANT COMMUNITIES DOMINATED BY TREES

SUBALPINE FOREST   A usually sparse conifer forest found on the often rocky slopes of the highest ridges and tallest peaks. It features trees up to medium height, some low and stunted, with an understory limited to scattered shrubs and herbs, or else barren. Not found north of the Lake Tahoe area.

MONTANE FOREST   A dense forest of tall pines and firs, often with a heavily shaded understory. Occurs on mid-elevation slopes south of Lake Tahoe and at lower elevations north of Tahoe.

EASTERN SIERRA RIPARIAN WOODLANDS   Lush woodlands of deciduous, medium-sized, broad-leaved trees and shrubs, often with a dense understory. These woodlands border mountain and desert streams where soil moisture

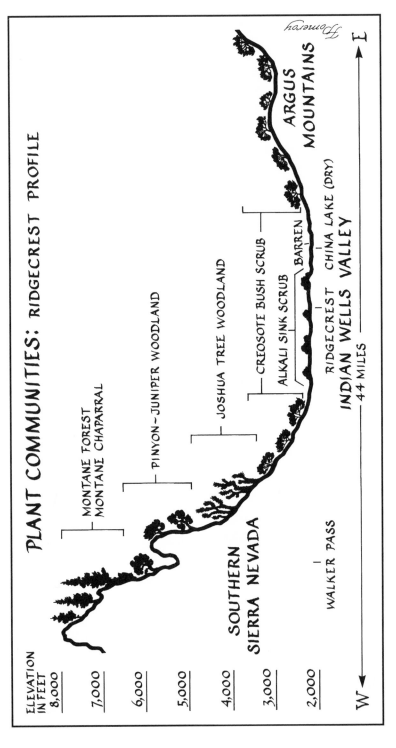

Fig. 5.1.

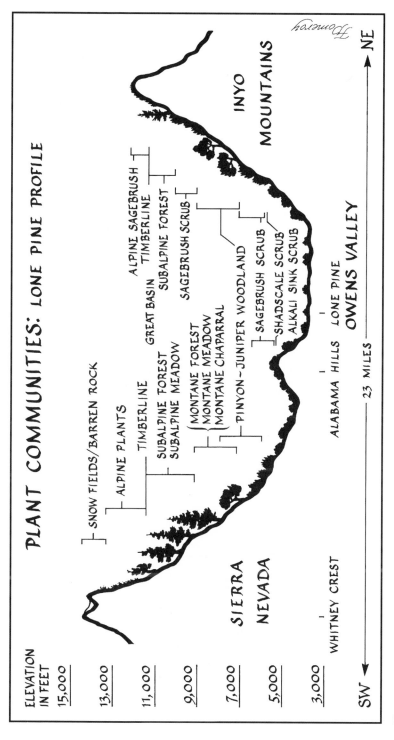

Fig. 5.2.

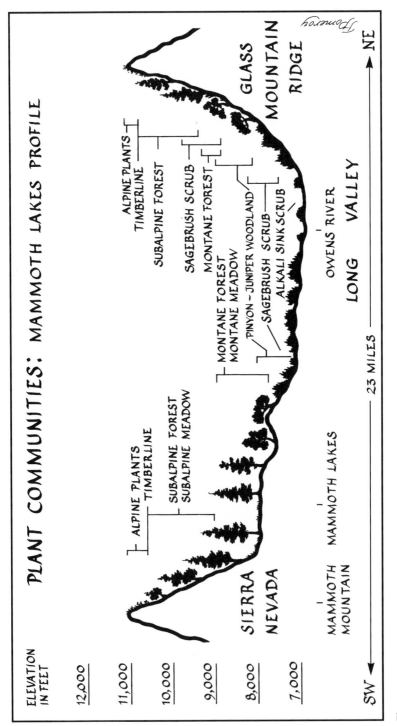

Fig. 5.3.

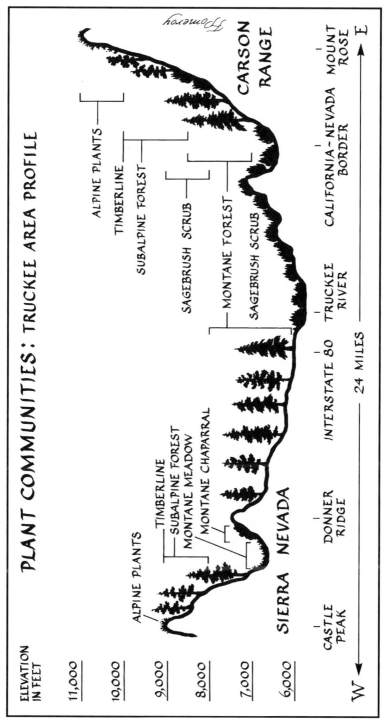

ELEVATION IN FEET

PLANT COMMUNITIES: TRUCKEE AREA PROFILE

F Pomeroy

CARSON RANGE

ALPINE PLANTS
TIMBERLINE
SUBALPINE FOREST
SAGEBRUSH SCRUB
MONTANE FOREST
SAGEBRUSH SCRUB

11,000
10,000
9,000
8,000
7,000
6,000

ALPINE PLANTS
TIMBERLINE
SUBALPINE FOREST
MONTANE MEADOW
MONTANE CHAPARRAL

SIERRA NEVADA

CASTLE PEAK
DONNER RIDGE
INTERSTATE 80
TRUCKEE RIVER
CALIFORNIA-NEVADA BORDER
MOUNT ROSE

24 MILES

W   E

Fig. 5.4.

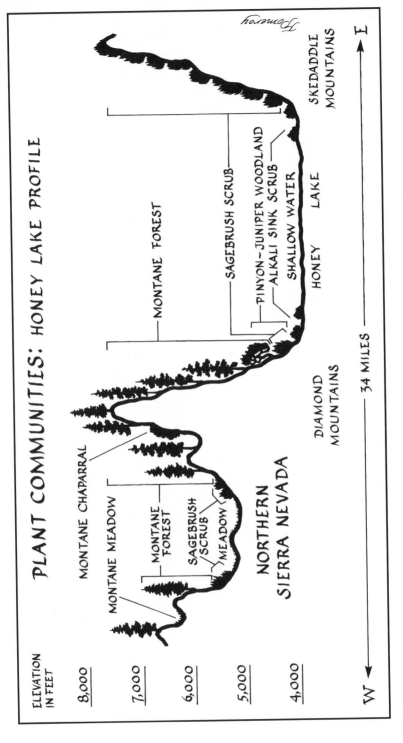

Fig. 5.5.

is consistently high. In our definition the term includes both Montane and High Desert riparian woodlands, as well as thickets of willows and aspens.

PINYON-JUNIPER WOODLAND   An open woodland of low, rounded, gray-green pinyons and junipers that occurs on the lower slopes of the Sierra Nevada and on mid-elevation slopes of mountain ranges to the east. The understory consists of sagebrush and other shrubs of the Sagebrush Scrub community.

JOSHUA TREE WOODLAND   A desert woodland with dense to scattered stands of Joshua "trees" with an understory of plants mainly of the Creosote Bush Scrub community. Patchy distribution in the southern part of Owens Valley and on the lower slopes of the southernmost Sierra Nevada, as at Walker Pass.

PLANT COMMUNITIES DOMINATED BY SHRUBS

MONTANE CHAPARRAL   A mountain community of dense, low, stiff-branched, mainly evergreen shrubs, found on warm, dry, steep, mid- to higher-elevation slopes. Often surrounded by Montane or Subalpine forest.

SAGEBRUSH SCRUB   A High Desert community of soft-woody, gray-green low shrubs. Sagebrush Scrub is widely distributed near the base of the Sierra on lower slopes and moraines, and on low to mid-elevation slopes of mountain ranges to the east. It occurs in a dwarf form on the Sierra's high ridges and on other high mountain ranges to the east. Frequently grades into Pinyon-Juniper Woodland.

SHADSCALE SCRUB   A High Desert community that looks similar to Sagebrush Scrub, but whose dominant shrubs are different species, lower, and sparser. Occurs on the lower slopes of alluvial fans at the base of desert mountains and on poorly drained flats with slightly alkaline soils.

ALKALI SINK SCRUB   A desert community of widely spaced, low to medium-sized shrubs and perennial herbs. Occurs in poorly drained alkaline or saline soils, often with a white-crusted surface, in valley bottoms and dry lake beds.

CREOSOTE BUSH SCRUB   A widely distributed, low-elevation desert community dominated by straggly, olive-green Creosote Bush, with other, lower gray-green shrubs. Its understory has much bare ground and, during wet years, may be carpeted with wildflowers in spring. Occurs on well-drained slopes and fans from the southern Owens Valley south to the Mojave Desert.

PLANT COMMUNITIES DOMINATED BY HERBS
(WILDFLOWERS, GRASSES, GRASSLIKE PLANTS)

ALPINE ROCK AND MEADOW COMMUNITIES   High-mountain communities of low-growing herbs and dwarf shrubs, with much rocky open ground, often above treeline.

MOUNTAIN MEADOWS   Includes both Montane and Subalpine meadows, with dense, lush, mainly low-growing sedges, rushes, grasses, and wildflowers. Species vary greatly from meadow to meadow. Mountain Meadows occur in basins with moist to wet soils, often near streams, within the elevation range of Montane and Subalpine forests.

ALKALI MEADOW   A High Desert community that looks similar to Mountain Meadows but has more alkaline soil and different plants. Found in the valleys east of the Sierra, often near streams.

DESERT MARSHLANDS   Fresh, brackish, or saline marshes with water-loving herbs ranging in height from a few inches to many feet. Usually in saturated soil, often near standing water, in valleys east of the Sierra.

HOW PLANTS ARE CLASSIFIED

Many kinds of plants are found in the communities described above. In the simplest classification, they are trees, shrubs, and herbs. Trees are woody plants, usually with one main trunk and commonly over 10 feet (3 m) tall when mature. Shrubs are also woody, but they usually branch at or near the base. Shrub height is variable but generally doesn't exceed 25 feet (8 m). Shrubs and trees in the western United States belong to one of two major groups: flowering plants or conifers. Flowering plants produce seed-bearing fruits from the female parts of their flowers and are often broad-leaved. They include trees such as aspen, willow, maple, and sycamore, and most shrubs and herbs. Conifers almost always have needlelike or scalelike leaves and bear woody cones that produce seeds. Conifers include the pines, firs, hemlocks, and junipers.

Herbs, or herbaceous plants, usually don't have hard, woody tissues. They include wildflowers, grasses, grasslike plants such as rushes, sedges, tules, and cattails, and also ferns and horsetails. They vary from tiny, short, compact, mat and rosette plants to tall, spreading, many-branched forms—from 2-inch-high (5-cm) buckwheats to 5-foot-tall (1.5-m) corn lilies. The life cycle of an herb may be annual, biennial, or perennial. Annuals, such as California Poppies, complete their entire life cycle—from seed to mature, seed-producing adult—within one year or less (some take no more than six

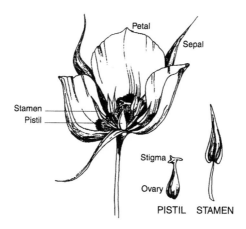

Fig. 5.6.   Flower parts of a Mariposa Lily. (Modified from *Sagebrush Country, Wild-flowers 2,* by permission of Ronald J. Taylor)

weeks). Perennials, such as penstemons and some lupines, live longer than two years. They usually die back at the end of each growing season, but an underground root or stem remains alive through the winter and puts forth new sprouts the following spring. Far less common, biennials, such as Scarlet Gilia, require parts of two consecutive growing seasons to mature. Some plants are neither completely woody nor herbaceous, but something in between. These are *suffrutescent* perennials (somewhat woody at the base with herbaceous stems above), a growth form quite common among plants in mountain and High Desert communities.

Most herbaceous plants of the Eastern Sierra are flowering plants that reproduce from seeds, such as daisies, buttercups, lilies, and violets; nonflowering herbs such as ferns and horsetails reproduce from spores. By far the majority of plants in eastern California (and in the world) are flowering plants, even though the dominant plants of several plant communities, such as Montane and Subalpine forest, are conifers.

PLANT NAMES

Each plant species has only one scientific name, shared with no other species of plant. Scientific names, always in Latin, are understood throughout the world. A botanist from China would know Quaking Aspen by its scientific name, *Populus tremuloides,* although probably not by the common name we use.

A scientific name consists of the name of the genus, such as *Lilium* (lily) or *Abies* (fir), plus a species name. The complete scientific name consists of at least two words, for example, *Lilium kelleyanum* or *Abies magnifica.* The generic name alone refers to the genus as a whole, for example, all of the species of *Lilium* or *Abies*. Many scientific names have descriptive meanings useful in remembering them. In the name *Stachys alba*, White Hedge-nettle, the species name *alba* means "white," the color of the plant's flower. Some scientific names commemorate the person who first collected the species, as in *Pinus jeffreyi*, Jeffrey Pine, collected by John Jeffrey in northern California in 1852; others indicate where the species grows, as in *Ranunculus californicus*, California Buttercup. In many species, several subspecies or varieties are also recognized. In these cases the scientific name includes the name of the variety (var.) or subspecies (ssp.), for example, *Rosa woodsii* var. *ultramontana*, Interior Wild Rose.

In addition, some plants have common names that are easier to remember, but common names are not as precise as scientific names. One species may be called by five or more common names, or several different species may be referred to by the same common name. Common names like rose, sunflower, or gooseberry refer to a group of species, usually within the same genus. The majority of plants have no common names. Common names in this book are derived from a variety of sources; they are commonly used and recognized by California and Nevada botanists and the general public. Scientific names in this book follow those used in *The Jepson Manual* (1993), the most recent identification manual for the California flora, with earlier scientific names (those used in Philip A. Munz's *A California Flora*, 1968) provided after an "equals" sign. For example, Short-rayed Daisy's new and old scientific names would be given as *Trimorpha lonchophylla (=Erigeron lonchophyllus)*. Changes in scientific names may result from new studies that cause taxonomists to revise earlier ideas about relationships between plant species.

EXPLANATION OF TERMS

Plants that occur only on the eastern slope of the Sierra Nevada (rather than on both slopes) are noted by an asterisk (*). The *High Desert* includes the Great Basin plant communities above 4000 feet (1219 m) elevation—in our discussion, Pinyon-Juniper Woodland, High Desert Riparian Woodlands, Sagebrush Scrub, Shadscale Scrub, Alkali Meadow, Desert Marshlands, and Alkali Sink Scrub. The lower, *Mojave Desert* communities at the southern end of the Eastern Sierra include Joshua Tree Woodland, Creosote Bush Scrub, Alkali Meadow, Desert Marshlands, and Alkali Sink Scrub. A plant with a *circumboreal* distribution occurs throughout the high-latitude regions of the Northern Hemisphere.

The plants discussed in this chapter do not by any means constitute an exhaustive list of species occurring in the Eastern Sierra. We include those that are most conspicuous, widespread, or interesting.

Fig. 5.7. Alpine Rock and Meadow Communities

ALPINE ROCK AND MEADOW COMMUNITIES

Above treeline the climate is harsh and plant life is sparse. Alpine winters are cold, snowy, and windy. For up to five months at a time, temperatures may stay below freezing. Snowbanks may persist into late summer, for they melt slowly in the cool air. Even in midsummer it can seem like winter above 10,000 feet (3048 m), since freezing temperatures can threaten anytime. In July and August, although intense sunlight floods the snow-free peaks, gusty, cool winds often keep daytime temperatures in the 60s. Summer storms engulf the high ridges in clouds and drench them with rain, hail, or even snow. Consequently the growing season for alpine plants is extremely short, only six to ten weeks.

In spite of these harsh conditions, however, nearly 600 plant species inhabit the alpine regions of the Sierra Nevada. About 200 of them, the true mountaineers of the plant world, are rarely found below treeline. Even on dry alpine slopes that look barren from a distance, tiny tufts and mats of green, scattered among the rocks, form natural rock gardens. In moister sites, on benches and lakeshores, carpets of dense low sedges surround miniature willows and blueberries just a few inches high.

Mountaintops are rocky places, where porous, quick-drying gravels with little humus often serve as "soil." Good soils do not develop because there are few level places for small particles to accumulate and because the water, warm temperatures, and tiny organisms needed for decomposition aren't present. But even in these austere surroundings, favorable microsites provide enough shelter for plants to survive. Crannies in rocks serve as windbreaks and also collect bits of debris that form soil of a sort. A sunny nook may be several degrees warmer than open scree. Cracks in the rocks often lead to subsurface moisture that remains available to plants even in late summer, long after surface moisture has dried up.

Alpine plants have developed special features in response to their rigorous environment. Since small size minimizes many hardships, most alpine plants are just a few inches tall—even those with much larger low-elevation relatives. Arctic Willow's woody stems, creeping along the base of a sunny boulder, form prostrate shrubs 2 inches (5 cm) high that bear catkins at ground level. No energy is wasted in producing tall stems that winter storms could easily snap. Alpine Everlasting forms a groundhugging mat; Cushion Phlox and Dwarf Daisy are compact cushion plants with many-branched stems and dense foliage of small, tough leaves. These plants have low branches that trap soil particles and keep food-producing leaves near the ground, where warmer temperatures allow more efficient photosynthesis. Other features counter the severe water stress that alpine plants must endure, because after the snow melts many alpine areas dry out rapidly. Pussypaws has succulent leaves that store moisture, while Alpine Paintbrush and Silver Raillardella are covered with a dense layer of pale hairs that reflects sunlight. This hairy covering insulates stems and leaves, and may also lower leaf temperatures in summer, reducing water loss. Many alpine plants, like Timberline Phacelia, have large woody taproots that penetrate deeply into water-holding rock fractures.

Most high-living plants have a conservative lifestyle—that is, they take few risks in the game of surviving from one year to the next. Nearly all are perennials, so if snow covers them before their seeds ripen, they will try again the following year. In addition, most have developed ways of reproducing without seeds. Often this takes place as in Timberline Phacelia, whose leaves and stems arise from a ground-level root-crown. In older plants, as the crown becomes branched, each branch produces a separate flowering clump, united belowground. Eventually the underground connections break, leaving the satellites on their own, truly separate. Other alpine perennials send out spreading subterranean stems or roots that grow to the surface, sprout leaves and flowers, and eventually lose their ties to the parent. The few alpine annuals grow in the warmest, sunniest sites, where they sprout, flower, and set seed in record time, some in only four or five weeks. Most of these, like Sierra Monkeyflower *(Mimulus mephiticus =M. coccineus),* have desert relatives that survive intense heat and lack of water through similarly speedy life cycles.

Alpine habitats are widespread but often patchy. They occur in the Lake Tahoe area on Pyramid and Dick's peaks above 9000 feet (2743 m) and on Mount Rose above 10,000 feet (3048 m). On peaks like Mount Dana near Tioga Pass and south toward Mammoth Crest, they begin at 10,300–10,800 feet (3140–3290 m). Alpine habitats are extensive above 10,800 feet (3290 m) through the Whitney Crest. Olancha Peak, where they occur above 11,300 feet (3440 m), marks their southern limit in the Sierra. The neighboring Sweetwater, Inyo, and White mountains also have alpine habitats, at higher elevations than in the Sierra Nevada. In these eastern ranges, the alpine flora includes mainly species that have Great Basin affinities.

DWARF WOODY PLANTS

Dwarf woody plants occur on the margins of alpine meadows and in other moist sites, especially near lakes. Many extend downslope into Subalpine Forest.

ALPINE LAUREL, *Kalmia polifolia*   Heath Family. Plate 9c. Low shrub with many branched stems. Leaves oblong, in opposite pairs, dark green above, 0.5–1 in (1.2–2.5 cm) long. Flowers rose-purple, about 0.5 in (1.2 cm) wide, in few-flowered clusters. *Flowers:* June–Aug.   *Distribution*   Widespread. Klamath and Cascade ranges, Sierra Nevada in California; to Alaska; Rocky Mountains; eastern North America.

ARCTIC WILLOW, *Salix arctica*   Willow Family. Fig. 5.8d. Prostrate, creeping, mat-forming woody plant. Leaves narrow, less than 1.5 in (3.8 cm) long, yellow-green when young, purplish black when older. Catkins 0.35–2.5 in (0.8–6.3 cm) long. *Flowers:* July–Aug. A related species, **Snow Willow (Salix reticulata ssp. nivalis =S. nivalis),** occurs in alpine cirques in Mono Co. *Distribution*   Cascade Range, Sierra Nevada in California; Nevada; to northern Canada.

BUSH CINQUEFOIL, *Potentilla fruticosa*   Rose Family. Fig. 5.8b. Dwarf shrub, branchlets very leafy. Leaves divided into 3–7 leaflets, each less than 1 in (2.5 cm) long. Flowers about 1 in (2.5 cm) wide, each with 5 yellow petals. *Flowers:* June–Aug.   *Distribution*   Bush Cinquefoil belongs to a small group of California alpine plants that range throughout the arctic regions of

Fig. 5.8.   Alpine dwarf woody plants. (a) White Mountain Heather, *Cassiope merten-siana.* (b) Bush Cinquefoil, *Potentilla fruticosa.* (c) Sierra or Dwarf Bilberry, *Vaccinium caespitosum.* (d) Arctic Willow, *Salix arctica*

Alaska and Labrador, and the high mountains of Eurasia. Widespread. Klamath and Cascade ranges, Sierra Nevada in California; northern Nevada; to Washington; Idaho; circumboreal.

RED MOUNTAIN HEATHER, *Phyllodoce breweri*   Heath Family. Low woody plant with spreading branches. Leaves needlelike, about 0.5 in (1.3 cm) long. Flowers cup-shaped, rose-purple to pink, about 0.5 in (1.3 cm) long, borne in clusters at ends of branches. *Flowers:* July–Aug.   *Distribution*   Cascade Range, Sierra Nevada, San Bernardino Mountains in California; northwestern Nevada.

SIERRA OR DWARF BILBERRY, *Vaccinium caespitosum (= V. nivictum)* Heath Family. Fig. 5.8c. In fall the changing leaves of Sierra Bilberry fashion a red and gold carpet around alpine lakes. Tufted, low-growing woody plant with underground stems. Leaves deciduous, oval, about 1 in (2.5 cm) long. Flowers oval, 0.25 in (0.6 cm) long, white to pink, nodding, solitary at the bases of the leaves. Berry blue-black, sweet-tasting. *Flowers:* June–July. **Western Blueberry (*Vaccinium uliginosum = V. occidentale*),** a larger plant with insipid berries, grows in Subalpine Forest.   *Distribution*   Mountains of California; eastern Nevada (Ruby Mountains); to Alaska, Montana; northeastern United States and adjacent Canada.

WHITE MOUNTAIN HEATHER, *Cassiope mertensiana*   Heath Family. Fig. 5.8a. A creeping plant with branches that spread gradually upward. Leaves very small, scalelike. Flowers bell-shaped, white to pinkish, 0.25 in (0.6 cm) long, solitary and nodding from slender stalks. *Flowers:* July–Aug.   *Distribution*   Klamath and Cascade ranges, Sierra Nevada in California; to Alaska, western Canada, Montana.

WILDFLOWERS OF DRY SITES

Wildflowers of dry sites grow mainly in rock crevices, boulder fields, and talus, and on cliffs and ledges. Some alpine wildflowers of moist sites are described under Mountain Meadows, since they are also common in meadows at subalpine altitudes.

Flowers White to Cream or Greenish

ALPINE EVERLASTING, *Antennaria media (= A. alpina)*   Sunflower Family. Mat-forming perennial herb with upright stems to 4 in (10 cm) high. Leaves spoon-shaped, covered with white hairs. Flowering heads gray-green with cream centers, in groups of 3–6, on stalks to 5 in (12.5 cm) tall. *Flowers:*

Fig. 5.9.   Alpine wildflowers. (a) Timberline Phacelia, *Phacelia hastata* ssp. *compacta.*
(b) Alpine Sorrel, *Oxyria digyna.* (c) Rosy Everlasting, *Antennaria rosea.* (d) Silver Raillardella, *Raillardella argentea*

July–Aug. **Rosy Everlasting** *(A. rosea)* (fig. 5.9c), with pinkish flower heads, also grows in alpine habitat. *Distribution* Mountains of northern California, northern Nevada; to southwest Canada, Montana, Colorado, New Mexico.

ALPINE PAINTBRUSH, *Castilleja nana* Figwort Family. Fig. 5.10b. A perennial plant of dry alpine barrens. Several erect stems to 4 in (10 cm) tall, from a perennial root-crown. Stems and leaves with long gray hairs. Leaves narrow, with 3–5 lobes. Flower bracts usually cream, sometimes dull red. *Flowers:* July–Aug. *Distribution* Sierra Nevada, northern Nevada.

CUSHION PHLOX, *Phlox condensata (= P. covillei)* Phlox Family. Fig. 5.10d. Perennial cushion plant of rocky sites, especially limestone. Very short branchlets, about 1 in (2.5 cm) long. Leaves tiny and narrow, about 0.2 in (0.5 cm) long, crowded on branchlets. Flowers white to pinkish, about 0.5 in (1.3 cm) wide. *Flowers:* June–Aug. *Distribution* San Bernardino Mountains, Sierra Nevada, mountains of eastern California; Nevada.

TIMBERLINE PHACELIA, *Phacelia hastata* ssp. *compacta (= P. frigida)* Waterleaf Family. Fig. 5.9a. Perennial rosette plant with several stems to 6 in (15 cm) tall, covered with short, stiff, gray hairs. Leaves 1.5–3 in (3.8–7.5 cm) long, narrow, but wider toward the ends. Flowers white to pale lavender, small, densely clothing the coiled branchlets of the flower stalks. *Flowers:* July–Sept. *Distribution* Cascade Range, Sierra Nevada, mountains of eastern California; western Nevada; to Washington.

Flowers Yellow to Orange

ALPINE COLUMBINE, *Aquilegia pubescens* Buttercup Family. Plate 9a. Delicate-looking flower common on talus slopes. Erect, many-stemmed perennial herb to 1.5 ft (0.5 m) tall. Leaves divided into many pale green, lobed segments, round in outline. Flowers large, tilted upward at the ends of long, slender stems, the petals forming distinctive long nectar-bearing spurs. *Flowers:* June–Aug. In some areas the flowers range from pale yellow (often with blue spurs) to pink or white, a result of hybridizing with the lower elevation **Red Columbine** *(A. formosa)*. *Distribution* Sierra Nevada.

ALPINE GOLD, *Hulsea algida* Sunflower Family. Plate 9e. A conspicuous, sturdy perennial of gravelly slopes that ranges to 14,000 feet (4270 m). Rosette plant with disagreeable odor. Leaves narrow, to 6 in (15 cm) long, with small marginal teeth. Flowering heads bright yellow, about 2 in (5 cm) wide, each borne on a stalk about 5 in (12.5 cm) tall. *Flowers:* July–Aug.

Fig. 5.10. Alpine wildflowers. (a) Lemmon's Draba, *Draba lemmonii* var. *lemmonii.*
(b) Alpine Paintbrush, *Castilleja nana.* (c) Dwarf Daisy, *Erigeron pygmaeus.* (d) Cushion
Phlox, *Phlox condensata.* (e) Pussypaws, *Calyptridium umbellatum*

Distribution Sierra Nevada, mountains of eastern California; Nevada; to Oregon, Montana.

LEMMON'S DRABA, *Draba lemmonii* var. *lemmonii* Mustard Family. Fig. 5.10a. Compact perennial rosette plant with several flowering stems to 5 in (12.5 cm) tall. Leaves tufted, thick, oval, less than 1 in (2.5 cm) long, covered with short, stiff hairs. Flowers small, each with 4 yellow petals. Fruit an erect, twisted "pod." *Flowers:* July–Aug. *Distribution* Sierra Nevada; Mount Rose, Nevada.

SILVER RAILLARDELLA, *Raillardella argentea* Sunflower Family. Fig. 5.9d. Perennial from underground stems, with leaves in many small tufts. Leaves narrow, to 2.5 in (6.3 cm) long, covered with silky silver hairs. Flower heads yellow-orange, each on a stalk one to several inches high. *Flowers:* July–Aug. *Distribution* Cascade Range, Sierra Nevada, San Bernardino Mountains, mountains of eastern California; western Nevada; Oregon.

Flowers Pink to Red

ALPINE SORREL, *Oxyria digyna* Buckwheat Family. Fig. 5.9b. Perennial rosette herb with several slender erect flowering stalks 2.5–10 in (6.3–25 cm) high. Leaves slightly succulent, roundish, dark green, about 1 in (2.5 cm) wide. Flowers minute, with red and green parts, clustered on stalks that rise above the leaves. *Flowers:* July–Sept. Occurs in alpine and arctic habitats throughout the Northern Hemisphere, often in moist rock crannies. *Distribution* Mountains of California and Nevada; circumboreal.

PUSSYPAWS, *Calyptridium umbellatum* Purslane Family. Fig. 5.10e. Annual or perennial. Succulent rosette plant with prostrate flower stems. Leaves spoon-shaped, to 1.5 in (3.8 cm) long at high elevations. Flowers in dense roundish clusters, the parts pink and white. *Flowers:* June–Aug. *Distribution* Mountains of northern and central California; northern and central Nevada; to Montana, western Wyoming.

ROCK FRINGE, *Epilobium obcordatum* Evening-primrose Family. Plate 9h. Perennial herb, stems several, spreading, sometimes matted, forming a "fringe" on moist rocky ledges. Leaves in pairs, oval, crowded on stems, about 0.5 in (1.3 cm) long, with tiny teeth along the edges. Flowers 1–1.5 in (2.5–3.8 cm) wide, each with 4 rose-pink heart-shaped petals. *Flowers:* July–Sept. *Distribution* Cascade Range, Sierra Nevada, Modoc Plateau in California; northern Nevada; to eastern Oregon, central Idaho.

SIERRA PRIMROSE, *Primula suffrutescens* Primrose Family. Plate 9b. Perennial with creeping, branched woody stems; mat-forming. Leaves crowded,

0.5–1.5 in (1.3–3.8 cm) long, broadened outward and toothed along the outside margin. Flowers deep magenta, about 0.75 in (1.8 cm) wide, clustered at the tops of elongated stalks. *Flowers:* July–Aug. *Distribution* Klamath Range, Sierra Nevada in California.

Flowers Blue to Violet or Purple

ALPINE OR DWARF LUPINE, *Lupinus lepidus (= L. lyallii)* Pea Family. Plate 9f. Perennial with several semi-prostrate stems from a woody base. Plants covered with smooth, silky hairs. Leaves divided into oval leaflets about 0.5 in (1.3 cm) long, arranged as the fingers are to the palm (palmately). Pea flowers about 0.5 in (1.3 cm) long, various shades of blue, on short erect stalks. *Flowers:* July–Sept. Highly variable. *Distribution* Cascade Range, Sierra Nevada, desert mountains in California; northern Nevada; to British Columbia, Montana, Colorado.

CREEPING PENSTEMON, *Penstemon davidsonii* var. *davidsonii* Figwort Family. Plate 9d. Perennial creeping mat plant, with several spreading flower stems to 4 in (10 cm) high. Leaves oval, about 0.5 in (1.3 cm) long. Flowers large, to 2 in (5 cm) long, purple-violet, with a broad tube and 2 flaring lips. *Flowers:* July–Aug. *Distribution* Klamath and Cascade ranges, Warner and Sweetwater mountains in California; Sierra Nevada.

DWARF DAISY, *Erigeron pygmaeus* Sunflower Family. Fig. 5.10c. Compact cushion plant, perennial, with leaves about 1 in (2.5 cm) long, narrow, in dense rosettes. Flowering heads solitary on erect stems 1–2.5 in (2.5–6.3 cm) tall. Heads about 1 in (2.5 cm) wide, with purple marginal rays and yellow central disks. *Flowers:* July–Aug. Many other species of *Erigeron* are found in the Eastern Sierra. *Distribution* Sierra Nevada, High Desert mountains in California; west-central Nevada.

SKY PILOT, *Polemonium eximium* Phlox Family. Plate 9g. At home on gravelly slopes and ridges of the highest peaks, Sky Pilot occurs up to 14,000 ft (4270 m) on the Whitney Crest and above 13,000 ft (3960 m) on Mount Dana. Perennial with several erect stems to 6 in (15 cm) tall, from a woody base. Strong musky odor. Leaves divided into many tiny leaflets that closely clothe the stalks. Flowers about 0.5 in (1.3 cm) wide, blue and white, each with a slender tube and 5 flaring lobes, crowded at the ends of the flower stalks. *Flowers:* July–Aug. *Distribution* Sierra Nevada.

Fig. 5.11. Subalpine Forest

SUBALPINE FOREST

Few roads reach the rugged heights of the Subalpine Forest. Broad vistas open to the sky and superb views of distant peaks reward those who hike its sunny trails. Within this highest Sierran forest, stands of gnarled pines surround lakes that lie below towering masses of pale granite, ancient junipers command rocky ridges, and regal hemlocks sway on slopes too steep to hike. The conifers of this domain are usually no more than 40 feet (12 m) tall. Erect, single-stemmed trees are the most common, although spreading, candelabra-shaped, skirted, and even prostrate shrublike forms grow on craggy peaks. Rarely do the trees form a continuous cover. Only the dense hemlock groves that clothe moist north-facing slopes have the closed canopy and deeply shaded understory we usually associate with forest.

Subalpine Forest plants must withstand conditions nearly as harsh as those of alpine regions. Winters are long and cold, with blowing ice and snow. More snow accumulates in Subalpine Forest than on the exposed slopes above; 5–10 feet (1.5–3 m) can cover the ground for up to six months. Spring is slow to arrive, with snow remaining longer on shaded subalpine slopes than on sunny ones above treeline. However, summer weather is noticeably milder than in the alpine zone. The growing season is a little longer, eight to twelve weeks, and a little warmer. Rocky soils, often with more rock than soil, are common, although moist sites near lakes and streams have better developed soils.

Subalpine Forest trees include Lodgepole, Western White, Limber, Whitebark, and Foxtail pines, and Mountain Hemlock. Sierra Juniper occurs both here and at the higher elevations of Montane Forest. These trees sometimes grow in mixed stands; more commonly, a single kind dominates a local area. The understory is quite variable; its composition is influenced both by the soil and by the amount of sunlight that penetrates the canopy. Granitic soils, by far the most common, support many species of low shrubs, grasses, and herbaceous plants. In the vicinity of Mammoth Mountain, deep layers of pumice form a porous, nutrient-poor "soil" with widely scattered herbaceous plants. Subalpine Forest on the limestone rocks of the Convict Basin contains few understory species, especially on steep slopes subject to erosion.

In general, conifers are better adapted to the cold, dry conditions at high elevations than are broad-leaved deciduous trees. Pine leaves, or "needles," have a thick outer layer that resists water loss. In addition, they are evergreen and therefore able to begin photosynthesis as soon as the days warm up. A typical deciduous broad-leaved tree would barely finish growing a new set of leaves before winter was upon it. Aspens are the exception that proves the rule. In the subalpine region they leaf out early, often grow no larger than shrubs, and avoid drought stress by inhabiting moist sites.

On the Sierra's eastern slope, Subalpine Forest ranges from Horseshoe Meadows and the Cottonwood Lakes Basin (just south of the Whitney Crest, west of Lone Pine) to the Lake Tahoe region. Generally it occurs from about 9000 to 11,300 feet (2740–3440 m), at higher elevations in the south. South of the Cottonwood Lakes, Subalpine Forest is restricted to isolated stands on the highest mountains, such as Olancha Peak. North of the Lake Tahoe area it is all but absent.

The upper limit of Subalpine Forest is marked by treeline. Trees on the bleakest, highest crests have been twisted and pruned by furious winds into low, shrubby forms, called *krummholz* (German for "twisted wood"). Many researchers have speculated about the causes of treeline. One possibility is that harsh winters, especially extreme cold and wind, limit tree growth. While Sierra winter temperatures may occasionally reach exceptionally low levels, the scanty records suggest that moderately low temperatures, between zero and 25°F (−18 and −4°C), are typical. Trees tolerate much lower temperatures than these in areas farther north. Winds in the Sierra have been clocked at more than 100 mph (160 km per hour); they carry knife-edged ice particles that slice away the buds of woody branches protruding from the snow blanket, resulting in the "flagged" and other asymmetric shapes often observed near treeline. The weight of accumulated ice and snow can easily snap off brittle twigs, inhibiting growth. Another possibility is that poorly developed soils may limit tree growth. However, scientists now believe that it is the length of the growing season at high elevations that defines the upper limit of tree growth. A growing season that is sometimes as short as six weeks is too short for woody plants to produce enough new tissue to reach tree size. Under these conditions even small plants barely obtain the energy they need to maintain themselves, with none left over to make large woody parts.

TREES OF SUBALPINE FOREST

MOUNTAIN HEMLOCK, *Tsuga mertensiana*   Pine Family. Fig. 5.12a. Mountain Hemlock trees are easily recognized by their graceful, spirelike form, downcurved branches, and slender drooping tip at the top of even the largest trees. Tapering tree, 65–150 ft (20–46 m) tall. Skirted form and krummholz common. Needles about 1.5 in (3.8 cm) long, borne individually on tiny pegs on the slender branches. Cones oblong, 1–3 in (2.5–7.5 cm) long, deep brown. Widespread in the northern Sierra, where it forms dense forests. In the central Sierra it is restricted to moist sites near lakes and streams, north-facing slopes, and heavily shaded ravines that hold snow late in the season. It is not common south of Mammoth Lakes and reaches its southern limit near Bishop Creek. *Distribution*   Central Sierra to northern California mountains; to Alaska, Rocky Mountains.

WHITEBARK PINE, *Pinus albicaulis*   Pine Family. Fig. 5.12b. Trees erect to about 50 ft (15 m), or low, dwarfed, or prostrate. Trunks usually multiple, often twisted. Needles 1.5–3 in (3.8–7.5 cm) long, dark green, in bundles of 5. Cones egg-shaped, 1.5–3 in (3.8–7.5 cm) long, dark purple at maturity. The bark is beige, in spite of what the common name suggests. Often grows at treeline, in varied forms: candelabralike clumps with upward-reaching branches and gnarled, twisted trunks; skirted forms with dense, shrubby bases; or low mats creeping over boulders on windswept ridges. Whitebark Pines occur most often on rocky slopes and benches where glacial scouring has left a soil-free surface. Cones are distinctive: The scales remain fleshy at maturity instead of drying and opening up, as in other pines. The fleshy scales of the ripe, deep purple cones are whacked open in late summer by Clark's Nutcracker, a bird that harvests and caches the seeds underground, to be retrieved and eaten later. Cached seeds missed by the birds may germinate to form new stands of Whitebark Pine. *Distribution*   Sierra Nevada to

Fig. 5.12.   Subalpine Forest trees. (a) Mountain Hemlock, *Tsuga mertensiana*.
(b) Whitebark Pine, *Pinus albicaulis*. (c) Western White Pine, *Pinus monticola*.
(d) Lodgepole or Tamarack Pine, *Pinus contorta* ssp. *murrayana*

northeastern California; western and northern Nevada; Alberta, British Columbia, Idaho, Wyoming.

WESTERN WHITE PINE, *Pinus monticola* Pine Family. Fig. 5.12c. Erect tall tree, 50–165 ft (15–50 m), with reddish brown plated bark and branches that curve upward at the ends. Needles 2–4 in (5–10 cm) long, in bundles of 5. Cones cylindric, 4–8 in (10–20 cm) long, light brown, with thin spreading scales. Most common in the lower elevations of Subalpine Forest and almost as abundant farther downslope in upper Montane Forest. The tree and its cones look like a smaller version of Sugar Pine, a massive tree of Montane Forest that is common on the western Sierra slope and around Lake Tahoe (see Montane Forest). *Distribution* Sierra Nevada to northern California mountains; western Nevada; to British Columbia, Alberta, Montana, Idaho.

LODGEPOLE OR TAMARACK PINE, *Pinus contorta* ssp. *murrayana* (=*P. murrayana*) Pine Family. Fig. 5.12d. Usually a slender tree, 50–130 ft (15–40 m). Trunks very straight, with scaly yellowish gray bark and many thin branches. Needles 1–2 in (2.5–5 cm) long, yellow-green, in bundles of 2. Cones roundish, 1–2 in (2.5–5 cm) long, with prickly scales. Widespread in Subalpine Forest, where it grows in both moist and dry habitats. Often it forms extensive pure stands on benches and other more or less level terrain. It also borders lakes and invades meadow margins. Frequently it is the first tree to become established after fire. Lodgepole Pine grows within a broad elevational range, occurring in the Eastern Sierra from about 6000 ft (1830 m) in sites north of Lake Tahoe to over 11,000 ft (3350 m) in the central Sierra. *Distribution* Mountains of California; western Nevada; to Alaska, Rocky Mountains; Baja California.

FOXTAIL PINE, *Pinus balfouriana* Pine Family. Fig. 5.13a. Low tree, 20–50 ft (6–15 m) tall. Trunk often stout, with cinnamon-brown bark. Needles about 1 in (2.5 cm) long, in bundles of 5, curving inward and densely covering the ends of the pendulous branches. Cones egg-shaped, 1.5–3 in (3.8–7.5 cm) long, deep reddish brown; scales with tiny prickles. Foxtail Pine does not tolerate shade; the trees form open pure stands on rocky slopes and ridges. The trees usually grow erect, but a low, shrubby form occurs on Olancha Peak in exposed sites at treeline. *Distribution* Foxtail Pine grows only in two widely separated locations: the Klamath Mountains in northwestern California and the southern Sierra. In the Eastern Sierra it is limited to a small section of western Inyo Co., mainly above 9500 ft (2890 m), where it reaches its southern limit on Olancha Peak near Higgins Lake. It is common near Horseshoe Meadows and in the Cottonwood Lakes Basin,

Fig. 5.13.   Subalpine Forest trees. (a) Foxtail Pine, *Pinus balfouriana.* (b) Limber Pine, *Pinus flexilis.* (c) Western Bristlecone Pine, *Pinus longaeva*

south of Mount Langley, and extends north to just beyond Kearsarge Pass. Here, Foxtail Pine is only 20 air miles (30 km) from the nearest stands, in the White Mountains, of its longer-lived and famous cousin, the Western Bristlecone Pine, which it closely resembles.

LIMBER PINE, *Pinus flexilis*   Pine Family. Fig. 5.13b. Erect tree 35–65 ft (11–20 m) tall, with short trunk and spreading crown. Needles about 2 in (5 cm) long, in bundles of 5. Cones egg-shaped to cylindric, 3–8 in (7.5–20 cm) long, light golden brown, the scales thick with rounded ends. *Distribution*   Restricted in the Eastern Sierra to Inyo and Mono counties, in scattered sites on steep, rocky slopes. (Four small stands occur on the western slope in Kern and Tulare counties.) Grows with Western Bristlecone Pine in Inyo, White, and Panamint mountains. Also mountains of southern California and Nevada; to Basin Ranges, Rocky Mountains; Oregon; Arizona.

WESTERN BRISTLECONE PINE, *Pinus longaeva*   Pine Family. Fig. 5.13c. Resembles Foxtail Pine. Cones and longevity distinguish the species. Bristlecone Pine cones, similar to those of foxtail in size, shape, and color, differ in having scales with prominent prickles up to 0.25 in (0.6 cm) long, whereas the prickles of Foxtail cones are small and inconspicuous.

Although bristlecones do not grow in the area this book covers, we find it impossible to ignore them since they are world-famous and occur in the adjacent White-Inyo Mountains. Bristlecone Pines are renowned for their great age and stark beauty. Called the world's oldest living things, they certainly hold that title as far as individual trees are concerned, since several trunks have been dated at more than 4600 years old. (See Creosote Bush Scrub for a discussion of the much older creosote "rings.") In California these unusual trees grow in a few sites on the highest peaks of the White, Inyo, Last Chance, and Panamint mountains. In Nevada they are scattered throughout the state in isolated groves on the highest peaks. They are often restricted to areas underlain by dolomite, a whitish carbonate rock that weathers to a porous, nutrient-poor soil in which few other plants can survive; there bristlecones have almost no competition. Bristlecone Pines live under very cold, dry conditions. As a result, their growth is exceedingly slow; a 3-foot-tall (1-m) "youngster" may be more than 100 years old. The trees add new wood in annual increments no thicker than a toothpick, and the needles may remain alive on the branches for 20 years or more. The oldest trees occupy the most forbidding sites, where icy winds of thousands of winters past have carved patterns in their twisted, reddish gold trunks. Frequently these gnarled patriarchs have lost most of their living tissue; a narrow strand of bark snaking along a reclining trunk is all that keeps the tree alive.

Fig. 5.14.   Montane Forest

MONTANE FOREST

Tall conifers, mainly pines and firs, typify Montane Forest. However, it is the most variable of the east slope's forests and woodlands. Some stands include five or more species of conifers, while others include just two or three, and some have only one. Eleven conifers grow here, including the graceful Incense Cedar, robust Jeffrey, Ponderosa, Western White, and Sugar pines, towering Red and White firs, and craggy Sierra Juniper. Under natural conditions that included frequent fires, Montane Forest was an open forest marveled at by early travelers who rode their horses through these resin-scented, sun-dappled woods. Where fires have been suppressed, the undergrowth has flourished, so that today some stands are heavily shaded tangles of fallen trees and overgrown shrubs. Cleared or burned sites tend more toward saplings, grasses, and flowering herbs. Unforested sites within the Montane Forest elevation zone support stands of Montane Chaparral, Sagebrush Scrub, Mountain Meadow, and Montane Riparian Woodland.

Compared to most other montane communities, Montane Forest has a relatively mild climate, comfortable for plants and people alike. Although cold and snowy in winter, it has more protection from wind and less snow-fall than at higher elevations—conditions enjoyed by both residents and visitors. Summers are warm and comparatively long, with a growing season of four and a half to five months. Summer evenings are cool, but usually not freezing, an important factor in determining the variety of plant life that dwells here.

Montane Forest extends most of the length of the Eastern Sierra—from the Susanville–Honey Lake area in the north to Whitney Portal (below the Whitney Crest, west of Lone Pine) in the south, with scattered outlying stands as far south as Olancha Peak. In the northern Sierra (Lake Tahoe northward) Montane Forest ranges from 5500 to 8000 feet (1680–2440 m). In the southern Sierra (south of Lake Tahoe) it ranges from about 7000 to 9000 feet (2130–2740 m). North of Lake Tahoe, east slope Montane Forest is similar to that on the Sierra's west slope. South of Tahoe, where the Sierra Crest is well defined and often above 12,000 feet (3660 m), Montane Forest occurs at higher elevations, is more open, contains fewer tree species, and is less extensive.

Around Lake Tahoe, Montane Forest varies considerably in species composition from place to place. Lower-elevation sites on the west shore contain Jeffrey, Ponderosa, and Sugar pines, Incense Cedar, and White Fir. Most of this forest, especially the Ponderosa Pine stands, was clear-cut before 1900 to provide lumber for Virginia City and the mines of the Comstock Lode. Today the slopes around Lake Tahoe are forested with trees that are largely second-growth. Higher elevation Montane Forest on the west shore is dominated by Jeffrey, Western White, and Lodgepole pines, Red Fir, and Sierra Juniper. On the east side of the lake, Montane Forest consists mainly of the more drought-tolerant Jeffrey Pine and White Fir. An isolated stand of the rare Washoe Pine grows on the east slope of Mount Rose, in Washoe County, Nevada. It was almost eliminated by logging in the late 1800s, more than fifty years before it was recognized as distinct. North of Tahoe, Montane Forest includes, in addition to species mentioned above, California Black Oak and, in a few sites, Douglas Fir. Clear-cutting in this region has resulted in a dense cover of Montane Chaparral in places that were previously forested.

South of Lake Tahoe, arid conditions and the steep eastern escarpment of the Sierra Nevada limit the distribution and the species diversity of Montane Forest. Here, the forest contains only Jeffrey, Western White, and Lodgepole pines, Red Fir, Sierra Juniper, and some scattered stands of White Fir in shaded ravines and other moist sites. The driest Montane Forest sites consist of pure stands of Jeffrey Pine with a sagebrush and chaparral understory, as around Markleeville (Alpine County) and on the well-drained pumice

Fig. 5.15.   Forest regeneration after fire. Fires are a natural part of the forest cycle; however, crown fires have become more common in the Sierra as a result of 100 years of fire suppression. On dry slopes of the Eastern Sierra, forest trees killed by crown fires may be replaced first by Montane Chaparral. It may take many decades for pines and firs to become reestablished.

soils north and east of Mammoth Lakes. This latter forest covers more than 100 square miles (259 sq km), even clinging to the eastern rims of the Mono Craters, south of Mono Lake. Here, in the High Desert east of the Sierra, one might expect to find a more drought-adapted community. It is likely that Jeffrey Pines flourish here because the area is actually more moist than neighboring areas. Low passes just north and south of Mammoth Mountain allow moisture-bearing winds to bring increased rainfall into the desert.

This Jeffrey Pine forest is interrupted by numerous low-lying pumice flats (locally called "sand flats") that are devoid of shrubs and trees. What accounts for these abrupt changes from forest to open flats? Cold air accumulation has been suggested. A more likely possibility is that these pumice flats, which appear dry and barren in summer, are actually too wet for trees. Their soils are saturated early in the season in wet years. Over time, this seasonal abundance of water could be enough to exclude Jeffrey Pine, a species that requires well-drained soil. Is it coincidence that a narrow band of Lodgepole Pine, a tree that tolerates moist soil conditions, encircles many of these pumice flats?

Jeffrey Pine stands are subject to periodic irruptions of the Pandora Moth (*Coloradia pandora*), a forest pest that has defoliated many square miles of trees. The adult moths are usually present in low numbers, but under favorable conditions the population may explode. In summer, the adult females deposit small chartreuse eggs on Jeffrey Pine trunks and branches. These hatch into caterpillars that eat pine needles before attaching to the branches and entering a dormant stage for the winter. To complete their two-year life cycle, the next spring they resume feeding and enlarge to finger size before crawling down the trunks, burrowing into the soil at the base of the trees, and transforming into a second dormant (pupal) stage. The caterpillars were called *piagee* (pronounced pee-ah-gee) by the local Paiutes, who harvested and ate them.

Near the town of Mammoth Lakes is one of the southernmost stands of well-developed Montane Forest—with Jeffrey Pine in the driest sites, Western White and Lodgepole pines in moister sites, Sierra Juniper on exposed ridges, and dense Red Fir on moist north-facing slopes. South of Mammoth Lakes Montane Forest is limited and patchy; Subalpine Forest grades, instead, into a narrow belt of Jeffrey Pine, which in turn is bordered by Pinyon Woodland. The southernmost outliers of Montane Forest, including a stand at Whitney Portal, consist of Jeffrey Pine, White Fir, and, rarely, Red Fir.

COMMON TREES OF MONTANE FOREST

CALIFORNIA BLACK OAK, *Quercus kelloggii*   Oak Family. Fig. 5.16c. Deciduous tree with broad crown, 35–80 ft (11–25 m) tall, with dark bark, deeply furrowed in old trees. Leaves broadly oval, deeply lobed, each pointed lobe with a bristle tip. Acorns about 1 in (2.5 cm) long, with thin scales covering the cup. *Flowers:* Apr–May. Common at lower elevations in Montane Forest in Plumas and Lassen counties. See also Eastern Sierra Riparian Woodlands. *Distribution* California mountains and hills, except deserts; Oregon, Baja California.

DOUGLAS FIR, *Pseudotsuga menziesii* Pine Family. Erect tree to 230 ft (70 m), with narrow crown. Main branches crowded on trunk, with long drooping secondary branches. Needles about 1 in (2.5 cm) long, borne individually on branches. Cones oval, 2–3 in (5–7.5 cm) long, dark brown, with 3-pointed bracts. An important timber tree of the Pacific Northwest. Scattered in moist sites north of Lake Tahoe, in Sierra, Plumas, and Lassen counties. *Distribution* Central and northern California mountains; Nevada; to southwestern British Columbia, Rocky Mountains, Texas, Mexico.

INCENSE CEDAR, *Calocedrus decurrens* Cypress Family. Fig. 5.16a. Erect tree, 80–160 ft (25–50 m), with graceful drooping branches closely covered with tiny scalelike leaves. Bark cinnamon-brown, furrowed and shredding in old trees. Cones about 1 in (2.5 cm) long, resembling a fleur-de-lis when dry. Occasional in shaded dry sites at lower elevations from Lake Tahoe north. Most lead pencils in the United States are made from this tree's wood. Preferred as fencing material because it is durable. *Distribution* Mountains of California, except deserts; western Nevada; Oregon, Baja California.

JEFFREY PINE, *Pinus jeffreyi* Pine Family. Fig. 5.16b. Erect tree, 65–195 ft (20–60 m), with mature bark in broad plates with pinkish to red or brown inner surface. Needles 5–11 in (12.5–27.5 cm), in bundles of 3. Cones broadly oval, reddish brown, 6–10 in (15–25 cm) long, with an inward-pointing prickle at the end of each scale. Common in open, dry, cold sites at higher elevations. The bark gives off a pleasant odor of vanilla or butterscotch. Most easily distinguished from Ponderosa and Washoe pines by its larger cones. *Distribution* Most California mountains; western Nevada; southwestern Oregon, northern Baja California.

LODGEPOLE PINE, TAMARACK PINE, *Pinus contorta* ssp. *murrayana* (= *P. murrayana*) Pine Family. For description see Subalpine Forest. Lodgepole Pine has the broadest ecological tolerances of any Eastern Sierra conifer. It tolerates both moist and dry habitats and occurs at elevations ranging from 6000 to 11,000 ft (1830–3353 m). Often occurs on cold benches.

PONDEROSA PINE, YELLOW PINE, *Pinus ponderosa* Pine Family. Fig. 5.16d. Erect tree, 50–230 ft (15–70 m), mature bark in broad plates with

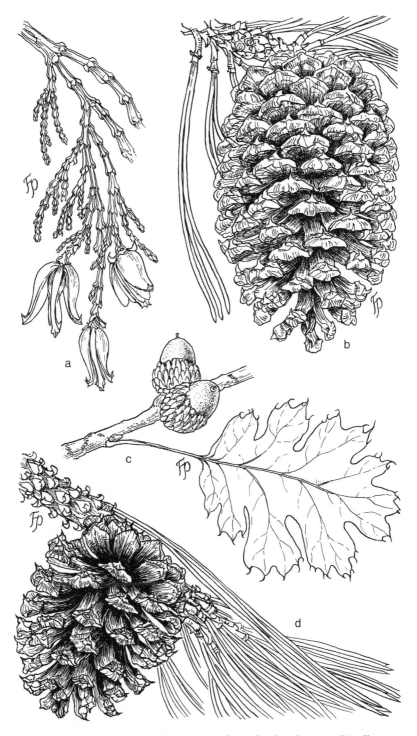

Fig. 5.16. Montane Forest trees. (a) Incense Cedar, *Calocedrus decurrens*. (b) Jeffrey Pine, *Pinus jeffreyi*. (c) California Black Oak, *Quercus kelloggii*. (d) Ponderosa Pine, *Pinus ponderosa*

yellow inner surface. Needles 5–10 in (12.5–25 cm) long, in bundles of 3. Cones ovoid, 3–6 in (7.5–15 cm) long, with an outward-pointing prickle at the tip of each scale. Common in open dry sites at lower elevations within Montane Forest. Scarce around Lake Tahoe, but more common to the north. *Distribution* Most mountains of California, except deserts; western Nevada; to British Columbia, Montana, Nebraska, northern Mexico.

RED FIR, SILVER-TIP FIR, *Abies magnifica* Pine Family. Fig. 5.17a. Erect tree, 65–195 ft (20–60 m), with narrow crown, trunk bearing unevenly spaced declining branches on old trees. Mature bark dark reddish brown, deeply furrowed. Needles about 1–1.5 in (2.5–3.8 cm) long, 4-sided, pale gray-green when young. Cones oblong-cylindric, 6–8 in (15–20 cm) long, erect, borne on uppermost branches; cones fall apart at maturity while still attached to the tree. Common on moist shaded sites in the Lake Tahoe region and in scattered localities north and south of Lake Tahoe. *Distribution* Mountains of northern and central California; west-central Nevada; southern Oregon.

SUGAR PINE, *Pinus lambertiana* Pine Family. Fig. 5.17b. Erect tree, 65–245 ft (20–75 m) tall, with flat-topped crown in mature trees. Main branches widely spaced, horizontal. Needles 3–4 in (7.5–10 cm) long, in bundles of 5. Cones light brown, to nearly 20 in (50 cm) long, hanging from the tips of the main branches. Sugar Pine, the largest of 100 pine species, is an important source of commercial timber. Widely scattered stands are found in moister, lower-elevation sites within Montane Forest from Alpine Co. north. *Distribution* Mountains of California; Oregon; Baja California.

WASHOE PINE, *Pinus washoensis* Pine Family. Trees resemble Ponderosa Pine. Cones distinguish the species. Washoe cones resemble Jeffrey Pine cones but are much smaller: 2.5–4 in (6.3–10 cm) for Washoe versus 5–10 in (12.5–25 cm) for Jeffrey. A rare tree throughout its range. Most closely related to Ponderosa Pine. "Hybrid swarms" of Washoe and Ponderosa pines, including many individuals with intermediate characteristics, are found in several valleys north of Lake Tahoe. Washoe Pines are not known to hybridize in nature with Jeffrey Pines, in spite of the resemblance of their cones. *Distribution* Cascade Range, northern Sierra Nevada, Warner Mountains in California; western Nevada; Oregon.

SIERRA JUNIPER, *Juniperus occidentalis* Cypress Family. Fig. 5.17d. Low spreading tree, 15–65 ft (5–20 m), with a thick trunk and shredding cinnamon-colored bark. Branchlets closely covered with tiny scalelike leaves.

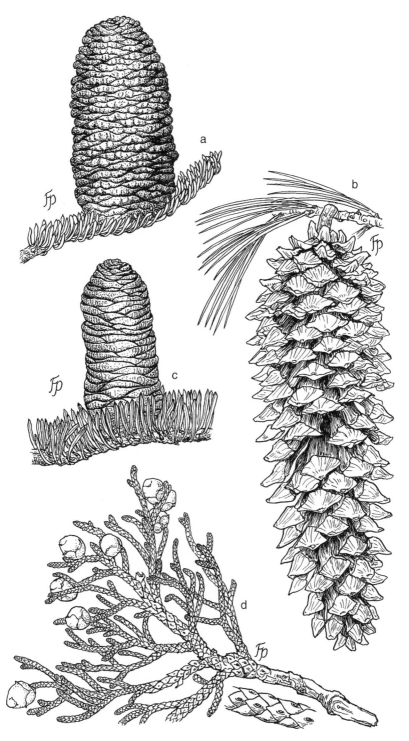

Fig. 5.17.   Montane Forest trees. (a) Red Fir, *Abies magnifica*. (b) Sugar Pine, *Pinus lambertiana*. (c) White Fir, *Abies concolor*. (d) Sierra Juniper, *Juniperus occidentalis*

Cones round, berrylike, about 0.35 in (0.8 cm) long, blue-black. Widely distributed in exposed dry sites. *Distribution* Higher mountains of California, Modoc Plateau; western Nevada; to Washington, Idaho.

WESTERN WHITE PINE, *Pinus monticola* Pine Family. For description see Subalpine Forest. Found in moist sites in scattered localities in the higher elevations of Montane Forest.

WHITE FIR, *Abies concolor* Pine Family. Fig. 5.17c. Erect tree, 50–230 ft (15–70 m), with narrow crown and short stiff branches. Mature bark gray, furrowed. Needles about 1.2–2.4 in (3–6 cm), flat, bluish green. Cones oblong-cylindric, 3–5 in (7.5–12.5 cm) long, erect, borne on upper branches, falling apart on tree when mature. Common in drier, lower elevation sites than Red Fir, although mixed stands are common. They can be difficult to tell apart. Both have gray bark when young; mature bark is distinctive. A Red Fir's new needles are grayish green, silvery; while those of White Fir are yellow-green. *Distribution* Most mountains of California and Nevada; to Baja California, Rocky Mountains.

COMMON UNDERSTORY HERBS OF MONTANE FOREST

Flowers White to Cream or Greenish

NUDE BUCKWHEAT, *Eriogonum nudum* Buckwheat Family. Fig. 5.18a. So named because all the leaves are at the base of the plant, leaving the flowering stems leafless, or "nude." Perennial rosette plant with flowering stems to 1.5 ft (0.5 m) tall. Leaves broadly oval, 0.5–1.5 in (1.3–3.8 cm) long, white-hairy on the lower surface. Flowers small, each with 6 petal-like parts, cream with pink veins. Flowers aggregated into globular clusters at the ends of the flowering stems. *Flowers:* June–Aug. Common in sunny to semi-shaded sites. Size, flower color, and leaf characteristics are highly variable. *Distribution* Much of California, western Nevada; to Washington, northwestern Mexico.

SIDEBELLS, ONE-SIDED WINTERGREEN, *Orthilia secunda* (=*Pyrola secunda)* Heath Family. Named for the arrangement of the small, bell-shaped flowers, which hang from one side of the stalk. Low perennial herb with slender underground stems that connect small colonies of seemingly distinct individuals. Leaves mainly at the base, bright green, oval, 1–2 in (2.5–5 cm) long. Flowering stalks 3–4 in (7.5–10 cm) tall, with small greenish flowers attached along one side. *Flowers:* July–Sept. Prefers moist, shaded spots beneath conifers. *Distribution* Widespread. High mountains of California, northern Nevada; circumboreal, subarctic.

Fig. 5.18.   Montane Forest understory herbs. (a) Nude Buckwheat, *Eriogonum nudum.*
(b) Western Wallflower, *Erysimum capitatum* ssp. *perenne.* (c) White-veined Shinleaf,
Pyrola picta. (d) Sierra Soda Straw, *Angelica lineariloba*

SIERRA SODA STRAW, SIERRA ANGELICA, *Angelica lineariloba* Carrot Family. Fig. 5.18d. Named for its long, straight, hollow stems. Robust perennial herb, 1.5–4.5 ft (0.5–1.5 m) tall, with leaves 4–14 in (10–35 cm) long, divided into a few very narrow segments. Flowers small, each with 5 tiny white petals, aggregated into large rounded clusters about 4 in (10 cm) across. *Flowers:* June–Aug. Common on dry, well-drained slopes. *Distribution* Sierra Nevada, eastern California, western Nevada.

WHITE-VEINED SHINLEAF, *Pyrola picta* Heath Family. Fig. 5.18c. Common in moist, shady places. Many rosettes connected by slender underground stems. Perennial. Leaves oval, 1–3 in (2.5–7.5 cm) long, gray-green, veined or mottled with white. Flowers greenish to cream, cup-shaped, about 0.5 in (1.3 cm) wide, on elongated flower stalks 4–8 in (10–20 cm) long. *Flowers:* June–Aug. "Shinleaf" is the common name for the genus; English peasants used a shinleaf in preparing plasters for healing bruises or sores on the legs. *Distribution* Mountains of California, except deserts; northern Nevada; to southwestern Canada, New Mexico.

Flowers Yellow to Orange

PINE-WOODS LOUSEWORT, *Pedicularis semibarbata* Figwort Family. Fig. 5.19d. Low perennial rosette plant with flowers borne at ground level on short stalks. Leaves green with a purple tinge, 2–6 in (5–15 cm) long, divided into many irregularly toothed segments. Flowers pale yellow with purple tips, less than 1 in (2.5 cm) long, 2-lipped, the upper lip forming a curved beak. *Flowers:* May–July. *Distribution* High mountains of California, western and southern Nevada.

SINGLE-STEMMED GROUNDSEL, *Senecio integerrimus* Sunflower Family. Fig. 5.19a. Perennial with leaves mainly at the base. Single flowering stem to 2 ft (0.7 m) tall. Leaves 2–6 in (5–15 cm) long, oval, with thin soft hairs. Flowering heads few, each with yellow marginal ray and central disk flowers, aggregated into a flat-topped open cluster. *Flowers:* May–Aug. *Distribution* Mountains of northern and central California, Nevada; to southwestern Canada, Wyoming, Colorado.

WESTERN WALLFLOWER, *Erysimum capitatum* ssp. *perenne (=E. perenne)* Fig. 5.18b. Mustard Family. Erect perennial herb with single flowering stalk to 1 foot (30 cm) tall. Leaves mainly at the base, narrow, to 2 in (5 cm) long. Flowers yellow, 4-petaled, about 0.75 in (2 cm) wide, developing into long "pods." *Flowers:* June–Aug. Common on dry slopes.

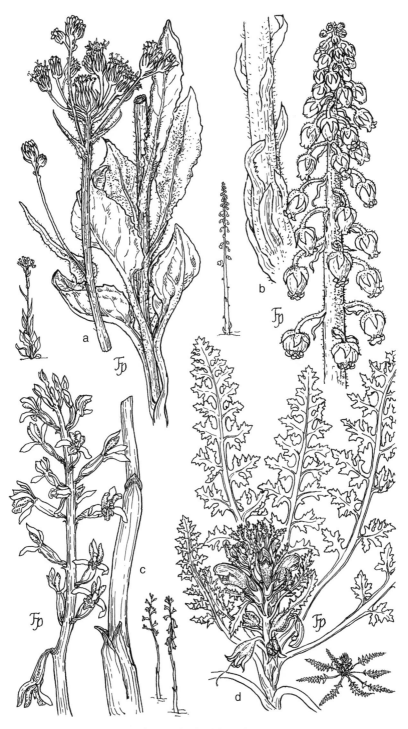

Fig. 5.19. Montane Forest understory herbs. (a) Single-stemmed Groundsel, *Senecio integerrimus*. (b) Pinedrops, *Pterospora andromedea*. (c) Spotted Coralroot, *Corallorhiza maculata*. (d) Pine-woods Lousewort, *Pedicularis semibarbata*

Distribution   Central Sierra to mountains of northern California; Oregon; Nevada.

Flowers Pink to Red

PINEDROPS, *Pterospora andromedea*   Heath Family. Fig. 5.19b, plate 10h. Stout, erect root-parasite, 1–3 ft (0.3–1 m) tall, stems and scalelike leaves purple-brown. Flowers urn-shaped, 0.35 in (0.8 cm) long, reddish yellow, scattered along the flowering stalk. *Flowers:* June–Aug. Pinedrops is a plant parasite. It lacks the ability to conduct photosynthesis, the process by which most plants convert the sun's energy into compounds necessary to make living tissues. Instead, it derives its nutrition from other plants, mainly pines, through root connections. *Distribution*   Mountains of northern and central California, western and northern Nevada; to British Columbia, Mexico; eastern North America.

SNOWPLANT, *Sarcodes sanguinea*   Heath Family. Plate 10f. Blood-red Snowplants begin to emerge from decaying forest duff as the last dregs of winter snow recede. California's only totally red plant. Stout, erect *saprophyte* (derives nutrients from decaying plant matter in the soil), 6–12 in (15–30 cm) tall, with thick stem buried in conifer needle litter. No leaves. Flowers about 0.5 in (1.3 cm) long, urn-shaped; many in a compact inflorescence. *Flowers:* May–July. *Distribution*   San Jacinto Mountains, Sierra Nevada, high mountains of northern California; Oregon, Baja California.

SPOTTED CORALROOT, *Corallorhiza maculata*   Orchid Family. Fig. 5.19c. This saprophytic orchid of heavily shaded sites derives its nutrients from decaying pine and fir litter. Erect, 8–16 in (20–40 cm) tall; branched, coral-like underground stems. Stems and scalelike leaves yellowish brown. Flowers about 0.5 in (1.3 cm) wide, reddish purple, the largest petal spotted and veined with crimson. Flowers loosely arranged along stalk. *Flowers:* June–Aug. *Distribution*   High mountains of California, San Francisco Bay area, northern Nevada; to British Columbia, eastern United States, New Mexico, Mexico, Guatemala.

MONTANE CHAPARRAL

Cross-country hikers dread Montane Chaparral, a community of dense, low shrubs that forms impenetrable thickets on hot, dry slopes. The dominant shrubs—including early-flowering species of manzanita and ceanothus, Huckleberry Oak, Bitter Cherry, Bush Chinquapin, and Curly-leaved Mountain Mahogany—are 3–6 feet (1–1.8 m) high and mainly evergreen. They tolerate extreme heat and drought, as well as winter cold. Their small, thick,

Fig. 5.20. Montane Chaparral

leathery leaves, with resinous or waxy coatings, minimize water loss. Several species, such as Snowbush, have stiff, spiny branches that overlap with those of neighboring shrubs, forming a dense cover that leaves little room for an understory. Occasional breaks in the canopy are filled with grasses and wildflowers that often poke out from the bases of the shrubs.

In the Eastern Sierra, Montane Chaparral occurs mainly from 6000 to 9500 feet (1830–2900 m), as patches within the Montane and Subalpine forests, where it occupies some of the hottest, driest sites. Usually these are south-facing slopes—steep and rocky with little soil, where greater exposure to the sun results in early snowmelt. Winters when snowfall is sparse keep the shrubs pruned low, since branches that protrude above the insulating snow blanket don't survive the desiccating effects of wind and cold.

Montane Chaparral occurs at higher elevations and is composed of lower-growing shrub species than either California's Coastal Chaparral or the Foothill Chaparral of the Sierra's west slope, which are fire-adapted communities that require frequent burnings to maintain them over the long term. In Montane Chaparral fires occur less frequently, and their role is not as clearly understood. However, one of the most common and widespread Montane

Chaparral shrubs, Greenleaf Manzanita, is adapted to fire. Its root-crown (burl), protected by thick, fire-resistant bark, can sprout new twigs and leaves within a month after fire. A deep root system that anchors it to the slopes and taps underground water probably contributes to its rapid sprouting ability. On dry forested slopes that have been burned or clear-cut, Montane Chaparral shrubs are often the first woody plants to appear. They may dominate such disturbed sites for decades, eventually being replaced by drought-tolerant forest trees like Jeffrey Pine.

COMMON SHRUBS OF MONTANE CHAPARRAL

BITTER CHERRY, *Prunus emarginata*   Rose Family. Plate 12c. Deciduous spreading shrub, 3–9 ft (1–3 m) tall, with shining red twigs. Leaves oval, 1–2 in (2.5–5 cm) long. Flowers about 0.5 in (1.3 cm) wide, with 5 white petals, borne in small clusters among the leaves. *Flowers:* Apr–May. The small round fruits, bright red cherries that sparkle invitingly, appear in late summer. The temptation to sample them is best resisted, for they are exceedingly bitter. *Distribution*   Mountains of California, western and southern Nevada; to British Columbia, Idaho, Montana, Wyoming.

BUSH CHINQUAPIN, *Chrysolepis sempervirens*   Oak Family. Fig. 5.21c. Spreading round-topped evergreen shrub, 2–7 ft (0.7–2.3 m) tall. Leaves oblong, 1–3 in (2.5–7.5 cm) long, with golden hairs on the lower surface. Female flowers very small, inconspicuous. Male flowers in erect catkins, about 1.5 in (3.8 cm) long. *Flowers:* July–Aug. Although Bush Chinquapin is related to oaks, its fruit, a distinctive spiny bur containing 1–3 seeds, bears little resemblance to an acorn.   *Distribution*   In California, San Jacinto and San Gabriel mountains, Sierra Nevada, northern Coast Ranges; western Nevada; to southern Oregon.

CURLY-LEAVED MOUNTAIN MAHOGANY, *Cercocarpus ledifolius*   Rose Family. Fig. 5.21d. Shrub 6–30 ft (2–9 m) tall. Branches spreading; bark pale when young, later reddish brown and furrowed. Leaves narrow, about 0.5–1 in (1.3–2.5 cm) long, curled under along the edges. Flowers small, greenish, borne singly or in small clusters along the branches. *Flowers:* Apr–June. In late summer thousands of silvery plumes—the "tails" of the fruits, which aid in seed dispersal—ornament the branches of these shrubs. Also occurs in Pinyon-Juniper Woodland.   *Distribution*   In California, Sierra Nevada, northern and desert mountains; Nevada; to Washington, Montana, Arizona, Baja California.

GREENLEAF MANZANITA, *Arctostaphylos patula*   Heath Family. Plate 12a. Spreading evergreen shrub 3–6 ft (1–2 m) tall, with several branches from

Fig. 5.21. Montane Chaparral shrubs. (a) Pinemat Manzanita, *Arctostaphylos nevadensis*. (b) Huckleberry Oak, *Quercus vaccinifolia*. (c) Bush Chinquapin, *Chrysolepis sempervirens*. (d) Curly-leaved Mountain Mahogany, *Cercocarpus ledifolius*

an enlarged root-crown (burl). Bark smooth, bright reddish brown. Leaves bright deep green, oval to roundish, 1–1.5 in (2.5–3.8 cm) long. Flowers urn-shaped, about 0.35 in (0.8 cm) long, light pink, in pendant clusters at the ends of the branches. *Flowers:* Apr–June. The fruit is a small, slightly flattened berry, resembling a tiny apple (the Spanish word *manzanita* means "little apple"). Another common manzanita of Montane Chaparral is **Pinemat Manzanita *(A. nevadensis)*** (fig. 5.21a), a prostrate shrub with narrow leaves that lacks a burl. The species sometimes grow together, but Pinemat Manzanita extends to higher elevations. *Distribution* Sierra Nevada, northern Coast Ranges; Nevada; to Oregon, Utah.

HUCKLEBERRY OAK, *Quercus vaccinifolia* Oak Family. Fig. 5.21b. Spreading, sometimes prostrate evergreen shrub to about 5 ft (1.7 m) tall. Leaves narrowly oval, to about 1.25 in (3 cm) long. Female flowers very small, inconspicuous. Male flowers in pendant catkins. Acorns nearly round, about 0.5 in (1.3 cm). *Flowers:* May–July. Extends nearly to treeline in dry sites. *Distribution* Sierra Nevada, northern California, western Nevada; Oregon.

SNOWBERRY, *Symphoricarpos rotundifolius (= S. vaccinoides)* Honeysuckle Family. Named for its fruits, fleshy white berries that mature in late summer. Spreading deciduous shrub to 5 ft (1.7 m) tall, with curving wand-like branches. Leaves oval to roundish, less than 1 in (2.5 cm). Flowers bell-shaped, pink, about 0.35 in (0.8 cm), borne individually or in pairs at the base of the leaves. *Flowers:* June–Aug. *Distribution* Sierra Nevada and desert mountains of California; Nevada; to Washington, Wyoming, Colorado, western Texas.

SNOWBUSH, *Ceanothus cordulatus* Buckthorn Family. In favorable years masses of white flowers, like a mantle of snow, cover these shrubs. Intricately branched evergreen shrub, 3–6 ft (1–2 m) tall, with spiny twigs and gray-green bark and leaves. Leaves oval, less than 1 in (2.5 cm) wide, white, in small dense clusters. *Flowers:* May–July. Fire enhances seed germination. *Distribution* Most California mountains; Nevada; to Oregon, northern Baja California.

TOBACCO BRUSH, *Ceanothus velutinus* var. *velutinus* Buckthorn Family. Plate 12b. On warm sunny days Tobacco Brush gives off a heavy, sweet fragrance, like certain pipe tobaccos. Round-topped, much-branched evergreen shrub, 3–6 ft (1–2 m) tall. Leaves oval, 1–3 in (2.5–7.5 cm) long, bright green and shining on top, dull beneath, resinous. Flowers small, white, about 0.25 in (0.6 cm) wide, in small dense clusters. *Flowers:* Apr–July. Tobacco Brush and Snowbush are closely related species that often grow to-gether. *Hybrids* (plants that result from cross-breeding between parents of different species) between the two are common in the wild. These offspring, with characteristics between those of the parents, probably result from "mis-takes" in pollination. The flowers of the two species, both pollinated by bees, are similar and bloom at the same time. *Distribution* Sierra Nevada, mountains of northern California; Nevada; to British Columbia, South Dakota.

MOUNTAIN MEADOWS

Meadows are the flower gardens of the mountains. Changing patterns of color and form emerge—this is their special beauty—as waves of summer flowers bloom and then die. Early in the season there are crimson paint-brushes, blue lupines, and pink shooting-stars. In midsummer—when the meadows are buzzing with bees and adorned by butterflies—the yellow groundsels and monkeyflowers, delicate pink willow-herbs, and robust Corn Lilies and Cow Parsnips appear. As summer fades, lavender fleabanes and asters, diminutive blue gentians, and spreading goldenrods take their places. At least 150 species of wildflowers grow in east slope meadows; 30 or more may occur within a single acre. Meadows occupy shallow basins that re-main moist or even soggy through all or part of the growing season. These basins are often filled-in tarns and lakes, with soils high in organic mat-ter. Although some wildflowers, like sky-blue Camas Lily, may be scattered throughout meadows, other species are often more dense and diverse on the slightly higher ground along the margins or on raised hummocks, perhaps indicating a preference for drier soils. Lower-elevation meadows boast the tallest species, while higher-elevation meadows have mostly low-growing plants.

Meadows vary tremendously; in fact, a different mix of species occurs in nearly every one. Variations in soil moisture, elevation, and species compo-sition provide a basis for distinguishing different types of Mountain Mead-ows. Wet meadows differ from dry ones in the amount and duration of soil moisture, as well as in the kinds of plants growing in them. Mountain Meadows, as we use the term, include both Subalpine and Montane mead-ows. They differ in elevation (which affects growing season), temperature, moisture, and other variables—all of which determine which species grow where.

The most abundant plants of Mountain Meadows are not the brightly colored perennial herbs we call wildflowers; they are, instead, the grasses, rushes, and sedges. Their narrow, ribbony leaves make up the shaggy green

Fig. 5.22.   Subalpine Meadows

carpets and silky lawns of Mountain Meadows. Of special importance are
the sedges—grasslike plants that form the dense sods of meadows. Several
dozen species grow in east slope meadows, many of them shorter than the
neighboring grasses and wildflowers. Sedge species are difficult to identify,
but as a group they can be distinguished from grasses and rushes (another
type of grasslike plant) by rolling the stems between your fingers. Since
grasses and rushes have round stems, they feel smooth. But "sedges have
edges," to quote a familiar botanical refrain; the three angles of the triangular stem can be felt easily.

SUBALPINE MEADOWS

On the Sierra's east slope the largest meadows, the Subalpine Meadows, are
found mainly along the headwater reaches of the major creeks, generally between 9000 and 11,000 feet (2743–3353 m). Often they occur in exposed sites,
surrounded by slopes of bare rock or by scattered conifers of the Subalpine
Forest. You will find good examples in Little Lakes Valley west of Toms Place,

Fig. 5.23.   Montane Meadows

in the Cottonwood Lakes Basin above Lone Pine, and in Twenty Lakes Valley below Mount Conness. Most meadows at these elevations, although covered by a continuous blanket of snow in winter, have shallow soils that dry out by the end of the growing season. Most plants are short compared to those in meadows downslope. Dwarf woody plants, such as White Mountain Heather and Alpine Laurel, commonly grow along the margins. Lodgepole Pine saplings may encroach upon the edges during dry years, then be killed back when wetter conditions return. In some high-altitude meadows, the "shorthair meadows," Shorthair Sedge *(Carex exserta),* is abundant; its bronze leaves impart a tawny hue in late summer.

MONTANE MEADOWS

Smaller, wetter meadows that occur within the Montane Forest zone, from 6000 to 9000 feet (1830–2740 m), are known as Montane Meadows. These meadows are typically in more protected sites, surrounded by Montane Forest or by groves of Quaking Aspen. Soils are dark and rich in organics, and quite soggy in early summer after winter snows have melted. Many are wet throughout the growing season, fed by flows from streams and springs. Wildflowers thrive as in a well-watered garden: luxuriant purple Meadow Lupine, crimson Meadow Paintbrush, erect stalks of Corn Lily, broad-leaved Cow Parsnip, nodding pink willow-herbs, all in great profusion. You will find good examples scattered in and near the Lake Tahoe Basin, in Hope Valley (south of Tahoe), in the Mammoth Lakes area, at Onion Valley (west of Independence), at Horseshoe Meadows (southwest of Lone Pine), and at Olancha Pass.

Almost all of California's plant communities have been grazed by livestock during some part of the last 150 years; Mountain Meadows are no exception. Many High Sierra meadows were severely overgrazed by sheep until the land was included in one of the national parks or national forests. In some areas, destructive grazing practices were unchecked until the 1920s and 1930s. Precisely what was lost we probably never will know. Comparisons with early photographs show that some meadows have not regained their original condition even though grazing ended 50–80 years ago. The impact of sheep grazing around Mount Whitney was described by mountain climber Thomas Magee in 1885 in the following passage: "Mountain meadows are abundant, but the sheepherder and his flocks have more largely worked their ruin in the Whitney region than anywhere else in the Sierra that I have visited. Each of these meadows is yearly cropped several times by various flocks of sheep, and the result is that, even where there was a genuine mountain meadow, there are now only shreds and patches. The sod and the verdure are gone—eaten and trodden out; gravel is now in the ascendant." Mountain Meadows within national forest boundaries are still seasonally grazed by cattle and pack stock, but this practice does not appear to be as destructive as the unrestricted sheep grazing of years past.

COMMON WILDFLOWERS OF MOUNTAIN MEADOWS

Flowers White to Cream or Greenish

ALPINE GENTIAN, *Gentiana newberryi*   Gentian Family. Plate 10d. Low perennial herb, 2–4 in (5–10 cm) high. Leaves broadly spoon-shaped or narrower, 1–2 in (2.5–5 cm) long. Flowers broadly funnel-shaped, 1–1.5 in (2.5–3.8 cm), white with green spots inside, with darker green bands outside. Each plant with one to a few flowers at ground level. *Flowers:* Aug–Sept. Occasional in moist meadows and on streambanks below 12,000 ft (3658 m). *Distribution*   Sierra Nevada and mountains of eastern California; western Nevada; Oregon.

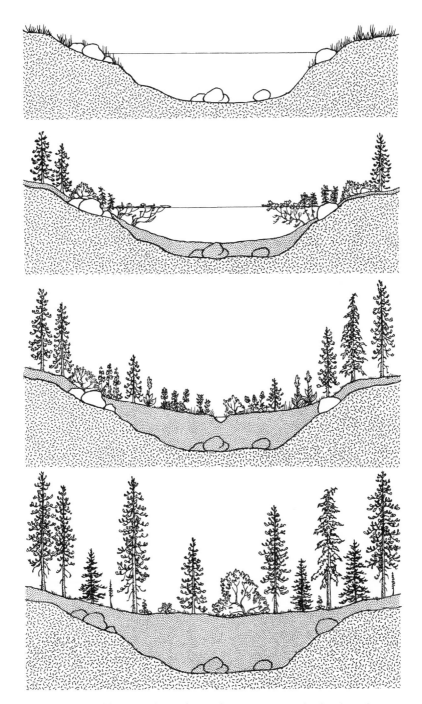

Fig. 5.24. From lake to meadow to forest. Plant communities develop through a process called *ecological succession*. Changes occur in stages, as shown in this example. A lake gradually fills with sediment, and aquatic life is replaced by a meadow community. Meadow plants invade the lake margins as soil develops, with species that prefer wet soils becoming established first. Much later, as the meadow fills in and dries, it will be replaced by a forest of conifers.

Fig. 5.25. Subalpine and Alpine meadow plants. (a) American Bistort, *Polygonum bistortoides*. (b) Primrose Monkeyflower, *Mimulus primuloides*

AMERICAN BISTORT, *Polygonum bistortoides* Buckwheat Family. Fig. 5.25a. Slender-stemmed perennial herb with underground rootstocks, to 2 ft (0.7 m) tall. Leaves mainly near base of stem, oblong, 4–10 in (10–25 cm). Flowers small, whitish, aggregated into elongated heads. *Flowers:* June–Aug. Common in wet meadows and along streams below 10,500 ft (3200 m). *Distribution* Mountains of California, northern Nevada; to Alaska, eastern North America.

CORN LILY, *Veratrum californicum* Lily Family. Plate 11a. A Midwesterner, upon seeing this plant, could easily believe that short, white-tasseled corn sprouts from Sierran meadows. However, in spite of their name, Corn Lilies are not edible. In fact, the roots and young shoots are poisonous. Robust herb to 6 ft (2 m) tall, the stems densely clothed with 1-foot-long (0.3-m), oval, pleated leaves that are pointed at the ends. Flowers about 1 in (2.5 cm) wide, each with 6 white petal-like parts, clustered densely in a spreading inflorescence 1–2 ft (0.3–0.7 m) long. *Flowers:* July–Aug. Common in wet meadows below 11,000 ft (3353 m). *Distribution* Mountains of California; Nevada; to Washington, Rocky Mountains, Mexico.

COW PARSNIP, *Heracleum lanatum (=H. sphondylium)* Carrot Family. Plate 11h. Giant perennial herb to 9 ft (3 m) tall, the stems sheathed by broad leaf bases. Leaves roundish, divided into 3 toothed leaflets, each up to 1 foot (0.3 m) long. Flowers about 0.5 in (1.3 cm) wide, each with 5 white petals, in dense, flat-topped clusters up to 1 foot (0.3 m) wide. Flowers on the margins of the clusters are larger than those in the center. *Flowers:* June–July. In moist meadows and wet places below 9000 ft (2743 m). *Distribution* Mountains and north coast, California; northern Nevada; to Alaska, eastern United States.

RANGER'S BUTTONS, SWAMP WHITEHEADS, *Sphenosciadium capitellatum* Carrot Family. Fig. 5.26a. Slender perennial herb to 5 ft (1.7 m) tall. Leaves up to 1.5 ft (0.5 m) long, divided into many oblong, toothed leaflets, each up to 5 in (12.5 cm) long. Flowers tiny, white, tightly aggregated into balls about 0.7 in (1.6 cm) wide. *Flowers:* July–Aug. Flourishes in meadow margins and other wet places, especially along streams, below 10,500 ft (3200 m). *Distribution* Mountains and north coast of California; western Nevada; Oregon; to Idaho, Baja California.

Flowers Yellow to Orange

ALPINE BUTTERCUP, *Ranunculus eschscholtzii* Buttercup Family. Plate 10a. Because many buttercups grow in moist places where frogs abound, the Roman scholar Pliny called them by the Latin name *ranunculus,* meaning "little frog." Slender-stemmed perennial herb, 2–6 in (5–15 cm) high, with leaves mainly at the base. Leaves round in outline, lobed and toothed, 0.5–1.5 in (1.3–3.8 cm) wide. Flowers about 1 in (2.5 cm) wide, each with 5 shiny, bright yellow petals, borne individually on slender stalks. *Flowers:* July–Aug. Grows in meadows and in rocky places from 8000 to 11,500 ft (2438–3505 m). *Distribution* Mountains of California, northern Nevada, Arizona; to Alaska, Rocky Mountains.

ARROWLEAF BUTTERWEED, *Senecio triangularis* Sunflower Family. Plate 11f. Several-stemmed perennial herb to about 4 ft (1.3 m) tall. Leaves arrowhead-shaped, toothed on the margins, up to 8 in (20 cm) long. Flowering heads yellow, aggregated into a flat-topped cluster up to several inches broad. *Flowers:* July–Sept. Common in wet meadows and along streambanks below 11,000 ft (3353 m). The related **Meadow Groundsel *(S. hydrophilus),*** with long, gray-green leaves and smaller flower heads that bloom late in the season, is also widespread in Mountain Meadows. *Distribution* Mountains of California, western and northern Nevada; to Alaska, Colorado.

COMMON MONKEYFLOWER, *Mimulus guttatus* Figwort Family. Plate 11b. Annual or perennial. Low succulent herb up to 1 foot (0.3 m) tall, with paired

Fig. 5.26.   Montane Meadow plants. (a) Ranger's Buttons, *Sphenosciadium capitellatum*.
(b) Spike Mallow, *Sidalcea oregana* ssp. *spicata*. (c) Wandering Daisy, *Erigeron peregrinus*.
(d) Monkshood, *Aconitum columbianum*

oval leaves up to 2.5 in (6.3 cm) long, irregularly toothed on the margins. Flowers look like a little monkey's face, yellow, 1–2 in (2.5–5 cm) long, 2-lipped, with maroon blotches and dots on lower lip. Flowers borne in pairs along the flowering stalk. *Flowers:* June–Aug. Thrives in wet meadows and other wet places below 10,000 ft (3048 m). **Lewis's Monkeyflower** *(M. lewisii),* a tall, pink-flowered perennial, is common on streambanks and around seeps and springs. *Distribution* Wide-ranging; throughout California, Nevada; to Alaska, Canada, Rocky Mountains, northern Mexico.

PRIMROSE MONKEYFLOWER, *Mimulus primuloides* Figwort Family. Fig. 5.25b. Diminutive perennial rosette plant, with flowering stalks to 5 in (12.5 cm) high. Leaves and stems covered with long, soft, white hairs. Leaves oblong, to 1.5 in (3.8 cm) long. "Monkeyface" flowers about 0.5 in (1.3 cm) wide, 2-lipped, yellow, with reddish dots on lower lip, borne individually on slender, naked stalks. *Flowers:* July–Aug. Flourishes in wet meadows and on grassy banks below 11,000 ft (3353 m). *Distribution* Mountains of California, Nevada; to Washington.

Flowers Pink to Red

ALPINE SHOOTING STAR, *Dodecatheon alpinum* Primrose Family. Plate 10e. Perennial rosette plant with flowering stalks to 6 in (15 cm) tall. Leaves narrow, up to 2.5 in (6.3 cm) long. Flowers about 0.75 in (1.8 cm) long, nodding, with 5 reflexed magenta petals exposing yellow, black, and maroon central parts. *Flowers:* July–Aug. Common in meadows below 12,000 ft (3658 m). The flowers of this group have inspired such whimsical names as roosters' heads, mosquitobills, prairie pointers, and mad violets! **Subalpine Shooting Star** *(D. subalpinum),* a similar species, occurs in meadows below 10,000 ft (3048 m). *Distribution* Mountains of California, northern Nevada; to Oregon, Utah, Arizona.

FIREWEED, *Epilobium angustifolium* Evening Primrose Family. Plate 11e. Perennial herb with flowering stalks up to 6 ft (2 m) tall. Flowers about 1 in (2.5 cm) wide, each with 4 deep pink to magenta petals. Tiny seeds with tufts of cottony hairs are borne aloft by the wind. *Flowers:* July–Aug. Common in open sites in meadows, near lakes, and along streams below 10,000 ft (3048 m). Especially common after fire and in other disturbed sites. *Distribution* Mountains of California, Nevada; to the Rocky Mountains; circumboreal.

ELEPHANT'S HEADS, *Pedicularis groenlandica* Figwort Family. Flowers resemble tiny elephants' heads with curled trunks. Perennial herb with stems to 2 ft (0.7 m) tall. Leaves at base and on lower stems, 4–6 in (10–15 cm) long, deeply divided, almost fernlike, with toothed divisions. Flowers pink, about 0.5 in (1.3 cm) long, in dense spikes on elongate flowering stalks. *Flowers:* June–Sept. Elephant's Heads are pollinated by bumble bees, whose fluttering wings provide the stimulus for pollen release. A similar plant, **Little Elephant's Heads** *(P. attolens)* (plate 10c), is about 1 foot (0.3 m) tall and has a fuzzy flowering stalk. Both occur in meadows below 11,000 ft (3353 m). *Distribution* Sierra Nevada and mountains of northern California, western and northern Nevada; to Alaska, eastern North America.

MEADOW PAINTBRUSH, *Castilleja miniata* Figwort Family. Plate 11d. Erect, several-stemmed perennial herb to 2 ft (0.7 m) tall. Leaves bright green, narrow, 1–2 in (2.5–5 cm) long, sometimes lobed. Flower bracts scarlet, enclosing narrow green flowers 1–1.5 in (2.5–3.8 cm) long. *Flowers:* May–Sept. Grows in meadows below 11,000 ft (3353 m) but is most common below 9500 ft (2896 m). More common in meadows at higher elevations is **Lemmon's Paintbrush** *(C. lemmonii),* a shorter plant with deep rose-pink flowers. *Distribution* Mountains of California, Nevada; to Alaska, Rocky Mountains.

SPIKE MALLOW, *Sidalcea oregana* ssp. *spicata* Mallow Family. Fig. 5.26b. Many-stemmed erect perennial herb to 2 ft (0.7 m) tall, stems and leaves covered with star-shaped hairs, visible with a hand lens. Leaves roundish in outline, deeply lobed and divided, to 3 in (7.5 cm) wide. Flowers like tiny hollyhocks, light pink, about 0.75 in (1.8 cm) wide, each with 5 white-veined petals, borne in dense spikes on elongate flowering stalks. *Flowers:* June–Aug. Inhabits meadows and other moist places below 8500 ft (2591 m). *Distribution* Mountains of northern and central California, western Nevada; Oregon.

SWAMP ONION, *Allium validum* Lily Family. Onion-scented bulb plant with elongate, naked flower stalks to 3 ft (1 m) tall. Leaves long and narrow, erect, almost as tall as the flowering stalks. Flowers rose-pink, each with 6 petal-like parts, tightly grouped into clusters about 1 in (2.5 cm) wide. *Flowers:* July–Sept. Edible and flavorful. Grows in meadows and near lakes and streams below 11,000 ft (3353 m). *Distribution* Sierra Nevada and mountains of northern California, northern Nevada; to British Columbia, Idaho.

Flowers Blue to Violet or Purple

CAMAS LILY, *Camassia quamash* Lily Family. Plate 10b. Bulb plant with erect flower stems, 0.8–2.4 ft (2–8 dm) tall. Leaves from the base of the stem, long, narrow. Flowers with 6 petal-like parts, usually blue, ranging from purple to white, about 1 in (2.5 cm) wide; many on elongate flower stalk. *Flowers:* May–July. An important food plant of native peoples and early settlers, now rare in areas where they established large settlements. Does not persist in sites heavily grazed by livestock. Flower color in this lily's populations is extremely variable; this has been attributed to native peoples, trappers, and pioneers transporting bulbs from place to place. *Distribution* Mountains of northern California, Modoc Plateau, northern Sierra Nevada, northern Nevada; to southwestern Canada, Wyoming, Utah.

MEADOW LUPINE, *Lupinus polyphyllus* Pea Family. Plate 11c. Robust spreading perennial herb to 5 ft (1.7 m), with stout hollow stems. Leaves divided into 5–9 leaflets, dark green, oval, 2–3 in (5–7.5 cm) long, arranged as the fingers are to the palm. Pea flowers, blue-purple, about 0.5 in (1.3 cm) long, loosely scattered along elevated flowering stalks. *Flowers:* June–Aug. Common in meadows and wet places below 8500 ft (2591 m). *Distribution* Mountains of California, northwestern Nevada; to British Columbia.

MONKSHOOD, *Aconitum columbianum* Buttercup Family. Fig. 5.26d. Robust perennial herb with several erect or spindly, climbing stems to 6 ft (2 m). Leaves roundish in outline, 2–5 in (5–12.5 cm) wide, deeply cleft into 3–5 toothed divisions. Flowers purplish blue, sometimes pale blue, 1–1.5 in (2.5–3.8 cm) high, the upper part forming a curved hood. Flowers in loose spikes up to 1.5 ft (0.5 m) long. *Flowers:* July–Aug. Monkshood varies in height, flower color, and flower shape from place to place. Grows in moist places below 8500 ft (2591 m), including meadow margins, often among willows. Monkshood contains an alkaloid, aconitine, that can impair vision and speech, and can cause nervousness, nausea, vertigo, and even death. Related species of *Aconitum* are grown as a source of several medicinal drugs. *Distribution* Sierra Nevada and mountains of northern California, northern Nevada; to British Columbia, South Dakota, New Mexico.

SIERRA GENTIAN, *Gentianopsis holopetala (=Gentiana holopetala)* Gentian Family. Plate 11g. The largest gentian of the Sierra Nevada; grows in wet meadows below 11,000 ft (3353 m). Several-stemmed perennial herb to 1 foot (0.3 m) tall. Lower stems with narrow leaves to 2 in (5 cm) long. Flowers deep blue-purple, funnel-shaped, up to 2 in (5 cm) long, borne individually at the ends of naked stalks. *Flowers:* July–Sept. *Distribution* Central and southern Sierra Nevada; western Nevada.

SIERRA PENSTEMON, *Penstemon heterodoxus* Figwort Family. Plate 10g. Slender-stemmed perennial to 8 in (20 cm) tall; most leaves in a rosette at the base of the flowering stalk. Leaves narrow, deep green, 1–2 in (2.5–5 cm) long. Flowers tubular, deep blue-purple, about 0.7 in (1.6 cm), extending straight out from the flowering stalk. *Flowers:* July–Aug. Common in meadows and rocky places, 8000–12,000 ft (2438–3658 m). *Distribution* Sierra Nevada, mountains of northern and eastern California, western Nevada.

WANDERING DAISY, *Erigeron peregrinus* Sunflower Family. Fig. 5.26c. Leafy-stemmed perennial herb to 2 ft (0.7 m). Leaves narrow, 2–4 in (5–10 cm) long. Flowering heads with purple marginal ray flowers and yellow central disk flowers, 1–1.5 in (2.5–3.8 cm) wide, one or a few per flowering stalk. *Flowers:* June–Aug. Common in meadows below 10,500 ft (3200 m). A close relative, **Short-rayed Daisy** *(Trimorpha lonchophylla =E. lonchophyllus),* with white to lavender ray flowers only 0.1 in (0.25 cm) long, blooms late in the season in moist meadows below 11,500 ft (3505 m). *Distribution* Wide-ranging; high mountains of California, northern and western Nevada; to Alaska, Montana, Colorado, New Mexico, eastern Asia.

EASTERN SIERRA RIPARIAN WOODLANDS

Lush woodlands of cottonwood, birch, and willow flourish on the banks of Eastern Sierra streams, following them from the mountains down into the arid desert scrublands. Although they intergrade, two types of streamside woodlands can be distinguished: Montane Riparian Woodland and High Desert Riparian Woodland. Both types can be found along the same stream; Montane Riparian Woodland borders the higher, mountain reaches, and High Desert Riparian Woodland borders the lower, desert stretches of the same stream. Although both consist of deciduous broad-leaved trees and shrubs with a luxuriant herbaceous understory, in general, the two woodlands contain different plant species.

Montane Riparian Woodland borders larger streams and many of the lower lakes of the Eastern Sierra. The predominant woody plants include Quaking Aspen, Mountain Alder, and Black Cottonwood. Some stands may include Mountain Maple, Narrow-leaved Cottonwood, and Lodgepole or Jeffrey pines (the latter two are also important trees of the surrounding Montane Forest). Willows and Creek Dogwood are abundant; their wand-like branches overarch the stream and sway with every breeze. The sun-dappled understory is crowded with bright-colored wildflowers, moisture-loving rushes and sedges, coarse-stemmed horsetails, tall, ribbon-leaved grasses, and lacy ferns. Many of these inhabit Mountain Meadows as well. Where canyon bottoms are flat and streams meander, this community ex-

Fig. 5.27. Montane Riparian Woodland

tends away from the immediate streambanks, intermingling with trees of the Montane Forest. In steeper terrain, Montane Riparian Woodland hugs the streambanks and consists of just a narrow strip of trees nearly hidden by the surrounding conifers. Although the neighboring forest may be hot and dry in summer, the shaded streambank woodland remains cool, thanks to the rushing water. On the Sierra's east side, Montane Riparian Woodland ranges from 6000 to 9000 feet (1830–2740 m). Good examples are found along Mammoth, Lee Vining, Mill (in Lundy Canyon), and Virginia creeks, around Twin Lakes (near Bridgeport), and along the Truckee and Carson rivers and their tributaries.

Below 7000 feet (2130 m), especially south of Mammoth Lakes, different plants grow along the streams, although some are the same as those found higher. This woodland, High Desert Riparian Woodland, borders streams of high-elevation desert habitats. Trees include Black and Fremont cottonwoods, Copper Birch, and several species of willow. In summer the tree canopy forms a rich ribbon of green—a striking contrast to the dull browns and gray-greens of the surrounding desert scrub. Three trees more typical of Montane Forest also occur here: Jeffrey and Ponderosa pines, and California

Black Oak. Interestingly, the latter two, common members of west slope Montane Forest, don't occur in east slope Montane Forest south of Lake Tahoe. They are restricted to just a few Eastern Sierra streams, a sign of the east slope's greater aridity. Under the trees grow thorny thickets of Interior Wild Rose, sticky shrubs of Sierra Nevada Currant, and dense tangles of water-loving herbs. In some sites, trailing vines of Virgin's Bower climb into the tree canopy. Even though many Eastern Sierra streams are small, compared to rivers of the west slope, in most years they maintain a perennial flow that provides abundant moisture for plants growing along their banks. Ample moisture, coupled with constant sunshine and warm temperatures that come as early as April, creates a favorable growing season of about six months—one of the longest in the Eastern Sierra, with plenty of time for large deciduous trees and shrubs to grow an entire new set of leaves each year. Because they aren't moisture-limited, riparian plants reach their peak of growth in summer as temperatures climb into the triple digits. On dry fans, literally a stone's throw away, desert shrubs have long since become drought-dormant. Most of the streams and some of the lakes from Convict Creek (in Mono County) south are bordered by High Desert Riparian Woodland from about 4000 to 7000 feet (1220–2130 m) elevation. This community also borders the lower stretches of the Walker, Carson, and Truckee rivers, and the Susan River near Susanville.

WOODY PLANTS OF MONTANE RIPARIAN WOODLAND

AMERICAN DOGWOOD, CREEK DOGWOOD, *Cornus sericea (=C. stolonifera)*   Dogwood Family. Fig. 5.28b. Creek Dogwood's eye-catching deep red leaves and clusters of white fruits enhance the Eastern Sierra's fall color. Spreading shrub, 6–15 ft (2–5 m) tall, with reddish purple twigs. Leaves dark green, 2–4 in (5–10 cm) long, arranged in pairs opposite each other. Flowers small, each with 4 white petals, in small dense clusters about 1.5 in (3.8 cm) wide, at the ends of the branches. *Flowers:* May–July.   *Distribution*   Wide-ranging. Streambanks throughout California and Nevada; to Alaska; eastern North America; Mexico.

MOUNTAIN ALDER, *Alnus incana* ssp. *tenuifolia (=A. tenuifolia)*   Birch Family. Fig. 5.28c. Shrub or small tree, 3–25 ft (1–7.6 m) tall, with smooth gray or reddish brown bark. Leaves somewhat heart-shaped, 1–2.5 in (2.5–6.3 cm) long, toothed along the margin. Male catkins clustered, female "cones" about 0.5 in (1.3 cm) long, both hanging from the branches. *Flowers:* Apr–June.   *Distribution*   Mountains of California, western and northern Nevada; to Alaska, western Canada, Wyoming, New Mexico.

Fig. 5.28.   Montane Riparian Woodland shrubs. (a) Mountain Red Elderberry, *Sambucus racemosa* var. *microbotrys*. (b) American Dogwood, *Cornus sericea*. (c) Mountain Alder, *Alnus incana* ssp. *tenuifolia*. (d) Prickly Currant, *Ribes montigenum*

MOUNTAIN RED ELDERBERRY, *Sambucus racemosa* var. *microbotrys* *(= S. microbotrys)* Honeysuckle Family. Fig. 5.28a. Shrub to 6 ft (2 m) tall, leaves divided into 5–7 leaflets, each 1–3 in (2.5–7.5 cm) long, coarsely toothed on margins. Flowers small, each with 5 cream-colored petals, in flat-topped clusters. Berry red, poisonous. *Flowers:* June–Aug. **Blue Elderberry** *(S. mexicana =S. caerulea),* with blue berries, is also common in Montane Riparian Woodland and other moist sites. *Distribution* Higher mountains of California, Nevada; to Colorado, Arizona.

PRICKLY CURRANT, *Ribes montigenum* Gooseberry Family. Fig. 5.28d. Straggling low shrub to 2 ft (0.7 m) tall, with prickly twigs. Leaves roundish, up to 1 in (2.5 cm) wide, deeply 5-lobed. Flowers small, saucer-shaped, greenish purple. Berries small and red. *Flowers:* June–July. Prickly Currant extends up into alpine elevations, in both moist and dry sites. Shrubs in the genus *Ribes* are an intermediate host for White Pine Blister Rust (*Cronartium ribicola*), a serious pest of Sugar Pine and other commercial timber species.

Relatives found in Montane Riparian Woodland include: **White-stemmed Gooseberry** *(R. inerme =R. divaricatum* var. *inerme),* with greenish purple flowers and dark, reddish purple berries; **Golden Currant** *(R. aureum),* with yellow flowers and red to black berries; and **Mountain Pink Currant** *(R. nevadense),* with rose to red flowers and blue-black berries. *Distribution* Higher mountains of California and Nevada; to British Columbia, Idaho, Nevada, Arizona.

QUAKING ASPEN, *Populus tremuloides* Willow Family. Plate 14b. Eastern Sierra fall color is largely a gift of the aspen, whose shimmering golden leaves foretell winter's approach. Slender tree 6–70 ft (2–20 m) tall, smooth greenish white bark, many narrow branches. Leaves roundish, 1–2.5 in (2.5–6.3 cm) long, bright yellow-green, on slender stalks that allow them to tremble in the slightest breeze. Catkins 1–2.5 (2.5–6.3 cm) long, drooping, appear before the leaves. *Flowers:* Apr–June.

Dense groves of Quaking Aspen are not restricted to streambank sites. They surround springs and seepy areas, and extend over broad slopes where abundant moisture lies near the surface. For example, the aspen-covered slopes north of Conway Summit and below Monitor Pass are far above the nearest stream. The presence of large, old aspens in seemingly dry spots may indicate that the area previously was wetter or that water is present near the surface. Aspens reproduce mostly from lateral roots that send up numerous sucker shoots, forming extensive clones that may be thousands of years old. Different clones on the same hillside may show slight differences in flowering and leafing-out times, and in autumn leaf color. *Distribution* Widespread. Mountains of California, Nevada; to Alaska, eastern North America, Mexico.

TEA-LEAFED WILLOW, *Salix planifolia* Willow Family. Erect shrub, 1–6 ft (0.3–2 m) tall, with yellow-green to reddish brown twigs. Leaves narrowly oval, pointed, 1–1.5 in (2.5–3.8 cm) long. Catkins about 1 in (2.5 cm) long, appearing with the leaves. *Flowers:* June–Aug. Other willows found in Montane Riparian Woodland include: **Shining Willow** *(S. lucida* ssp. *caudata = S. caudata)* and **Drummond's Willow** *(S. drummondiana)*, both shrubs 3–12 ft (1–4 m) tall. *Distribution* Central Sierra to northern and eastern North America.

WOODY PLANTS OF HIGH DESERT RIPARIAN WOODLAND

ARROYO WILLOW, *Salix lasiolepis* Willow Family. Fig. 5.30a. Shrub or small tree, 6–35 ft (2–10 m) tall. Leaves narrow, 2–4 in (5–10 cm) long, dark green. Catkins 1–3 in (2.5–7.5 cm) long, made up of many tiny flowers, all either male or female. Male and female catkins borne on different plants. *Flowers:* Feb–Apr. Two other tree-sized willows found in High Desert Riparian Woodland are: **Goodding's Black Willow** *(S. gooddingii)*, with yellowish twigs, and **Shining Willow** *(S. lucida* ssp. *lasiandra =S. lasiandra)*, with reddish twigs. *Distribution* Widespread. Streams throughout California and Nevada; to Washington, Idaho, Texas, Mexico.

BLACK COTTONWOOD, *Populus balsamifera* ssp. *trichocarpa (=P. trichocarpa)* Willow Family. Fig. 5.30c. Deciduous spreading tree with rounded crown, 90–195 ft (28–60 m) tall. Leaves broadly oval with pointed tip, 1.5–4 in (3.8–10 cm) long, dark green on top, pale beneath. Catkins 1.5–3 in (3.8–7.5 cm) long, made up of many small flowers, all either male or female. Male and female catkins borne on different trees. *Flowers:* Feb–Apr. Black Cottonwood also grows around meadows in the Bishop area and near Coleville, where many stands have been planted. It hybridizes with Fremont Cottonwood. *Distribution* Much of California and Nevada except deserts; to Alaska, northern Rocky Mountains, Utah, northern Baja California.

CALIFORNIA BLACK OAK, *Quercus kelloggii* Oak Family. For description see Montane Forest. In the southern part of the Eastern Sierra this oak grows only on the banks of Independence and Oak creeks.

CALIFORNIA COFFEEBERRY, *Rhamnus californica* Buckthorn Family. Fig. 5.30d. Upright shrub to 6 ft (2 m), with reddish bark on the young twigs. Leaves persistent, oval, about 1–2 in (2.5–5 cm) long. Flowers small, green-

Fig. 5.29.   High Desert Riparian Woodland

ish, inconspicuous. Fruit a reddish brown berry. *Flowers:* Apr–July.   *Distribution*   Throughout most of California; southwestern Oregon, Arizona, New Mexico.

COPPER BIRCH, WATER BIRCH, *Betula occidentalis (=B. fontinalis)*
Birch Family. Plate 14a. Large, spreading deciduous shrub to about 25 ft (8 m) tall, with several main trunks. Trunks and branches with shining copper-colored bark. Leaves broadly oval, about 1 in (2.5 cm) long, toothed on the margins and covered with sticky secretions. Catkins about 1 in (2.5 cm) long. *Flowers:* Apr–May.   *Distribution*   Widespread. Mountains of California, Nevada; scattered throughout western North America.

FREMONT COTTONWOOD, *Populus fremontii* ssp. *fremontii (=P. fremontii)*   Willow Family. Fig. 5.30b. Deciduous tree with broad open crown, 30–100 ft (10–31 m) tall. Leaves triangular in outline, 1.5–3 in (3.8–7.5 cm) long, light green on both surfaces, with teeth along the margins. Catkins about 2 in (5 cm) long, made up of many small flowers, either all male or all female. Male and female catkins borne on different trees. *Flowers:* Mar–Apr. Fremont Cottonwood is restricted mainly to the Eastern Sierra's desert valleys; its range overlaps that of Black Cottonwood, but the latter extends up

Fig. 5.30. High Desert Riparian Woodland shrubs and trees. (a) Arroyo Willow, *Salix lasiolepis*. (b) Fremont Cottonwood, *Populus fremontii* ssp. *fremontii*. (c) Black Cottonwood, *Populus balsamifera* ssp. *trichocarpa*. (d) California Coffeeberry, *Rhamnus californica*

some stream courses to nearly 9000 ft (2743 m). *Distribution* Most of California, Nevada; to central Rocky Mountains, northern Mexico.

INTERIOR WILD ROSE, *Rosa woodsii* var. *ultramontana* Rose Family. Plate 14d. Erect, pleasant-smelling prickly shrub with sprawling branches, 3–9 ft (1–3 m) tall. Leaves divided into 5–7 oval leaflets, 0.5–1.5 in (1.3–3.8 cm) long. Flowers 1–2 in (2.5–5 cm) wide, typical "wild roses" with 5 pink petals and many yellow stamens. Rosehips bright red. *Flowers:* June–Aug. *Distribution* Sierra Nevada and mountains of southern California, Nevada; to British Columbia, Montana.

JEFFREY PINE, *Pinus jeffreyi* For description see Montane Forest. Jeffrey Pine occurs in some higher elevation sites of High Desert Riparian Woodland—for example, along Convict Creek above 6500 ft (1980 m) and along Lower Rock Creek near 6000 ft (1830 m).

MOUNTAIN PINK CURRANT, *Ribes nevadense* Gooseberry Family. Open deciduous shrub, 3–6 ft (1–2 m) tall. Leaves roundish in outline, 1–3 in (2.5–7.5 cm) wide, with 3–5 lobes. Flowers about 0.35 in (0.8 cm) long, rose to deep red, drooping, in loose, elongate clusters. Fruit a small, blue-black berry. *Flowers:* May–July. Currants and gooseberries are both members of the genus *Ribes*. Gooseberries have spiny twigs and berries, whereas currants lack spines. All the Sierran currants and gooseberries are edible, although some are more palatable than others. *Distribution* Mountains of California, western Nevada; southern Oregon.

PONDEROSA PINE, *Pinus ponderosa* For description see Montane Forest. Ponderosa Pine is absent from the Eastern Sierra south of Alpine Co., except for a few sites in High Desert Riparian Woodland along Lower Rock, Pine, Oak, Big Pine, Bishop, and Independence creeks, where it occurs below 6500 ft (1980 m). The tall pine that the community of Lone Pine was named after, which fell in 1900 when undermined by high water, was almost certainly a Ponderosa Pine.

VIRGIN'S BOWER, *Clematis ligusticifolia* Buttercup Family. Woody vine that climbs into crowns of trees and shrubs, sending trailing branches to the ground. Leaves divided into 5–7 oval, pointed leaflets with large teeth along the margins. Flowers about 1 in (2.5 cm) wide, petal-like parts cream-colored. *Flowers:* Mar–Aug. The stems and foliage are sometimes so dense that they blanket the branches of supporting trees. Early settlers used a potion made from Virgin's Bower to treat cuts and sores on horses. *Distribution* Throughout California and Nevada; to British Columbia, South Dakota, New Mexico, northwestern Mexico.

Fig. 5.31.   Pinyon-Juniper Woodland

PINYON-JUNIPER WOODLAND

Squat gray-green conifers, more shrubby than treelike, dominate the Eastern Sierra's Pinyon-Juniper Woodlands. The understory is decidedly desertlike, sparsely populated with scattered low shrubs, perennial wildflowers, and grasses. These woodlands cover slopes between 5000 and 9000 feet (1520 – 2740 m), where soils are gravelly and well drained. Yearly rainfall, mainly winter snow with occasional summer thundershowers, ranges from 10 to 20 inches (25 – 50 cm). Pinyon-Juniper Woodland extends southward along the eastern flank of the Sierra in a nearly continuous band from Alpine County to Kern County; farther east, it covers most of the low hills and mountain slopes of eastern California and western Nevada. Fine examples are found along Highway 395 south of Carson City and on Sherwin Grade north of Bishop. Throughout its range, this community is closely associated with Sagebrush Scrub. On Sierran slopes and uplands immediately to the east, stands of pinyon and juniper commonly occupy higher, steeper slopes with rockier soils, higher rainfall, and cooler temperatures, while Sagebrush Scrub occurs just below in sites that are less steep, less rocky, hotter, and drier. In desert ranges farther east this pattern holds with one exception—a second band of Sagebrush Scrub grows above the Pinyon-Juniper belt.

In the Eastern Sierra, most Pinyon-Juniper Woodlands are dominated by just one kind of tree, Single-leaf Pinyon; along the eastern base of the Sierra and on some slopes of the desert ranges to the east, extensive woodland stands lack junipers entirely. In some areas junipers grow with the pinyons to form mixed woodlands. On Sierran slopes it is the Sierra Juniper, a tree more at home in the upper reaches of the Montane Forest, that keeps company with the pinyons. East of the Sierra most mixed woodlands have Utah Juniper, a large bushy tree that extends across the Great Basin but stops short of Sierran slopes. The understory shrubs in these woodlands are, for the most part, the same plants found in the Sagebrush Scrub community, including Big Sagebrush, Antelope Bitterbrush, and Rubber Rabbitbrush. The wildflowers of Pinyon-Juniper Woodland are mainly perennials, including vividly colored species of penstemon and paintbrush; most of these also occur in Sagebrush Scrub. Woodlands composed entirely of Sierra Juniper are conspicuous along Highway 395 from just north of Hallelujah Junction (intersection of Highways 395 and 70) to Honey Lake Valley.

The effects of fire on Pinyon-Juniper Woodlands, though largely unstudied, appear to include a dramatic, long-term loss of trees, especially pinyons. Burned sites remain treeless for decades after fire; pinyon seedlings establish slowly. One can only speculate on the reasons for this lengthy recovery period. Perhaps the few pinyon seeds that have escaped being eaten by Pinyon Jays, Clark's Nutcrackers, Golden-mantled Ground Squirrels, and other wildlife are killed by fire. The habitat, especially after it has been cleared by fire, is hot and dry. Possibly pinyon seeds can germinate only when these conditions are ameliorated by unusually cool, wet weather. Also, fire can change the understory from native shrubs and herbs to a sea of grass—often nonnative grass—including the annual Cheat Grass *(Bromus tectorum),* and the perennial Crested Wheatgrass *(Agropyron cristatum),* the latter a native of Russia and Turkey that has been planted throughout the Great Basin to improve forage quality. Unfortunately, both of these grasses can compete with pinyon seedlings, slowing woodland recovery.

SHRUBS AND TREES OF PINYON-JUNIPER WOODLAND

SINGLE-LEAF PINYON, *Pinus monophylla*   Pine Family. Fig. 5.32a. Small rounded tree, 15–50 ft (5–15 m) tall, with divided trunk. Needles gray-green, incurved, about 1–1.5 in (2.5–3.8 cm) long, borne singly on the branches. Mature female cones nearly spherical, 1.5–2.5 in (3.8–6.3 cm) long, golden brown, with spreading scales, very resinous. Single-leaf Pinyon is the only pine of about 100 species worldwide with single needles; all others have them in bundles of 2, 3, 4, or 5. Each fall Pinyon Jays and Clark's Nutcrackers cache quantities of the large nutritious seeds, storing food for winter and spring. Seeds in unrecovered caches may germinate. This is the main way that pinyon pines are dispersed.

The seeds ("pine nuts") of these trees were an important staple food of the native Paiutes. In autumn family groups harvested the nuts from groves that they "owned." Unopened cones were knocked from the trees and roasted in fire pits to release the seeds. The nuts were eaten plain or ground into a mush, to which other foods, such as meat, seeds, or insects, were added. Mono Lake Paiutes carried pine nuts—along with obsidian, salt cakes, and dried brine fly pupae—over the Sierra Crest to barter with west slope groups for acorns, bead money, and manzanita berries. Some pine nuts sold in stores today are harvested by hand by local Native Americans, although most are imported from China. Single-leaf Pinyon is the state tree of Nevada, where mechanical harvesting of the seeds is prohibited to protect the traditional collecting rights of the native peoples. *Distribution* Mountains of central California, Sierra Nevada, California High Desert, and desert mountains; Nevada; to southeastern Idaho, northern Baja California.

*UTAH JUNIPER, *Juniperus osteosperma* Cypress Family. Fig. 5.32b. Small tree or large shrub, 9–20 ft (3–6 m) tall, with short unbranched trunk covered with gray bark. Crown rounded but fairly open. Leaves tiny, scale-like, closely pressed to the branches. Female cones ("berries") about 0.35 in (0.8 cm) long, reddish brown beneath a thin whitish waxy coating. Berries are dispersed by birds. *Distribution* San Gabriel and San Bernardino mountains, High Desert, and desert mountains in California; Nevada; to Montana, New Mexico.

CURLY-LEAVED MOUNTAIN MAHOGANY, *Cercocarpus ledifolius* For description see Montane Chaparral.

GREEN EPHEDRA, MORMON TEA, *Ephedra viridis* Ephedra Family. Fig. 5.32d. Shrub to 4.5 ft (1.5 m) high, with many stiff, erect, bright green branches. Male and female "cones" on separate plants. Female "cones" oval, about 0.35 in (0.8 cm) long, with a pair of brown seeds in each. Male "cones" similar in size and shape, but with many protruding yellow stamens. Desert travelers used twigs of this plant to brew a supposedly stimulating (but rather blah-tasting) tea. *Distribution* Central and eastern California mountains, High Desert in California; Nevada; to Colorado.

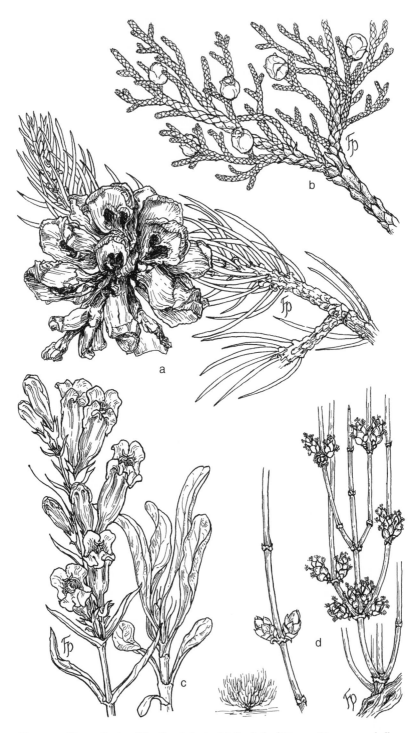

Fig. 5.32.   Pinyon-Juniper Woodland plants. (a) Single-leaf Pinyon, *Pinus monophylla*.
(b) Utah Juniper, *Juniperus osteosperma*. (c) Showy Penstemon, *Penstemon speciosus*.
(d) Green Ephedra, *Ephedra viridis*

SIERRA JUNIPER, *Juniperus occidentalis*   For description see Montane Forest.

WILDFLOWERS OF PINYON-JUNIPER WOODLAND

Most wildflowers described below under Sagebrush Scrub extend at least into the lower ranges of Pinyon-Juniper Woodland.

PENSTEMON, BEARD-TONGUE, *Penstemon* species   Figwort Family. Erect perennial herbs with leaves in opposite pairs along the stems. Flowers elongate, often tubular or funnel-shaped, with upper and lower lips, often blue to purple, sometimes pink or red, rarely other colors. *Flowering times:* May–Aug. The name "beard-tongue" refers to the sterile stamen (male flower part) that, in some species, is densely "bearded" with hairs. Descriptions follow for a few of the more distinctive or widespread penstemons found in Pinyon-Juniper Woodland.

*INYO MOUNTAIN PENSTEMON, *Penstemon scapoides*   Stems few, to 1.5 ft (0.5 m) tall. Leaves mainly at the base, oval to roundish, to 0.75 in (1.8 cm) long. Flowers blue with pale throat, about 1.35 in (3.3 cm) long. Sterile stamen with yellow "beard" of hairs. *Flowers:* June–July.   *Distribution*   On dry hills surrounding the Owens Valley, and in Mono Co.

*ROSE PENSTEMON, *Penstemon floridus*   Plate 13g. Erect stems to 3.5 ft (1.2 m) tall. Stems and leaves blue-green. Flowers rose-pink, about 1 in (2.5 cm) long, with strongly inflated throat. *Flowers:* May–June.   *Distribution*   East slope of Sierra, in and near the White-Inyo Mountains, southern Nevada.

SCARLET PENSTEMON, *Penstemon rostriflorus (=P. bridgesii)*   Stems erect, to 3 ft (1 m) tall, with yellow-green leaves. Flowers scarlet to vermilion, about 1 in (2.5 cm) long. *Flowers:* June–Aug. Pollinated by hummingbirds. Also in Montane Chaparral.   *Distribution*   Sierra Nevada, desert mountains in California, southern Nevada; to Colorado, New Mexico.

SHOWY PENSTEMON, *Penstemon speciosus*   Fig. 5.32c. Stems in clumps, to 2.5 ft (0.8 m) tall, sometimes grayish green. Flowers showy, blue-violet, 1–1.5 in (2.5–3.8 cm) long. *Flowers:* May–July. Also in Sagebrush Scrub and Montane Forest.   *Distribution*   Widespread. Most mountains in California, Nevada; to Washington, Idaho, Utah.

Fig. 5.33. Sagebrush Scrub

SAGEBRUSH SCRUB

Sagebrush, a rather drab little shrub, forms a shadowy gray-green mantle over some of the Eastern Sierra's most dramatic scenery. Sagebrush Scrub is the treeless community of low shrubs that stretches across much of the High Desert. It seems to go on forever, and it does range across that enormous expanse of high country between the Rockies and the Sierra—the Great Basin—where rivers drain into landlocked basins instead of into the sea. Some find the landscapes of this region monotonous, others find them soothing. Regardless, the smell of wet sagebrush is unforgettable, bringing to mind crashing summer thunderstorms that split the sky with lightning, then grace it with double rainbows.

Most common shrubs of Sagebrush Scrub, such as Big Sagebrush, Antelope Bitterbrush, and Rubber Rabbitbrush, grow 3–6 feet (1–2 m) high and have small gray-green leaves present throughout the year. While these shrubs grow densely enough to hide many a jack rabbit, it is possible to wander through them unimpeded. The understory is often sparse. In early summer annual herbs flower between the shrubs; the native perennial grasses mature later.

The climate is harsh and consists mainly of two seasons: long, cold winters and hot, dry summers. Precipitation is scarce, ranging from 5 to 15 inches (12.5–37.5 cm), most of it falling as snow in winter, some coming from summer thundershowers. Most plants living in the arid sagebrush climate grow and flower during a brief period in early summer when the soil is still moist and daily temperatures are moderate. For example, Rubber Rabbitbrush completes at least 60 percent of new stem growth during a two-week surge in May. Grasses and other herbs follow a similar pattern, although in favorable years, when warm-wet conditions are prolonged, wildflowers grow more vigorously and flower well into summer. Most shrubs, such as Antelope Bitterbrush, Spiny Hop-sage, Desert Peach, and Cotton Thorn, flower from late May to early June. Interesting exceptions are species of sagebrush and rabbitbrush, which flower from August to October; this may be advantageous for pollination. Since most other insect-pollinated plants have finished flowering by late summer, many beetles, bees, and butterflies are looking for new sources of nectar and pollen at about the time that rabbitbrush bursts into golden bloom.

From eighty to a hundred years of livestock grazing throughout the range of Sagebrush Scrub have changed dramatically the composition of the understory. Now absent, except in ungrazed plots such as old fenced cemeteries, the understory of pregrazing times was dominated by dense clumps of perennial bunch grasses. Cattle and sheep, preferring them to other plants, decimated them. Introduced annual grasses, especially Cheat Grass or Downy Brome, have largely replaced the native perennials, a shift that has reduced the grazing value of the rangelands. Since the growth of annual grasses is highly variable, depending upon local rainfall, and since they are palatable for no more than a few weeks, they are not as reliable a source of forage as the longer-lived perennials. Annual grasses, which become tinder-dry in summer, are also more fire-prone. After fire, the shrub composition of burned stands often changes, since Cotton Thorn, Desert Peach, and rabbitbrush species sprout back after fire, whereas sagebrush and most bitterbrush plants do not. Once-dominant bunch grasses still found scattered in the understory include Great Basin Wildrye *(Leymus cinereus =Elymus c.)*, Idaho Fescue *(Festuca idahoensis)*, Bluebunch Wheatgrass *(Pseudoroegneria spicata =Agropyron s.)*, and several kinds of needlegrass *(Achnatherum* species *=Stipa)*.

In eastern California, Sagebrush Scrub extends from northern Inyo County to the Oregon border. It covers the slopes of alluvial fans and moraines, extending down into valley bottoms in some sites. It occurs on a wide variety of soils—from gravelly to fine-textured, although usually well drained. The typical community ranges from about 4000 to 9000 feet (1220 – 2740 m) in elevation. On higher Sierran slopes Sagebrush Scrub occurs within Montane Forest. Even higher up, a low-growing, high-altitude form of Sagebrush Scrub, sometimes dominated by Alpine Sagebrush, extends up to 12,000 feet (3659 m).

The Sagebrush Scrub found on higher Sierran slopes is far richer in wild-flowers than the valley-bottom community. In July, brilliant Scarlet Gilia and Desert Paintbrush, sunshine-yellow Mules Ears and Sulphur Buckwheat, purple and blue lupines, penstemons, and locoweeds cover the slopes. Per-haps the good drainage or cooler weather favors these flowers, or perhaps they have escaped the ravages of livestock grazing. Good examples of this "enriched" Sagebrush Scrub occur on Monitor Pass and at Minaret Summit west of Mammoth Lakes.

COMMON SHRUBS OF SAGEBRUSH SCRUB

BIG SAGEBRUSH, GREAT BASIN SAGEBRUSH, *Artemisia tridentata* ssp. *tridentata* Sunflower Family. Fig. 5.34b. Rounded aromatic shrub 2–6 ft (0.7–2 m) high. Leaves narrow, 0.5–1.5 in (1.3–3.8 cm) long, with 3 rounded teeth. Flowers very small, yellow, in clusters 0.25 in (6 mm) long on elongate flowering stems. *Flowers:* Aug–Oct. Big Sagebrush occurs throughout the range of this community and is the most common sagebrush of the Eastern Sierra.

Big Sagebrush provides nesting sites for the Black-throated Sparrow, Sage Sparrow, gnatcatchers, and Costa's Hummingbird. The Gray Vireo often hangs its cup-shaped nest—of grasses and shredded bark, ornamented with spoon-shaped sagebrush leaves—from the branches of Big Sagebrush. The distribution of Big Sagebrush also determines the range of the Sage Grouse, a chicken-sized game bird that hides, breeds, and feeds amid Big Sagebrush. Sagebrush leaves and flowers make up 75 percent of its diet. Native peoples used sagebrush to cure various ailments, especially as a cold remedy. Some boiled green leaves for a tea, taken several times a day, or burned branches and inhaled the fumes for head colds, or mashed green leaves as a poultice for chest colds. Others used an infusion of the leaves as a tonic, a wash for hair and eyes, an antiseptic for wounds, and a cure for stomach ache.

Also occurring in the Eastern Sierra, and looking much alike, are: **Silver Sagebrush (*A. cana* ssp. *bolanderi*),** which grows in moist depressions or on meadow margins, sometimes in more alkaline sites than Big Sagebrush; **Dwarf or Low Sagebrush (*A. arbuscula*),** found in rocky soils on slopes and ridges in northeast California; **Black Sagebrush (*A. nova*),** found in similar

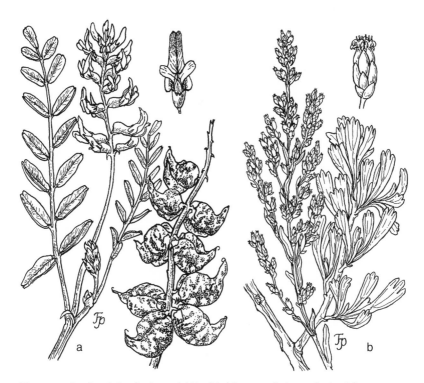

Fig. 5.34. Sagebrush Scrub plants. (a) Freckled Locoweed, *Astragalus lentiginosus.*
(b) Big Sagebrush, *Artemisia tridentata* ssp. *tridentata*

habitats to the south; and **Alpine Sagebrush** *(A. rothrockii),* a large-flowered
species found from 6500 ft (1980 m) to above treeline, usually on dry slopes.
Distribution Northern and eastern California, Nevada; to Washington,
Montana, New Mexico.

ANTELOPE BITTERBRUSH, Purshia tridentata Rose Family. Plate 12d.
Antelope Bitterbrush is an important browse plant both for livestock and for
native herbivores, such as Mule Deer and Pronghorn. Tall and spreading to
prostrate shrub, height exceedingly variable, 1–15 ft (0.3–5 m). Leaves wedge-
shaped, 3-lobed, 0.25–1 in (6–25 mm) long. Flowers 0.5 in (13 mm) wide,
each with 5 pale yellow petals. Fruit 0.5–1 in (13–25 mm) long, spindle-
shaped, leathery. *Flowers:* May–July. Bitterbrush reproduces from seed as
well as by sending out branches that develop roots and eventually form new
shrubs. Fires are hard on this species. In some localities bitterbrush plants
can resprout from root-crowns after fire and grow back quickly. However,
even with artificial reseeding, most bitterbrush stands require at least several
decades to reestablish after fire. A variety of this species, ***Desert Bitterbrush** *(P. tridentata* **var.** *glandulosa),* grows with Antelope Bitterbrush
from the Mono Lake region south along the Sierra's east slope. *Distribu-*

tion Mountains and deserts of northern and central California; Nevada; to British Columbia, Montana, New Mexico.

COTTON THORN, Tetradymia axillaris Sunflower Family. Plate 12h. The common name is derived from two characteristics: new leaves gradually are transformed into slender rigid spines, and the bristles of the flowering heads fluff up into cottony clusters as they dry out. Shrub 2–5 ft (0.7–1.7 m), with rigid branches covered with dense white hairs. Leaves very narrow, bright green, about 0.5 in (13 mm) long, in small clusters. Flower heads small, each with 6–7 yellow disk flowers. *Flowers:* Apr–May. Also in Sagebrush Scrub: ***Spineless Horsebrush** *(T. canescens)* and ***Littleleaf Horsebrush** *(T. glabrata).* **Distribution** Higher deserts of California, southern Nevada; to southwestern Utah.

DESERT PEACH, Prunus andersonii Rose Family. Plate 12e. Fruit resembles a tiny green to brownish peach. Deciduous shrub 3–6 ft (1–2 m), with diffuse spiny branches. Leaves oval, 0.5–1 in (1.3–2.5 cm) long, clustered along the branches. Flowers 0.75 in (1.8 cm) wide, each with 5 light pink petals. *Flowers:* Mar–Apr. **Distribution** Eastern Sierra Nevada, High Desert in California; western and central Nevada.

RUBBER RABBITBRUSH, *Chrysothamnus nauseosus* Sunflower Family. Plate 12g. Rounded, strongly scented shrub 1–6 ft (0.3–2 m). Twigs covered with dense felty hairs. Leaves very narrow, 1–3 in (2.5–7.5 cm) long. Flower heads small, yellow, in oval clusters grouped at the ends of the branches. *Flowers:* Aug–Oct. A wide-ranging High Desert species with at least 9 subspecies in California and Nevada. Abundant in roadside gulleys, washes, and other low-lying disturbed places, which it highlights in golden yellow during its fall bloom. During World War II the University of California investigated this shrub as a source of emergency rubber. The latex was found to contain high-quality rubber that vulcanized easily. Also in the Eastern Sierra: ***Yellow or Sticky-leaved Rabbitbrush** *(C. viscidiflorus),* a shrub about 1 foot (0.3 m) tall, with white stems and twisted bright green leaves; and ***Parry's Rabbitbrush** *(C. parryi),* similar to Rubber Rabbitbrush except for its wider leaves and different arrangement of the flower heads. *Distribution* Eastern and southern California, Nevada; to British Columbia, Montana, Colorado, Baja California.

SPINY HOP-SAGE, Grayia spinosa Goosefoot Family. Plate 12f. Intricately branched shrub 1–3 ft (0.3–1 m) tall. Twigs with black and white lines. Leaves feel as if covered with fine meal, oblong, 0.5–1.5 in (1.3–3.8 cm).

Flowers green, in dense terminal spikes. Fruits enclosed in 2 greenish bracts that turn red with age. *Flowers:* Mar–June. Male and female flowers produced on different plants. Female plants are distinctive in summer, covered with large clusters of red bracts that enclose the fruits. Even though the twigs are spiny, sheep, goats, and deer browse them avidly, fattening on their nutritious fruits. Able to tolerate many conditions, Spiny Hop-sage occurs in Creosote Bush, Shadscale, and Sagebrush Scrub, and in Pinyon-Juniper Woodland. *Distribution* Eastern Sierra, central California mountains, California deserts; Nevada; to Washington, Montana, New Mexico.

COMMON WILDFLOWERS OF SAGEBRUSH SCRUB

Although wildflowers are generally less abundant here than in low desert communities, favorable seasons bring bright-colored displays, especially on moraines and higher gravelly slopes. Most species listed below extend into lower elevations of Montane Forest in dry open sites.

Flowers White to Cream

*FRAGRANT EVENING PRIMROSE, *Oenothera caespitosa* Evening-primrose Family. Plate 13e. The delicately scented flowers open at dusk and are pollinated at night by large hawkmoths. Each flower remains open only one night, closing the following morning. Semi-prostrate perennial. Leaves narrow with wavy margins, 1–4 in (2.5–10 cm) long, on petioles of similar length. Flowers 1.5–3 in (3.8–7.5 cm) wide, each with 4 white petals that age to pink. *Flowers:* Apr–Aug. *Distribution* Eastern California, Nevada, western United States.

*PRICKLY POPPY, *Argemone munita* Poppy Family. Plate 13a. The large white flowers with deep yellow centers have inspired an alternative common name: Fried Egg Flower. Erect annual or perennial, 1.5–4.5 ft (0.5–1.5 m) tall. Stems and leaves gray-green, densely covered with prickles. Flowers 2–5 in (5–12.5 cm) wide, each with 6 crinkled white petals. *Flowers:* June–Sept. Common along roadsides. Stems, when broken, exude a sticky yellow latex that is reported to be toxic. *Distribution* Coast and deserts of California; Nevada; northern Baja California.

Flowers Orange to Yellow

BLAZING STAR, *Mentzelia laevicaulis* Sandpaper Plant Family. Plate 13f. Large showy flowers open in the evening and close late the following morning. Erect biennial 1–6 ft (0.3–2 m) tall. Stems shining white. Leaves deeply lobed, 0.5–4 in (1.3–10 cm) long, with sandpapery surface. Flowers 4–6 in

(10–15 cm) wide, each with 5 satiny yellow petals. *Flowers:* June–Oct. Blazing Star inhabits steep gravelly slopes of canyon walls and roadcuts. *Distribution* California, except Central Valley and low desert; Nevada; to British Columbia, Montana.

MULES EARS, *Wyethia mollis* Sunflower Family. Plate 13c. Rounded perennial with many leaves and few flowering stems from an enlarged base, 1–3 ft (0.3–1 m) tall. Stems and leaves gray-green, densely covered with felty hairs. Leaves that look and feel like mules' ears, oval, 8–16 in (10–40 cm) long. Flower heads about 3 in (7.5 cm) across, resembling the common sunflower. *Flowers:* May–Aug. *Distribution* Klamath and Cascade ranges, Sierra Nevada; western Nevada, southeastern Oregon.

SULPHUR BUCKWHEAT, *Eriogonum umbellatum* Buckwheat Family. Plate 13d. Low, branched perennial, generally 1 foot (0.3 m) or less in height. Branches with leafy tips. Leaves narrowly oval, 0.5–1 in (1.3–2.5 cm) long, green on upper surface and white-hairy on lower. Flowers small, bright yellow, in dense rounded clusters. *Flowers:* June–Aug. Many varieties. One of many native wild buckwheats that can be cultivated successfully in California gardens. *Distribution* California, Nevada; to western Canada, Colorado, New Mexico.

Flowers Pink to Red

*DESERT PAINTBRUSH, *Castilleja angustifolia (= C. chromosa)* Figwort Family. Low, many-stemmed perennial to about 1 foot (0.3 m) tall. Leaves long and narrow, sometimes with 1–2 pairs of thin lobes. Flower bracts scarlet to orange-red. Flowers about 1 in (2.5 cm) long, narrow, greenish, hidden by bracts. *Flowers:* Apr–Aug. The reddish bracts attract hummingbirds, which transfer pollen on their bills while searching the flowers for nectar. *Distribution* Higher deserts of California, Nevada; to Oregon, Montana, Wyoming, Colorado, New Mexico.

SCARLET GILIA, *Ipomopsis aggregata* Phlox Family. Plate 13b. Supplies nectar for both hummingbirds and hawk moths. Erect biennial or perennial, 1–2.5 ft (0.3–0.8 m) tall. Leaves 1–2 in (2.5–5 cm) long, dissected into many narrow segments. Flowers bright red to pale pink, trumpet-shaped, about 1.5 in (3.8 cm) long, each with 5 pointed lobes. *Flowers:* June–Sept. Scarlet Gilia forms small colonies in gravelly soils. The plants grow a rosette of leaves until they flower once and die. *Distribution* Klamath and Cascade ranges, Modoc Plateau, Sierra Nevada in California; Nevada; to British Columbia, Colorado, Mexico.

Flower Color Various

LOCOWEED, RATTLEPOD, MILK-VETCH, *Astragalus* species. Pea Family. Mainly perennial herbs, erect to prostrate with leaves divided into numerous oval to narrow leaflets. Flowers pink, red, blue, violet, white, or yellow. Fruit is a pod, often enlarged and inflated when mature. *Flowering times:* spring through summer.

A large group with more than 30 species in the deserts and mountains of the Eastern Sierra, some widely distributed, others narrowly endemic (with a limited natural distribution). The common name "locoweed" is applied to those species suspected of being poisonous to livestock. Representatives include: *Scarlet Locoweed (A. coccineus),* with brilliant red flowers, an uncommon inhabitant of rocky places; **Freckled Locoweed (A.** *lentiginosus)* (fig. 5.34a), a wide-ranging, complex species with red-spotted stems, all varieties poisonous to stock; **Woolly Rattlepod (A.** *purshii),* low tufted plants with gray hairs and fuzzy pods, often in well-drained soils. Two of many rare species of *Astragalus* that are protected in California include: *Mono Milk-vetch (A. monoensis),* restricted to pumice flats of Mono Co.; and *Long Valley Milk-vetch (A. johannis-howellii),* found only in alkaline soils of Long Valley in Mono Co.

SHADSCALE SCRUB

A community of low, gray-green shrubs, Shadscale Scrub occurs throughout the Great Basin in areas with cold, snowy winters and hot, dry summers. Although it shares some characteristics with Sagebrush Scrub, the two are distinct communities with few plant species in common.

Perhaps the most notable feature of Shadscale Scrub is the uniform appearance of its many kinds of small, sparsely scattered shrubs. Shadscale, a spiny saltbush valued as forage, occurs in most stands. Its associates include Budsage, Blackbush, Winter Fat, Spiny Hop-sage, California Matchweed, Spiny Menodora, Nevada Ephedra, and Four-wing Saltbush—all small-leaved, drought-tolerant plants with inconspicuous flowers. Most have spiny branchlets and are 1–3 feet (0.3–1 m) tall, generally shorter than shrubs of Sagebrush Scrub. Bright-colored wildflowers like those found in some stands of Sagebrush Scrub are absent from Shadscale Scrub. Herbaceous plants of the latter community are few and inconspicuous. One notable example is the rare Nevada Oryctes *(Oryctes nevadensis),* a tender yellow-flowered herb related to tomatoes and eggplants. Known from a handful of sites in California and Nevada, it has been found only in years of above-average rainfall.

Shadscale Scrub averages only 6 inches (15 cm) of rain annually, com-

Fig. 5.35. Shadscale Scrub

pared with nearly double that amount for Sagebrush Scrub; Shadscale Scrub
areas are also somewhat warmer both in summer and in winter. Shadscale
Scrub often grows on heavy saline or alkaline soils; in fact, the community is
considered by some to be an indicator of subsurface salinity. Although it also
can grow on deeper, better drained soils, these are usually occupied by Sagebrush Scrub.

Most Shadscale Scrub sites have an 80- to 100-year history of livestock
grazing; several of the dominants, like Four-wing Saltbush and Winter Fat,
provide excellent forage. Heavy use of these rangelands in parts of Nevada
and Utah has brought about an undesirable change in the community—
invasion by a Central Asian weed called Halogeton *(Halogeton glomeratus)*.
Halogeton was introduced to the United States in the 1930s. Since then it has
become established in many areas with saline soils that were previously occupied by Shadscale Scrub. Where Halogeton is common it changes the soil
chemistry and causes a crust to form on the surface so that water does not
penetrate; then native plants, including valuable forage species, can no longer
grow there. Halogeton can be toxic to livestock due to its high concentration
of oxalates. Ingestion of a few ounces can be lethal to sheep, and even lim-

ited consumption is suspected of inhibiting growth and reproductive ability in other range animals. No practical means has been found to eliminate Halogeton, once it is firmly established. In the late 1970s and early 1980s Halogeton was found in several sites along Highway 395 near Mammoth Lakes and Bishop, and it continues to be found occasionally along Highway 168 toward the Nevada border. The California Department of Food and Agriculture eliminates these invasions with herbicides and regularly monitors susceptible areas for new colonies.

COMMON SHRUBS OF SHADSCALE SCRUB

*SHADSCALE, *Atriplex confertifolia*   Goosefoot Family. Fig. 5.36a. Stiff-branched, rounded shrub with spiny branchlets, to 3 ft (1 m) tall. Leaves round to oval, less than 1 in (2.5 cm) long, covered with tiny scales, like fine meal. Flowers small, green, inconspicuous, male and female on different plants. *Flowers:* Apr–July. Shadscale is the most common and widespread shrub of the Shadscale Scrub community, and the most frequently encountered saltbush of the High Desert. The common name comes from the resemblance of the leaves to the scales of the shad, a herringlike fish.   *Distribution*   Mojave and High Deserts of California and Nevada; to Oregon, north-central United States, northern Mexico.

*BLACKBUSH, *Coleogyne ramosissima*   Rose Family. Sometimes called Blackbrush in Nevada. Intricately branched shrub to 6 ft (2 m), with spiny blackish branchlets. Leaves variable, about 0.5 in (1.3 cm) long. Flowers about 0.5 in (1.3 cm) wide, yellow, scattered among the leaves. *Flowers:* Apr–June. Blackbush is especially common at the lower border of the Shadscale community, where it interfaces with Creosote Bush Scrub. Blackbush often forms pure stands, noticeable from a distance due to the smoky hue of the branches.   *Distribution*   Deserts of California and Nevada; to Colorado, New Mexico.

*BUDSAGE, *Artemisia spinescens*   Sunflower Family. Fig. 5.36b. Low, semi-prostrate shrub, to 1 foot (0.3 m) tall, with spiny branches. Leaves small, divided into 5–7 lobed parts, crowded on the branches. Flowering heads yellow, about 0.5 in (1.3 cm) wide, on stems that become spines the following year. *Flowers:* Apr–May.   *Distribution*   San Bernardino Mountains, Mojave and High Deserts in California and Nevada; to Oregon, Montana, New Mexico.

*DESERT ALYSSUM, *Lepidium fremontii*   Mustard Family. Fig. 5.36c. Rounded suffrutescent (slightly woody) perennial to 1.5 ft (0.5 m) tall, with

Fig. 5.36. Shadscale Scrub shrubs. (a) Shadscale, *Atriplex confertifolia*. (b) Budsage, *Artemisia spinescens*. (c) Desert Alyssum, *Lepidium fremontii*. (d) Four-wing Saltbush, *Atriplex canescens*

many branching gray-green stems. Leaves very narrow, up to 4 in (10 cm) long. Flowers small, each with 4 minute white petals, clustered on elongate flowering stems. *Flowers:* Mar–May. *Distribution* Deserts of California and Nevada; to southern Utah, western Arizona.

*FOUR-WING SALTBUSH, *Atriplex canescens* Goosefoot Family. Fig. 5.36d. Named for its fruits, each with 4 conspicuous wings, and for its tolerance of saline soils. Much-branched shrub to 6 ft (2 m) tall, with spreading branches. Leaves variable, but usually more or less long and narrow, to 2 in (5 cm) long, covered with tiny scales like fine meal. Flowers small, inconspicuous, green, male and female on different shrubs. *Flowers:* June–Aug. The extended papery wings of the seeds promote dispersal by the wind. *Distribution* Transverse, Peninsular, and interior South Coast ranges, and deserts of California and Nevada; to western Canada, South Dakota, northern Mexico.

*INDIGO BUSH, *Psorothamnus fremontii (=Dalea fremontii)* Pea Family. Much-branched aromatic shrub to 6 ft (2 m), branches gray-green. Leaves up to 1.5 in (3.8 cm) long, divided into 3–5 small oval leaflets. Leaves and branches dotted with tiny orange glands. Sweetpealike flowers about 0.5 in (1.3 cm) long, dark blue-purple. *Flowers:* Apr–May. One of the very few Shadscale Scrub dominants that has showy flowers. Most other shrubs of this community have small blooms that come and go without notice. *Distribution* Desert mountains of California, southern Nevada; to southern Utah, Arizona.

*STICKY SNAKEWEED, CALIFORNIA MATCHWEED, *Gutierrezia microcephala* Sunflower Family. Fig. 5.37a. Often occurs along roadsides, on abandoned roadways, and in other places with disturbed soils. Many-stemmed low shrub to 2 ft (0.7 m) tall. Branchlets very slender, covered with soft, narrow leaves to 2 in (5 cm) long. Flowering heads tiny, yellow, in small clusters on slender stalks raised above the body of the shrub. *Flowers:* July–Oct. *Distribution* Southern California deserts, mountains; southern Nevada; to Colorado, central Mexico.

*NEVADA EPHEDRA, *Ephedra nevadensis* Ephedra Family. Fig. 5.37b. Sometimes called Nevada Joint Fir. Erect to spreading shrub, to 4 ft (1.3 m) tall, with stiff gray-green, twiggy, jointed branches. No true leaves. Male and female "cones" on separate plants. Female "cones" oval, about 0.5 in (1.3 cm) long, each containing a pair of brown seeds. Male "cones" similar to female in size and shape, but with many protruding yellow stamens. Nevada Ephe-

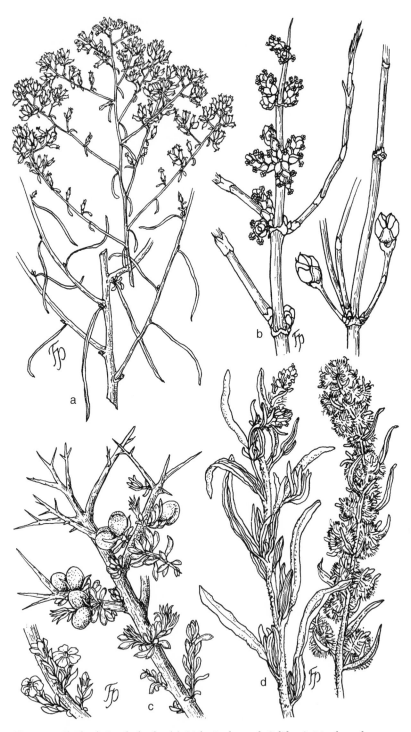

Fig. 5.37.   Shadscale Scrub shrubs. (a) Sticky Snakeweed, California Matchweed, *Gutierrezia microcephala*. (b) Nevada Ephedra, *Ephedra nevadensis*. (c) Spiny Menodora, Twinfruit, *Menodora spinescens*. (d) Winter Fat, *Krascheninnikovia lanata*

dra is similar to **Mormon Tea** *(E. viridis)*, which occurs at higher elevations in Pinyon-Juniper Woodland and Sagebrush Scrub. These weird desert shrubs are not flowering plants; they are more closely related to pines and firs. The seeds are naked, not enclosed in and protected by a fruit, as in flowering plants. Antelope Ground Squirrels harvest the nutritious seeds. Desert settlers used a tea made from the twigs to treat canker sores. *Distribution* Deserts of California and Nevada; to Oregon, Utah.

*SPINY HOP-SAGE, *Grayia spinosa* For description see Sagebrush Scrub.

*SPINY MENODORA, TWINFRUIT, *Menodora spinescens* Olive Family. Fig. 5.37c. Low or semi-prostrate shrub to 2 ft (0.7 m) high. Branches green, very rigid, irregularly divided; branchlets stoutly spine-tipped. Leaves narrow, fleshy, to 0.5 in (1.3 cm) long. Flowers white, about 0.5 in (1.3 cm) long, developing into fruits divided into 2 oval parts. *Flowers:* Apr–May. *Distribution* Higher deserts of California, southern Nevada; northwest Arizona.

*WINTER FAT, *Krascheninnikovia lanata (=Eurotia lanata)* Goosefoot Family. Fig. 5.37d. The most important winter forage plant of the Great Basin. Soft-woody shrub to 2 ft (0.7 m) high, young branches and long, narrow leaves covered with long, soft, light gray hairs. Flowers small and inconspicuous. Fruits covered with spreading tufts of silvery hairs. *Flowers:* Mar–June. Pure Winter Fat stands sometimes extend over many acres of valley bottomland, as in Deep Springs and Little Cowhorn valleys. They form silver-gray patches clearly visible from above. A pure Winter Fat stand provides more nutrition to livestock than a comparable area of mixed-species Shadscale Scrub. *Distribution* Mojave and Great Basin deserts, southern San Joaquin Valley and adjacent hills in California; Nevada; to Washington, north-central United States, New Mexico, northern Mexico.

ALKALI MEADOW

A postcard-perfect scene of lush pastoral beauty will greet you, should you drive into Bishop on a warm June day: green and flowery meadows grazed by sleek horses, bordered by cottonwoods, and framed by lofty Sierran peaks. The Alkali Meadows of Bishop and Round Valley occur where runoff from the Sierra's east slope (now commonly augmented by irrigation) continuously waters the ground, flowing from dozens of meandering natural and man-made brooklets. These Eastern Sierra Alkali Meadows vary considerably in aspect and composition. Well-watered Alkali Meadows in the Bishop area and in Bridgeport and Sierra valleys superficially resemble Mountain Meadows. They are composed of a verdant growth of grasses, rushes, sedges,

Fig. 5.38.   Alkali Meadow

and colorful herbs, including stands of pale blue Wild Iris and bright yellow
Alkali Cinquefoil, and their soils may be scarcely alkaline. In drier, more alkaline meadows, such as those of Long Valley, the vegetation is less dense,
contains fewer wildflowers, and, in the extreme case, may be dominated by
low mats of Saltgrass. Other common grasses of Alkali Meadows include Alkali Sacaton *(Sporobolus airoides)*, Squirrel Tail *(Hordeum jubatum)*, Alkali
Bluegrass *(Poa secunda =P. juncifolia)*, Scratchgrass *(Muhlenbergia asperifo-
lia)*, and Alkali Grass *(Puccinellia nuttalliana =P. airoides)*. Rushes, especially
Baltic Rush *(Juncus balticus)* (fig. 5.40d), are also common in Alkali Meadows. The Alkali Meadows of the Owens Valley harbor several rare plants, including the endangered Owens Valley Checkerbloom *(Sidalcea covillei)*, with
flowers like tiny pink hollyhocks.

Alkali Meadows inhabit fine-textured, permanently moist, sometimes alkaline soils (see Alkali Sink Scrub) that are occasionally saturated. They occur in valley bottoms and on lower alluvial slopes, sometimes near rivers and
streams, and around lake margins. They are found from the Owens Valley
north to Honey Lake, at altitudes ranging from 3500 to 7000 feet (1070 –
2130 m). You will find good examples just north of Bishop along both sides of

Highway 395, in the bottomlands near the Owens River and the West Walker River (near Coleville and Walker), and on the higher margins of Crowley and Mono lakes. Climatic conditions in this community vary from north to south, but include cold winters with some snow and hot, dry summers that are more moderate to the north. Average annual precipitation varies from about 6 inches (15 cm) near Bishop to nearly 15 inches (37.5 cm) in the vicinity of Susanville.

Alkali Meadows are largely unstudied and undescribed, although they occupy large areas in the Eastern Sierra. Local ranchers consider them important grazing lands. In some areas, where irrigation water is available, this community has been converted to alfalfa fields.

SOME FLOWERING HERBS OF ALKALI MEADOWS

Flowers White to Cream or Greenish

CRISPED THELYPODIUM, *Thelypodium crispum*   Mustard Family. Plate 14g. Annual or biennial rosette plant, usually with a single flowering stalk to 2 ft (0.7 m) tall. Leaves at base lyre-shaped, those on stem narrow, up to 2 in (5 cm) long. Flowers nearly erect on stalk, about 0.35 in (0.8 cm) long, each with 4 narrow, spoon-shaped white petals. Fruit an erect "pod" about 1 in (2.5 cm) long, with a beaded appearance. *Flowers:* June–July. Crisped Thelypodium also occurs around hot springs in Sagebrush Scrub and Pinyon-Juniper Woodland. *Distribution*   Sierra Nevada and High Desert in California; western Nevada.

Flowers Yellow to Orange

ALKALI CINQUEFOIL, *Potentilla gracilis* var. *elmeri* (=*P. pectinisecta*) Rose Family. Plate 14e. Perennial herb with several flowering stems to about 1 foot (0.3 m) tall. Leaves divided into 5–9 leaflets, these arranged as the fingers are to the palm of the hand, each one further subdivided into many narrow segments. Leaflets up to 2 in (5 cm) long, with silky hairs pressed closely to both surfaces. Flowers about 0.75 in (1.8 cm) wide, each with 5 yellow petals; borne in a loose inflorescence. *Flowers:* May–July.   *Distribution*   Sierra Nevada, mountains of southern California, High Desert in California and northern Nevada; to western North America.

*ALKALI CLEOMELLA, *Cleomella plocasperma*   Caper Family. Fig. 5.39e. Widely branched annual, to about 1 foot (0.3 m) high, with hairless, dull gray-green stems. Leaves divided into oblong leaflets about 1 in (2.5 cm) long. Flowers about 0.5 in (1.3 cm) wide, each with 4 pale yellow petals; densely aggregated along the flowering stems. *Flowers:* May–Oct. Alkali Cleomella

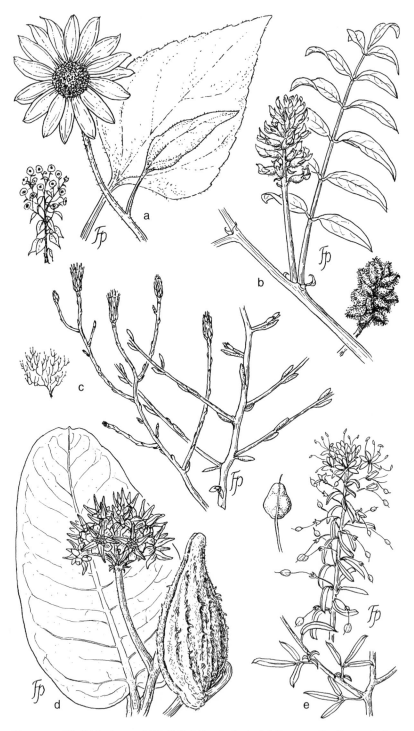

Fig. 5.39.   Alkali Meadow and Alkali Sink Scrub plants. (a) Common Sunflower, *Heli-anthus annuus.* (b) Wild Licorice, *Glycyrrhiza lepidota.* (c) Shrubby Alkali Aster, Broom Aster, *Machaeranthera carnosa.* (d) Showy Milkweed, *Asclepias speciosa.* (e) Alkali Cleo-mella, *Cleomella plocasperma*

often grows with the related *Small-flowered Stinkweed (C. parviflora), which has very narrow leaflets and tiny yellow flowers. *Distribution* Mojave and High Desert of California, Nevada; to Oregon, Idaho, Utah.

COMMON SUNFLOWER, *Helianthus annuus* Sunflower Family. Fig. 5.39a. Robust spreading annual herb to 6 ft (2 m), with stout, rough-hairy stems. Leaves to 6 in (15 cm) long, somewhat heart-shaped, dark green. Flower heads to 3 in (7.5 cm) wide, like typical sunflowers with a reddish central disk. *Flowers:* July–Oct. The cultivated sunflower, grown for seeds and oil, is a variety of Common Sunflower. *Distribution* Widespread. Throughout California and Nevada; to eastern North America.

GREAT VALLEY GUMPLANT, *Grindelia camporum* Sunflower Family. Branched suffrutescent perennial, with stems to nearly 4 ft (1.3 m) high. Leaves narrowly oblong, 1–3 in (2.5–7.5 cm), very resinous and sticky, irregularly toothed margins. Flower heads 1–1.5 in (2.5–3.8 cm), yellow marginal ray and central disk flowers. Bracts beneath the heads strongly recurved or hooked. *Flowers:* May–Oct. *Distribution* Central Valley, coastal California and California deserts; Baja California.

*MEADOW HAWKSBEARD, *Crepis runcinata* Sunflower Family. The "beard" in the name refers to the tuft of hairs attached to each seed. Perennial from a thickened root-crown with 1–3 flowering stems up to 1.5 ft (0.5 m). Leaves in a basal rosette, oblong, to nearly 1 foot (0.3 m) long, irregularly toothed along the margins. Flower heads 5–14, about 1 in (2.5 cm) long, with yellow ray flowers. Seeds with many white hairs at one end. *Flowers:* June–July. *Distribution* High Desert of California and Nevada; to Washington, south-central Canada, north-central United States.

WAND ASTER, *Pyrrocoma racemosa (=Haplopappus racemosus)* Sunflower Family. Fig. 5.40b. Perennial with slender stems curving upward from the base, to 2 ft (0.7 m) high. Leaves basal, long and narrow, to 1 foot (0.3 m). Flower heads about 0.5 in (1.3 cm) long, with yellow flowers obscured by the white-haired flower appendages. Flowers borne in small clusters along the flowering stems. *Flowers:* July–Oct. *Distribution* Much of California and Nevada; to Oregon, Utah.

WILD LICORICE, *Glycyrrhiza lepidota* Pea Family. Fig. 5.39b. Perennial with several spreading stems to 3 ft (1 m) long, from thickened rootstocks. Leaves to 4 in (10 cm) long, divided into 11–19 oval leaflets, each about 1 in (2.5 cm) long. Pea flowers yellowish white, about 0.5 in (1.3 cm) long, in dense spikes on short stalks attached at the leaf bases. Fruit a prickly pod resembling a cocklebur. *Flowers:* May–July. In California, cultivated **Lico-rice** *(G. glabra)* occurs occasionally, after having escaped from cultivation. *Distribution* Throughout California, Nevada; to central United States, Canada, Mexico.

Flowers Red to Pink

*MOJAVE THISTLE, DESERT THISTLE, *Cirsium mohavense* Sunflower Family. Fig. 5.40a. Erect biennial, 1.5–7.5 ft (0.5–2.5 m) tall, the single stem branching above, covered with white, cobwebby hairs. Leaves oval in outline, to nearly 2 ft (0.7 m) long, with large and small spiny teeth along the margins. Flower heads typical thistles, about 1.5 in (3.8 cm) long, with pink to reddish flowers. *Flowers:* June–Oct. *Distribution* Mojave and High Deserts of California, Nevada.

NORTHERN WILLOW-HERB, *Epilobium ciliatum* ssp. *ciliatum (= E. adeno-caulon)* Evening-primrose Family. This species and its relatives are called willow-herbs because their seeds, with tufted hairs that carry them aloft on the wind, resemble those of willows. Erect perennial herb to 3 ft (1 m), freely branched above. Leaves narrow, 1–2.5 in (2.5–6.3) long, with very small teeth along the margins. Flowers about 0.25 in (0.6 cm) wide, each with 4 pinkish petals; scattered along much-branched flowering stems. Seeds with a tuft of hairs at one end. *Flowers:* July–Sept. *Distribution* Widespread. Throughout California and Nevada; most of North America, eastern Asia, southern South America.

SHOWY MILKWEED, *Asclepias speciosa* Milkweed Family. Fig. 5.39d. Most plants in this family ooze milky juice when the stems or leaves are cut. Robust perennial herb with stems to 4 ft (12 dm) tall. Most plant parts often covered with woolly hairs. Leaves in pairs, opposite each other, oval to oblong, 3–6 in (7.5–15 cm) long. Flowers in large clusters, about 3 in (7.5 cm) wide. Each flower with rose-purple petals, about 0.5 in (1.3 cm) wide. Fruit a large, hairy teardrop-shaped capsule, about 3 in (7.5 cm) long, splitting open to release brown seeds with large tufts of snow-white hairs. Flowers: May–July. *Distribution* Much of California; to British Columbia, central Canada, Texas.

Flowers Blue to Violet or Purple

CAMAS LILY, *Camassia quamash* For description see Mountain Meadows.

WESTERN BLUE FLAG, WILD IRIS, *Iris missouriensis* Iris Family. Fig. 5.40c. Perennial from stout underground stems, flower stalks to 2 ft (0.7 m)

Fig. 5.40. Alkali Meadow plants. (a) Mojave Thistle, Desert Thistle, *Cirsium mohavense.*
(b) Wand Aster, *Pyrrocoma racemosa.* (c) Western Blue Flag, Wild Iris, *Iris missouriensis.*
(d) Baltic Rush, *Juncus balticus*

high. Leaves light dull green, narrow, to 1.5 ft (0.5 m) long. Flower stems branched, bearing showy pale blue to pale purple iris flowers 2–3 in (5–7.5 cm) wide. *Flowers:* May–June. Wild Iris also grows in Montane Meadows. According to the U.S. Forest Service, it has become more common in grazed Alkali Meadows in the Eastern Sierra, probably because livestock find it unpalatable. *Distribution* Widespread. Mountains and High Desert of California, Nevada; much of western North America, northern Mexico.

DESERT MARSHLANDS

The term *Desert Marshland* may seem incongruous or contradictory, yet—surrounded by miles of arid scrubland—small, scattered, unexpected marshes do occur in the desert, in waterlogged bottomlands. Patches of dark olive-green on desert lake margins, adjacent to desert streams, or near springs often indicate Desert Marshland. Its plants are mainly water-loving herbs, ranging from a few inches to more than 10 feet (3 m) in height. Although desert lowlands are generally arid, a marsh makes its own microclimate, which is hot and muggy through the summer months. All types of Desert Marshlands feature abundant water supplies, derived from springs, streams, or other permanent sources. The soils are typically waterlogged for most of the year and are high in organic matter. Climatic conditions and average rainfall are the same as those of the surrounding communities, which may be Pinyon-Juniper Woodland, Sagebrush Scrub, Shadscale Scrub, Alkali Sink Scrub, Alkali Meadow, or Creosote Bush Scrub, depending upon elevation and locality.

Desert Marshlands provide excellent growing conditions for properly adapted plants and a favorable environment for many animals. Birds and amphibians breed here in abundance, attracted by the dense vegetation and a myriad of insects, including dragonflies, mosquitoes, bees, wasps, true bugs, beetles, and biting flies. Many a summer visit to a desert marsh has been cut short by the persistent attentions of its six-legged inhabitants. But, even though insects abound in desert marshes, many of the dominant plants are wind-pollinated.

Desert Marshlands vary in alkalinity. Alkali marshes are dominated by low-growing, dense mats of Saltgrass, sedges, rushes, and other plants also found in Alkali Sink Scrub and Alkali Meadow. Often these marshes are associated with hot springs. Good examples are found near the Bridgeport hot springs and along the margins of Hot Creek, downstream from the Mammoth Lakes airport. Freshwater or brackish marshes commonly consist of dense stands of perennial herbs, sometimes with a few cottonwood or willow trees nearby. The Fish Slough area north of Bishop is perhaps the East-

Fig. 5.41.　Desert Marshlands

ern Sierra's best known freshwater marsh system. There tall grasses, sedges, rushes, and cattails form a thick, impenetrable border around small pools of standing water; lower-growing grasses and other herbs may form a broad soppy carpet that extends away from the open water. Sierra Valley north of Truckee also has many acres of freshwater marsh. Dense stands of tules, rushes, sedges, and grasses border the many open water channels—some with "islands" of Yellow Pond Lily *(Nuphar polysepalum)*—that eventually join to form the Feather River. Most marshes contain species of rushes and spikerushes (low, grasslike plants, often with wiry stems, that form dense turfs in the waterlogged soils bordering open water). Brightly colored wildflowers are uncommon in Desert Marshlands, although at Fish Slough and elsewhere they include a 10-foot-tall (3-m) native sunflower and a more modest crimson-tipped paintbrush.

Marsh plants exhibit interesting physiological and anatomical adaptations to their habitat. Most plants die if their roots are submerged for long periods because the low concentration of dissolved oxygen in shallow warm water does not permit them to "breathe." Most marsh plants, however, have developed chemical methods for making efficient use of low oxygen supplies. Some tall monocots, such as cattails and Common Tule, have stem and leaf tissue with many small pockets of air separated by thin walls. This tissue may allow extra oxygen to reach the roots by entering the leaves and traveling through these internal passageways.

When viewed from the perspective of geologic time, desert marshes are ephemeral habitats, short-lived when compared with dry uplands that change little over thousands of years. The brief lifetime of individual marshes has probably contributed to the development of the large ranges and wide dispersal capabilities seen in many marsh plants: They have had to move quickly to survive. Cattails, tules, rushes, and other marsh plants have exceedingly broad geographic distributions. Some range across North America, and some are found worldwide. Long-distance dispersal is often a factor in such wide distributions, and, in fact, many desert marsh species have seeds that are equipped to travel over long stretches of dry uplands. Cattails, for example, produce immense quantities of tiny, fuzzy seeds that are easily carried by the wind. Giant Reed, Common Tule, and Prairie Bulrush have small seeds that can be transported, internally or externally, by the waterfowl that rest and feed in marshes during migration. The ability of marsh plants to rapidly colonize isolated wetlands is apparent to anyone who has watched a newly dug stockpond fill with tules and cattails within a few years.

COMMON PLANTS OF DESERT MARSHLANDS

*ALKALI CORDGRASS, *Spartina gracilis*   Grass Family. Fig. 5.42d. Coarse perennial grass with many stems to 3 ft (1 m) tall, from creeping horizontal underground rootstocks. Leaf blades up to 8 in (20 cm), the edges inrolled when old. Flowering spikes 1–1.5 in (2.5–3.8 cm) long, 4–8 to a stalk. *Flowers:* June–Aug. Common in alkaline habitats. An important food source for waterfowl. Related species inhabit coastal salt marshes.   *Distribution* Widespread. California deserts, Nevada; to northern and eastern Canada, Kansas, New Mexico.

*ANNUAL INDIAN PAINTBRUSH, *Castilleja minor* ssp. *minor (=C. exilis)* Figwort Family. Plate 14h. Usually single-stemmed, rough-hairy annual to 2.5 ft (0.8 m) high. Stem leafy, leaves narrow, about 1.5 in (3.8 cm) long, pointed. Flower bracts leaflike, scarlet toward the tips, obscuring the inconspicuous dull yellow flowers that are about 0.7 in (1.6 cm) long. *Flowers:* June–Sept.   *Distribution*   High Desert in California, northern Nevada; to Washington, New Mexico.

BASIN OR SHOWY GOLDENROD, *Solidago spectabilis*   Sunflower Family. Plate 14c. Perennial herb with several stout, erect stems to 4 ft (1.3 m). Leaves

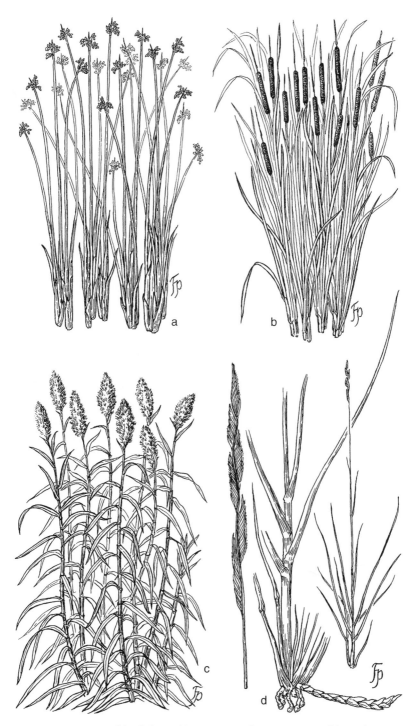

Fig. 5.42.   Desert Marshland plants. (a) Common Tule, *Scirpus acutus.* (b) Southern Cattail, *Typha domingensis.* (c) Giant Reed, *Phragmites australis.* (d) Alkali Cordgrass, *Spartina gracilis*

narrow, pointed, 4–11 in (10–27.5 cm) long, decreasing in length up the stem. Flowering heads tiny, with yellow marginal ray and central disk flowers, in dense terminal clusters, the clusters less than 4 in (10 cm) long. *Flowers:* July–Sept. *Distribution* Northern Sierra Nevada, High Desert in California, Nevada; to Oregon.

COMMON TULE, *Scirpus acutus* Sedge Family. Fig. 5.42a. Giant perennial herb with stems triangular in cross-section, to 15 ft (5 m) tall. Leaves inconspicuous, about 3 in (7.5 cm) long, sheathing the bases of the stems. Flowers minute, covered by papery brown scales, densely aggregated into ovoid heads that dangle from slender branchlets at the tips of the thick green stems. *Flowers:* May–Aug. Related species found in Desert Marshlands include: **Three-square** *(S. americanus),* with stems to about 3 ft (1 m) tall and flower heads in clusters of 1–7, and **Prairie Bulrush** *(S. maritimus),* with stems to 2 ft (0.7 m) tall and 3–10 flower heads in one dense terminal cluster. Both are found in marshy places across North America. *Distribution* Throughout California and Nevada; temperate North America.

GIANT REED, *Phragmites australis* Grass Family. Fig. 5.42c. Robust perennial reed-grass with stout erect stems to 12 ft (4 m), from creeping horizontal underground stems. Leaves up to 2 ft (0.7 m) long, 2 in (5 cm) wide. Inflorescence a spreading plume with many slender silky-haired branchlets. *Flowers:* July–Nov. Forms dense canelike thickets on the margins of Desert Marshlands and in other wet sites. *Distribution* Worldwide; it may have the most extensive range of any seed plant.

MEADOW SUNFLOWER, *Helianthus nuttallii* Sunflower Family. Robust perennial herb with erect stems to 9 ft (3 m). Leaves narrow-oval, pointed, up to 6 in (15 cm) long. Flowering heads are typical sunflowers, about 3 in (7.5 cm) wide, borne in loose clusters at top of main stems. *Flowers:* July–Sept. Found along lake margins and in moist, marshy places. *Distribution* San Gabriel and San Bernardino mountains, High Desert in California, northern Nevada; to Canada, New Mexico.

SOUTHERN CATTAIL, *Typha domingensis* Cattail Family. Fig. 5.42b. Robust perennial herb with many stalks and leaves to 9 ft (3 m) tall, from creeping horizontal stems. Leaves light yellowish green, about 0.5 in (1.3 cm) wide. Flowers minute, densely clustered in 2 elongate, erect cinnamon-brown spikes, one on top of the other, each 6–10 in (15–25 cm) long. The upper spike contains all male flowers; the lower, all female. *Flowers:* June–July. Native peoples used cattails in several ways. The Paiutes of Mono Co. ate the flower stalks before the pollen was produced, either raw or boiled. Other parts of the plant were used for torches and weaving materials; small boats were constructed from cattails and tules. The underground stems can be eaten raw or cooked like asparagus. Cattail pollen substitutes for flour in a modern dish known as Sunshine Pancakes. *Distribution* Much of California and Nevada; throughout tropics and temperate zone.

YERBA MANSA, *Anemopsis californica* Lizard-tail Family. Plate 14f. Low, spreading perennial herb with many horizontal woolly reddish stems. Flowering stalks to about 1 foot (0.3 m). Leaves oval, dark green, to 7 in (17.5 cm) long. Flower spikes resemble a single flower, but are actually made up of many small white flowers with several oval white bracts to about 1 in (2.5 cm) long at the base. *Flowers:* Mar–Sept. Folk medicine prescribed Yerba Mansa, which means "mild herb" in Spanish, for diseases of the skin and blood. The peppery-tasting, aromatic roots, made into tea or powder, were used to cure cuts and sores. The tea was also used to purify the blood, and a wash or poultice of the plant was applied to relieve rheumatism. The wilted leaves were pressed against the skin to reduce swelling. *Distribution* Coastal, valley, and desert marshes in California; Nevada; to Utah, western Texas, northwestern Mexico.

ALKALI SINK SCRUB

Alkali Sink Scrub is best described as the "barren wasteland" often pictured in cartoons of ill-fated desert prospectors. It is composed of dusty little shrubs, herbs, and grasses that are adapted to growing in the harsh, so-called alkali soils of the Owens, Panamint, and other desert valleys. Ranging from 3500 to 5000 feet (1070–1520 m) in the Eastern Sierra, true Alkali Sink Scrub is found from Kern and Inyo counties north to the Honey Lake Basin. Summers are hot and dry, winters are cold. Annual precipitation includes rain and snow, and generally does not exceed 6 inches (15 cm). Most plants can't tolerate these conditions, especially the blinding white alkali soils. Those that have evolved mechanisms for surviving in these parched bottomlands usually don't grow in more hospitable sites.

Although Alkali Sink Scrub soils are commonly called alkali and are referred to as alkaline or basic (as opposed to acidic), many of these soils are actually close to neutral. They are, however, highly saline, containing large concentrations of a variety of salts. These salts include, but by no means are limited to, table salt (sodium chloride). Salty soil creates a difficult problem for plants to solve: how to obtain water (and the dissolved nutrients it contains) when their roots are immersed in a salty solution that draws water out of most plants, causing them to wilt and die. Saltgrass and other salt-tolerant species have evolved the ability to take up salty water and concentrate the salts

Fig. 5.43.　Alkali Sink Scrub

in special glands. Excess salt is secreted in solution and hardens on the leaves and stems into tiny crystals that can be felt and even tasted. Seedling establishment is especially difficult in alkali soils that dry into a solid crust as summer approaches. The seeds of many Alkali Sink Scrub species are "planted" by ants that bury them as part of their food-collecting activities. Those that aren't eaten may germinate in the moister, more favorable conditions beneath the surface.

Alkali Sink Scrub soils occupy basins loaded with salts and other minerals that have been washed down from the uplands and concentrated over eons by evaporation. Shallow water fills many of these basins (called *playa lakes* or *alkali sinks*) during the wet season. Some of them once held lakes that have dried up naturally or, as in the case of Owens Lake, as a result of water diversions. The central depressions are filled with crusted white alkali inhospitable to plant growth. Even the salt-tolerant plants of Alkali Sink Scrub cannot colonize these glaring surfaces. The most salt-tolerant species include Iodine Bush, Inkweed, Alkali Pink, and Greasewood (all members of the Goosefoot Family), Alkali Bird's-beak, and Saltgrass. These species ring the central playa and dominate other saline sites. Dark green stands of Greasewood are characteristic of soils with varying degrees of alkalinity and salinity where groundwater lies within 15 feet (5 m) of the surface under natural conditions. Much of the floor of the Owens Valley and parts of the Honey Lake Basin are covered with this community, which has been called Greasewood Scrub. The perennial grass Alkali Sacaton and shrubs such as Nevada Saltbush, other saltbush species, and Rubber Rabbitbrush grow with Greasewood in areas with a high groundwater table.

COMMON PLANTS OF ALKALI SINK SCRUB

*ALKALI BIRD'S-BEAK, *Cordylanthus maritimus* ssp. *canescens* (= *C. canescens*) Figwort Family. Spreading, much-branched annual root-parasite to 1 ft (0.3 m) tall, stems hairy. Leaves gray-green, narrow-oval, pointed, to 1 in (2.5 cm) long. Flowers pale yellow, elongate, 2-lipped, the upper lip forming a beak. *Flowers:* June–Sept. *Distribution* Mojave and High Desert of California, northern Nevada; to southern Oregon, Utah.

ALKALI PINK, *Nitrophila occidentalis* Goosefoot Family. Low perennial herb with underground spreading stems and many lax branches to 1 ft (0.3 m) long. Leaves narrow, fleshy, borne in opposite pairs, up to nearly 1 in (2.5 cm) long. Flowers tiny, pink when fresh, borne at the leaf bases. *Flowers:* May–Oct. *Distribution* Deserts, Central Valley, and south coast of California; Nevada; to eastern Oregon, Utah, northern Mexico.

ALKALI SACATON, *Sporobolus airoides* Grass Family. Perennial bunchgrass in large tufts up to 3 ft (1 m) tall, with long, narrow, pointed leaves. Flowering stems very slender, diffusely branched, the flowers minute and enclosed in tiny bracts. *Flowers:* Apr–Oct. *Distribution* Much of southern California, Nevada; to eastern Washington, central and southern United States, Mexico.

BROOM ASTER, SHRUBBY ALKALI ASTER, *Machaeranthera carnosa* (= *Aster intricatus*) Sunflower Family. Fig. 5.39c. Rounded shrub to nearly 3 ft (1 m), with many slender, intricately divided, almost leafless branches. Leaves few on lower stems, narrow and less than 1 in (2.5 cm) long. Flowering heads yellow, about 0.5 in (1.3 cm) high, borne individually at the tips of slender branchlets. *Flowers:* June–Oct. *Distribution* San Joaquin Valley and higher deserts of California; southern Nevada; western and southern Arizona.

*GREASEWOOD, WORMWOOD, *Sarcobatus vermiculatus* Goosefoot Family. Fig. 5.44a. Much-branched spiny shrub to 6 ft (2 m) tall, young branches yellowish white. Leaves bright green, narrow, fleshy, to 1 in (2.5 cm)

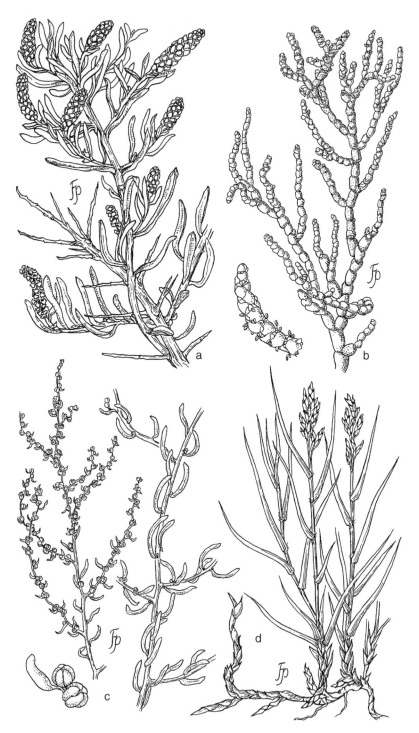

Fig. 5.44.   Alkali Sink Scrub plants. (a) Greasewood, *Sarcobatus vermiculatus.* (b) Iodine Bush, *Allenrolfea occidentalis.* (c) Inkweed, *Suaeda moquinii.* (d) Saltgrass, *Distichlis spicata*

long. Male flowers in conelike spikes, about 1 in (2.5 cm) long, at the ends of the branchlets. Female flowers about 0.5 in (1.3 cm) wide, green, borne singly at the bases of the leaves, each with a spreading green bract that turns red as the fruit matures. *Flowers:* May–Aug. *Distribution* Mojave and High Deserts of California, Nevada; to western Canada, Great Plains, northern Mexico.

INKWEED, BUSH SEEPWEED, *Suaeda moquinii (= S. torreyana)* Goose-foot Family. Fig. 5.44c. *Subshrub* (not totally woody) with many slender as-cending and spreading branches, to 2 ft (0.7 m) tall. Leaves narrow, about 1 in (2.5 cm) long at stem bases, decreasing in length toward stem ends. Flowers tiny, green, in clusters of 1–4 at leaf bases. *Flowers:* May–Sept. *Distribu-tion* San Francisco Bay, Central Valley, deserts of California; Nevada; to western Canada, Texas, Mexico.

IODINE BUSH, *Allenrolfea occidentalis* Goosefoot Family. Fig. 5.44b. Much-branched erect shrub or half-shrub to 6 ft (2 m) tall, with fleshy jointed stems and minute scalelike leaves. Flowers very tiny, inconspicuous, borne at stem joints. Resembles pickleweeds of coastal salt marshes. Fleshy stems have high salt levels. *Flowers:* June–Aug. *Distribution* Moist alkaline places in California deserts, San Joaquin Valley, and interior salt marshes in eastern San Francisco Bay region; to Oregon, Utah, northern Mexico, Baja California.

*NEVADA SALTBUSH, *Atriplex lentiformis* ssp. *torreyi (= A. torreyi)* Goosefoot Family. Erect, much-branched gray shrub to about 5 ft (1.7 m) tall. Twigs stiff and spiny. Leaves oval to arrowhead-shaped, 0.5–1.25 in (1.3–3.1 cm) long, covered with small gray scales. Flowers small, green, in loose, elongate clusters at the ends of the branchlets. *Flowers:* June–Oct. One of about 10 species of *Atriplex* found in the High Desert shrub communities east of the Sierra. Other saltbush species that occur in Alkali Sink Scrub are *Parry Saltbush (A. parryi),* a low, rounded shrub about 1 foot (0.3 m) tall with rounded to heart-shaped leaves, and *Wedgescale (A. truncata),* an an-nual herb with oval to arrowhead-shaped leaves. *Distribution* Mojave and High Deserts of California, Nevada; to southwestern Utah.

RUBBER RABBITBRUSH, *Chrysothamnus nauseosus* Sunflower Family. For description see Sagebrush Scrub.

SALTGRASS, *Distichlis spicata* Grass Family. Fig. 5.44d. A low, mat-forming grass with many creeping stems that root as they grow outward. Leaves narrow, diverging from the stems, up to 2 in (5 cm) long. Flowers tightly clustered in spikes about 1 in (2.5 cm) long, barely raised above the leaves. *Flowers:* May–Aug. Distributed widely in areas with both saline and alkaline soils. In many sites Saltgrass reproduces by spreading stems called *runners* and produces few seeds. It forms large, dense mats that extend over many acres and superficially resemble Bermuda Grass lawns. *Distribution* Widespread. California, Nevada; to southern Canada, United States.

JOSHUA TREE WOODLAND

Angular, tree-sized yuccas that embody the desert's strangeness and mystery dominate sparse, open stands of Joshua Tree Woodland. The upward-spreading limbs of these bizarre desert "trees" reminded early Mormon explorers of the biblical Joshua, his arms outstretched in prayer. Woodlands of Joshua Trees are scattered over the gravelly slopes and alluvial fans of the Mojave Desert at elevations where precipitation averages 8–10 inches (20–25 cm) per year. These slopes are a little cooler and moister than lower sites that are covered with Creosote Bush Scrub.

Joshua Trees are the only treelike plants of this desert woodland. Its associates are shrubs, other yuccas, a few cacti and low grasses, and many wildflowers. The familiar single-trunked Joshua Tree inhabits much of the Mojave Desert; however, a multi-trunked form is scattered along the eastern base of the Sierra Nevada. These Joshua Trees have multiple, clumped, often unbranched stems that sprout from scaly subterranean rootstocks. This clumped variety is fire-adapted: The underground stems sprout new shoots vigorously after fire. One of the best developed multi-trunked Joshua stands is in Freeman Canyon along Highway 178, east of Walker Pass, at about 4500 feet (1370 m). Here, in late May, the shrubby woodland understory is at its best. Tall, rounded shrubs of yellow-green California Juniper contrast with the lower mounds of Showy Goldenbush that are covered with bright yellow daisies, and with the spreading Blue Sage and its abundant blue-purple flower spikes. Penstemons, gilias, other phlox relatives, and poppies (including a white-flowered miniature only an inch high) further brighten the landscape.

In a community of sparse plant cover, the Joshua Tree is a source of food and shelter for many creatures. Birds seek refuge from predators by hiding their nests in its high branches. More than 20 species nest here, including the Ladder-backed Woodpecker and Northern Flicker, which excavate nest cavities in the soft wood. These hollows are used again by Ash-throated Flycatchers, House Wrens, and Western Bluebirds, hole-nesters that don't make their own holes. Among the Joshua's spiny leaves Scott's Oriole weaves a nest constructed mainly from yucca fibers. Nesting raptors, such as the Ameri-

Fig. 5.45.   Joshua Tree Woodland

can Kestrel and Red-tailed Hawk, perch on the highest branches to oversee their territories.

The lower branches often show evidence of Desert Wood Rats. These nocturnal rodents, in their never-ending search for nest materials, gnaw off quantities of tough, fibrous Joshua leaves. These leaves don't fall off readily. Instead, as they die, they fold back along the branch, dry, and turn brown, forming a dense thatch with the hard spiky leaf tips pointing downward. Desert Wood Rats seem undeterred by this armor. Their unruly nest piles— often heaped over the remains of fallen Joshuas—contain enormous numbers of detached leaves. Fallen Joshua Trees are home to the little Desert or Yucca Night Lizard *(Xantusia vigilis)*, a frail dark creature that seems out of place in the desert. About 2 inches (5 cm) long, it hides from enemies and the sun's heat beneath decaying Joshua stems, feeding on the termites and other insects that infest these downed logs.

Stands of Joshuas near their lower elevational limit, around 3000 feet (915 m), contain plants typical of the neighboring Creosote Bush Scrub, including a generous array of colorful wildflowers. At some higher sites the

Joshua's associates include Shadscale Scrub plants. The following species occur in Joshua Tree Woodland at 4000 feet (1220 m) and above.

SOME COMMON PLANTS OF
JOSHUA TREE WOODLAND

*JOSHUA TREE, *Yucca brevifolia*   Lily Family. Fig. 5.46e. Tree-sized yucca, 16–50 ft (5–15 m), branched 3–12 ft (1–4 m) above the ground. Stems and branches stout, branches angular. Leaves gray-green, 8–14 in (20–35 cm) long, rigid, each with a spine about 0.5 in (1.3 cm) long at the tip. Leaves clustered at the ends of branches. Flowers large, with 6 waxy, cream to greenish white petal-like parts, in dense clusters 1–1.5 ft (0.3–0.5 m) long, borne at the ends of branches. *Flowers:* Apr–May.

Until it reaches 5 ft (1.5 m) or more in height, the stem of the young Joshua remains erect and unbranched, covered with a dense brush of thick, leathery, bayonetlike leaves that protect it from nearly all invaders. The stem forks after it has flowered for the first time or after it has been attacked by the larva of the Yucca-boring Weevil *(Scyphophorus yuccae),* whose burrowing destroys the growing point and stimulates the production of side branches. In April large clusters of flowers form at the ends of the branches; the following season 2 or 3 heavy angular shoots will grow from each flowering site. Profuse flowering occurs irregularly; most years the trees produce just a few flower clusters. Joshua Tree flowers are pollinated only by the Joshua Tree Yucca Moth *(Tegeticula paradoxa).* The small pale female moth flies about at night visiting the large, white, musty-smelling blossoms. She collects pollen from several flowers, forms it into a compact ball, and places it on the stigma of a single flower. This act assures fertilization and abundant seed production in the fruit of that flower. The female moth also inserts one of her eggs within the young fruit of the flower she has pollinated. As the seeds mature, the moth egg hatches into a larva (caterpillar) that eats some (usually less than 20 percent) of the developing seeds before exiting from the fruit. The remaining seeds ensure the growth of new Joshua Trees. This relationship is a classic example of *mutualism* (a form of interaction between two species that benefits both). Mature Joshua Tree fruits remain on the tree until they are dry and brittle. Rarely do they open by themselves; eventually they are blown from the branches and broken open on the ground, the seed scattered by wind and rodents. *Distribution*   Southern Sierra Nevada, Mojave Desert in California and Nevada; to southwestern Utah, western Arizona.

*BLUE SAGE, *Salvia dorrii*   Mint Family. Plate 15f. Low, broad, much-branched shrub, 1–2 ft (0.3–0.7 m) tall, with hairy twigs. Leaves roundish to

Fig. 5.46.   Joshua Tree Woodland and Creosote Bush Scrub plants. (a) Gray California Buckwheat, *Eriogonum fasciculatum* var. *polifolium*. (b) Anderson's Boxthorn, *Lycium andersonii*. (c) Creosote Bush, *Larrea tridentata*. (d) Mojave Yucca, *Yucca schidigera*. (e) Joshua Tree, *Yucca brevifolia*

spoon-shaped, less than 1 in (2.5 cm) long. Flowers showy, 2-lipped, deep blue-purple to rose, about 0.5 in (1.3 cm) long, in spikes formed of several dense globular clusters. *Flowers:* May–July. *Distribution* Deserts of California, Nevada; to Washington, Idaho, Utah, Arizona.

CALIFORNIA JUNIPER, *Juniperus californica* Cypress Family. Large rounded shrub 3–13 ft (1–4 m) tall, with stout branches and ashy gray bark. Leaves yellowish green, minute, scalelike, closely covering the twigs. Berries blue with a pale waxy coating when young, reddish brown when older. *Distribution* Coast Ranges, Peninsular and Transverse Ranges, Sierra Nevada, desert mountains in California; southern Nevada; northwestern Arizona, Baja California.

*COTTON THORN, *Tetradymia axillaris* Sunflower Family. For description see Sagebrush Scrub.

GRAY CALIFORNIA BUCKWHEAT, *Eriogonum fasciculatum* var. *polifolium* Buckwheat Family. Fig. 5.46a. Low spreading shrub to about 1.5 ft (0.5 m) high, with stems to 4 ft (1.3 m) long. Leaves gray-green, narrow, about 0.5 in (1.3 cm) long, rolled under along the margins and borne in small clusters. Flowers small, each with 6 pink and white petal-like parts, in dense clusters on long, naked stalks. *Flowers:* Apr–Nov. *Distribution* Mountains of southern California, deserts in California; southern Nevada; to southwestern Utah, western Arizona, northwestern Mexico.

MOJAVE PENSTEMON, *Penstemon incertus* Figwort Family. Much-branched shrub to 2 ft (0.7 m) tall, the stems and leaves hairless and gray-green. Leaves long and narrow, to 2.5 in (6.3 cm). Flowers narrowly funnel-shaped, 2-lipped, about 1 in (2.5 cm) long, violet with a reddish cast, or purple. *Flowers:* May–June. *Distribution* Tehachapi and San Bernardino mountains, southern Sierra Nevada, desert mountains in California.

*MOJAVE YUCCA, *Yucca schidigera* Lily Family. Fig. 5.46d. Most common desert yucca. Woody plant with stout trunk 3–15 ft (1–4.5 m), sometimes branched above into a few short, broad stems. Leaves yellow-green, troughlike, to 4.5 ft (1.5 m) long, tipped with sharp spines and with coarse fibers curling from the margins. Flowers in a dense cluster to 3 ft (1 m) tall. Individual flowers with 6 cream or purple-tinged petal-like parts, about 2.5 in (6.3 cm) long. *Flowers:* Apr–May. The fruits, which ripen in mid-summer, have a bitter but sugar-rich outer covering that is often gnawed by Desert Wood Rats. *Distribution* San Diego Co., Mojave and Sonoran deserts in California; Nevada; Arizona, northern Baja California.

SHOWY OR INTERIOR GOLDENBUSH, *Ericameria linearifolia (=Haplopappus linearifolius)* Sunflower Family. Plate 15a. Spreading shrub to 4.5 ft (1.5 m) tall, the twigs resinous. Leaves very narrow, soft, up to 1.5 in (3.8 cm) long, sometimes borne in small clusters. Flower heads about 1.5 in (3.8 cm) wide, yellow, sunflowerlike, borne on stalks at the ends of the twigs. *Flowers:* Mar–May. *Distribution* Sierra Nevada, Central Valley and surrounding hills, Mojave and Sonoran deserts in California; central and southern Nevada, southwestern Utah, western Arizona.

CREOSOTE BUSH SCRUB

Creosote Bush's straggly black-ringed branches, olive-drab leaves, and yellow, pinwheel-shaped flowers are unmistakable, and its penetrating "telephone pole" odor—especially noticeable after rain—often scents the desert air. The Creosote Bush Scrub community includes other low shrubs, cacti, and herbaceous plants, but its most distinctive, common, and widespread plant is the Creosote Bush. Throughout the Mojave Desert this shrub dominates large areas. In the Eastern Sierra Creosote Bush Scrub ranges from Olancha in Inyo County south to Indian Wells Valley, covering hundreds of square miles, especially on slopes of alluvial fans, from about 3000 to 4500 feet (910–1370 m) elevation. It occurs on a wide range of soil types; however, Creosote Bush and most of its associates can't tolerate the finer-grained, poorly drained alkaline and salty soils of low-lying basins and wet areas, which usually support Alkali Sink Scrub.

In all of the Eastern Sierra, the area occupied by Creosote Bush Scrub has the warmest summers and winters and receives the least rainfall. Summers are long and dry, with hot days (over 90° F or 32°C) coming as early as May, and daytime temperatures commonly exceeding 100°F (38°C) from June through September. Rainfall seldom exceeds 5 inches (13 cm) a year, and snow only occasionally blankets the ground in winter. Yet more than 100 kinds of wildflowers grow in this community. Plants deal with the heat and drought in one of two ways: some, mainly short-lived wildflowers, avoid it by spending most of the year as seeds, sprouting and growing only during those periods when soil moisture is relatively high; others, the shrubs and cacti, are able to endure the long dry seasons because they possess a variety of mechanisms for conserving water.

During the dry season the desert ground is bare, but each spring masses of colorful wildflowers appear. They form bright rings around rocks and shrubs, outline moist gullies in vivid color, and transform drab hillsides into mosaics of pink, yellow, and blue. The amount of winter rain determines the extravagance of this display; wet winters are followed by spectacular carpets

Fig. 5.47. Creosote Bush Scrub

of color. Desert annuals are experts at making the most of scanty rainfall. They are referred to as "ephemerals" because their active growth period is relatively short. In a typical species, such as the bright pink Bigelow Monkeyflower *(Mimulus bigelovii),* the seeds germinate in fall or winter after rain. Annuals with this germination pattern are called *winter annuals.* Most species require about an inch or more of rain to germinate; it takes that much to leach from the seeds the chemicals that prevent germination. This mechanism keeps the plants from sprouting when the soil is too dry to sustain growth. Monkeyflower seedlings can grow rapidly and flower within two months, even if no more rain falls. They may be only an inch tall, with as few as one or two flowers, but they are capable of producing seed with just one watering. If more rain comes, the plants continue growing and flowering for up to four months, reaching 8 inches (20 cm) in height and producing dozens of flowers. As the soil dries out, the annuals set seed and die, leaving a scattering of seeds, protected by a tough seed coat, on the desert floor.

Another group, called *summer annuals,* exploits a highly unpredictable water resource: tropical downpours that occasionally reach the Mojave in summer. The seeds of summer annuals won't germinate in winter, no matter how much rain falls, because they are inhibited by low temperatures. Only summer rains can bring them forth. These seeds may remain dormant during several years of rainless summers.

Desert shrubs and cacti, which live for many years, survive long, hot summers by employing some amazing and effective methods of water conservation. Cacti are superbly adapted for storing and conserving water. The stems are thick, fleshy, and compact; their green surfaces conduct photosynthesis, eliminating the need for water-wasting leaves. Tissues inside the stems are specialized to store water for long periods. The shallow root system of a cactus extends over a large area, enabling it to absorb water quickly after rain. These plants also can suspend growth and most other metabolic activities for several years during a long drought. The stems may shrivel up and become wrinkled—some may die—but when rains return most recover their typical succulent forms. The spines of cacti, dense and overlapping in some species, may function as whitish hairs do in other desert plants, shading the stem surface and keeping it a little cooler. They also serve as a powerful deterrent to grazers.

Creosote Bush provides examples of adaptations found in many desert shrubs. The leaves are small and covered with resinous secretions, thereby limiting water loss through the leaf surface. They are also drought-deciduous: The leaves gradually fall from the twigs as dry conditions progress. This temporarily reduces the ability to conduct photosynthesis, but it also sharply limits water loss, permitting the shrub to survive prolonged drought. New leaves sprout rapidly when moisture becomes available. The shallow roots of the Creosote Bush, like those of cacti, allow the shrubs to benefit from even light rainfall. These shrubs are unusual in having the ability to reproduce without seed by cloning. In this process the parent shrub sends out a ring of branches, some of which eventually root and form new shrubs, while the central progenitor dies. The creosote rings thus formed can attain great size and age; one in the western Mojave Desert, nicknamed "King Clone," is more than 70 feet (21.5 m) wide and is estimated to be more than 12,000 years old, more than twice as old as the oldest Bristlecone Pine. Cloning is advantageous because it produces new shrubs directly from older shrubs, thus avoiding the risks and high mortality of the seedling stage of growth.

COMMON PLANTS OF CREOSOTE BUSH SCRUB

*CREOSOTE BUSH, *Larrea tridentata* Caltrop Family. Fig. 5.46c. Plate 15g. Evergreen (but drought-deciduous), aromatic, resinous shrub with many slender crooked branches, varying in height from 2 to 12 ft (0.7–4 m). Leaves in opposite pairs along the twigs, each consisting of 2 oval, olive-green leaflets to 0.5 in (1.3 cm) long. Flowers about 0.7 in (1.8 cm) wide, with 5 yellow, twisted petals. Fruit small, oval, covered with long white hairs. *Flowers:* Apr–May.

The size and spatial arrangement of Creosote Bush vary with local conditions. Shrubs may grow in regular, random, or clumped patterns and may be widely or closely spaced, as a result of competition for water. In very dry sites the shrubs are 1 or 2 ft (0.3–0.6 m) tall, widely and regularly spaced. In wetter areas, such as along paved roads where runoff accumulates, they are much larger and grow close together. Regular spacing is thought to be maintained by the failure of seedlings to survive within the circle occupied by the spreading root systems of existing shrubs. Seedling death is attributed both to chemical inhibition and to unsuccessful competition for water. Random distributions are found in arid areas, where competition with other shrub species is important. Clumped distributions occur in very old, undisturbed stands, where creosote rings have formed from individual shrubs. Creosote Bush was used for firewood and medicinal purposes by native peoples of the western deserts. Various concoctions of the leaves and stems were used in attempts to cure colds, chest infections, stomach cramps, cancer, gonorrhea, and tuberculosis; to heal wounds, draw out poisons, and prevent infections; and to deodorize the body and eradicate dandruff. *Distribution* Mojave and Sonoran deserts of California, southern Nevada; to southwestern Utah, Texas, central Mexico.

*ANDERSON'S BOXTHORN, *Lycium andersonii* Nightshade Family. Fig. 5.46b. Rounded, much-branched shrub to 6 ft (2 m) tall, with needlelike spines. Leaves pear-shaped, to 0.6 in (1.5 cm) long. Flowers whitish lavender, tubular, to 0.7 in (1.8 cm) long, 1 or 2 borne at the bases of some leaves. *Flowers:* Mar–May. The fruit is a small, red, fleshy berry, resembling a tiny tomato. (Tomatoes also are members of the nightshade family.) Desert quail relish both the fruits and leaves. A smaller shrub, *Cooper's Boxthorn (L. cooperi),* with greenish white flowers and a green berry, is also common in Creosote Bush Scrub. *Distribution* Southern San Joaquin Valley and surrounding hills, deserts in California; Nevada; to Utah, New Mexico, northwestern Mexico.

BEAVERTAIL CACTUS, *Opuntia basilaris* Cactus Family. Plate 15h. Low cactus to 1 foot (0.3 m) high, with several fleshy, broad, flattened stems. Stem sections gray-green, sometimes purple-tinged, covered with round spots made up of many short, slender, needlelike hairs. Leaves lacking. Flowers showy, about 3 in (7.5 cm) wide, each with many reddish purple to rose-pink petals; borne along the edges of the stem pads. *Flowers:* Mar–June. Native peoples ate the young fruits, after rolling them in the sand to remove the sharp hairs. Other cacti common in Creosote Bush Scrub include *Silver Cholla (O. echinocarpa),* *Old Man Cactus (O. erinacea),* *Hedgehog Cactus (Echinocereus engelmannii),* and *Mojave Mound Cactus (E. triglochidiatus =E. mojavensis).* Distribution Deserts and other dry sites in southern California, southern Nevada; to Utah, Arizona, Mexico.

*BLADDER SAGE, *Salazaria mexicana* Mint Family. Plate 15c. Also called Paperbag Bush because of the dry sacks that inflate around the fruit. Dense shrub with divergent branches, 2–3 ft (0.7–1 m) tall. Leaves oblong, about 0.5 in (1.3 cm) long. Flowers 2-lipped, the lips purple. Calyx becomes inflated and globular as the fruit matures. *Flowers:* Mar–June. Distribution Deserts of California, southern Nevada; to Utah, Texas, northern Mexico.

*BURRO-BUSH, *Ambrosia dumosa* Sunflower Family. The most frequent associate of Creosote Bush, but less noticeable due to its small stature and dull color. Low, intricately branched, rounded shrub, 8–24 in (0.2–0.7 m) high. Stems white and spiny. Leaves gray-green, to nearly 1 in (2.5 cm) long, oval in outline, deeply lobed. Flower heads yellow, about 0.25 in (6 mm) in diameter, in short erect spikes. *Flowers:* Feb–June, Sept–Nov. A preferred browse plant of burros and sheep. Distribution Deserts of California, southern Nevada; to southwestern Utah, Arizona, northwestern Mexico.

*BUSH SUNFLOWER, *Encelia virginensis* Sunflower Family. Plate 15d. Rounded, much-branched shrub, 2–4.5 ft (0.7–1.5 m) high, white stems, leaves deep green, oval, pointed. Flowering heads about 2 in (5 cm) in diameter; yellow sunflowers on long, naked flowering stalks that rise above the body of the shrub. *Flowers:* Apr–May. Prominent on gravelly slopes and in washes. Distribution Mojave Desert in California, southern Nevada; to southwestern Utah, northwestern Arizona.

*CHEESEBUSH, BURROBRUSH, *Hymenoclea salsola* Sunflower Family. Plate 15e. So named because the leaves emit a cheesy smell when crushed. Spreading, aromatic twiggy bush to 4 ft (1.3 m) tall. Leaves soft, 1–2 in (2.5–5 cm) long, very narrow, yellow-green. Flowering heads of 2 kinds: the female with one flower surrounded by several round whitish bracts, resembling a small pearlescent rose; the male with several very small yellow flowers in each. *Flowers:* Mar–June. Distribution South Coast, southern San Joaquin Valley, deserts of California; to southwestern Utah, Arizona, northwestern Mexico.

*DESERT OR APRICOT MALLOW, *Sphaeralcea ambigua*   Mallow Family.
Plate 15b. Perennial, woody at the base, with several erect stems to 3 ft (1 m)
high. Leaves thick, heart-shaped to roundish, gray-green, densely covered
with tiny star-shaped hairs. Flowers about 1 in (2.5 cm) wide, with 5 rose-
pink to peach-colored petals. Flowers borne on elongate spreading stalks.
Flowers: Mar–June. Exceedingly variable in size and number of flowers produced; ranges from below 500 ft (154 m) in the Sonoran Desert to above
9000 ft (2740 m) in the White Mountains. **Distribution**   Deserts of California, Nevada; to Utah, Arizona, Mexico.

*MOJAVE ASTER, *Xylorhiza tortifolia (= Machaeranthera tortifolia)*   Sunflower Family. Suffrutescent perennial with a woody base and many erect
stems 1–2 ft (0.3–0.7 m) tall. Leaves oval, elongate, toothed along the margins. Flower heads 1.5–2 .5 in (3.8–6.3 cm) wide; lavender sunflowers with
yellow centers, solitary at the ends of long, naked stalks. *Flowers:* Mar–May,
Oct.   **Distribution**   Mojave and Sonoran deserts in California, Nevada;
to southwestern Utah, western Arizona.

INTERMINGLING

As these plants from the Sierra Nevada, the Great Basin Desert, and the Mojave Desert intermingle, they bestow upon the Eastern Sierra an amazing
diversity. Desert species creep upward to mix with mountain flora; a few desert species even reach subalpine elevations in the southern Sierra. Streamside woodlands follow their streams out of the canyons and into the sagebrush. The green pine-fir forests of the Sierra merge with the gray shrubs of
the Great Basin Desert, and both blend with the Creosote Bush and other
widely spaced shrubs of the Mojave Desert. In spring, among the seemingly
monotonous expanses of gray shrubs, wildflowers bring riots of color. And
even in the most inhospitable places—in the alkali flats and on the highest
bare ridges—*something* manages to grow. Since altitude, climate, temperature, rainfall, and soil vary tremendously in the Eastern Sierra, due largely to
the abrupt mountain front, some plants endure months of no rain and withering heat, while others endure months of snow and bitter cold. Each plant
is adapted in wondrous ways to its particular environment, however harsh it
may be. Here, these extremes are not far from each other. Just a few miles
separate the sagebrush that covers the arid foothills from the alpine flowers
near treeline. All these factors have combined to fashion a region at the edge
of the Great Basin that is exceptionally rich in wildflowers, shrubs, and trees.

RECOMMENDED READING

Carville, Julie S. 1989, *Lingering in Tahoe's Wild Gardens.* Chicago Park, Calif.: Mountain Gypsy Press.

Graf, Michael. 1999. *Plants of the Tahoe Basin: Flowering Plants, Trees, and Ferns.* Berkeley: University of California Press.

Mozingo, Hugh M. 1987. *Shrubs of the Great Basin.* Reno: University of Nevada Press.

Munz, Philip. 1962. *California Desert Wildflowers.* Berkeley: University of California Press.

Niehaus, Theodore F. 1974. *Sierra Wildflowers.* Berkeley: University of California Press.

Taylor, Ronald. 1992. *Sagebrush Country.* Missoula: Mountain Press Publishing.

Weeden, Norman F. 1996. *A Sierra Nevada Flora.* Berkeley: Wilderness Press.

CHAPTER SIX

ARTHROPODS

EVAN A. SUGDEN

The arthropods are by far the most successful group of creatures on our planet, whether success is measured in terms of sheer numbers, in species diversity, or in the variety of their habitats. Of all the animal and plant species in the world, at least 60 percent are arthropods; most of these are insects. It has been estimated that out of every five multicelled organisms, one is a beetle. Around 1 million arthropod species have been identified; possibly another 10 million have not yet been named or even discovered. They occur in the ocean, the soil, and the air; in hot springs, oil pools, deserts, and ice caves; and inside plants and other animals. Their ecological importance is tremendous. Hardly any form of life is unaffected by arthropods, least of all our own species, *Homo sapiens.*

Since it is impossible to present all the members of even one family of arthropods in this chapter, we consider here a small but diverse sample of the roughly 10,000 arthropod species that inhabit the Eastern Sierra. Species described are common, important in their effects on humans, interesting because of some unusual adaptation, or important in their ecosystem. They are grouped according to their behavior and ecology: aquatic arthropods, decomposers, forest pests, human-biting arthropods, mimics, plant-feeders, pollinators, predators, parasites, and winter-active arthropods. Some species that belong to two or more groups have been chosen to illustrate species interrelations. For example, many aquatic arthropods are also winter-active, and some pollinators are mimics of others. Species in different groups interact, forming networks of interdependence. The Western Sand Wasp[†] for example, a predator and pollinator, preys on the Big Black Horse Fly[†], a human-biting arthropod. The horse fly attacks large mammals that also provide blood to the Bodega Gnat[†], another human-biting arthropod. Eggs of the Big

Black Horse Fly are parasitized by a scelionid wasp[†], and larvae of Edwards' Cuckoo Wasp[†] consume Western Sand Wasp larvae. The Cuckoo Wasp is in turn often parasitized by Sacken's Velvet Ant[†].

The dagger symbol ([†]) indicates species or groups that are discussed in more than one section of this chapter, illustrating how a single species may play several ecological roles or illustrate several different concepts.

Life histories of arthropods are varied, and many are complex. All begin as eggs and grow in a series of stages, which in insects are known as *instars.* Between instars, each individual sheds its confining exoskeleton and expands its soft exterior by inflating itself with air. The new exoskeleton then hardens around the enlarged body and becomes the supportive and protective armor during the next instar. Such molting occurs several times before adulthood. In general, the immatures of arachnids and other noninsects resemble adults in form and behavior throughout their development. This is also true of primitive nonwinged insects, such as the Snow-loving Springtail[†]. Some species undergo gradual development during which the immature forms (*nymphs* or *naiads*) possess undeveloped wings, which enlarge between instars (fig. 6.3). Most nymphs have a lifestyle similar to that of the adult form; this is the case, for example, with grasshoppers and true bugs. In some aquatic species, however, the nymph may have a completely different lifestyle than the fully winged adult—for example, dragonflies, damselflies, mayflies, and stoneflies. In most other species development is more complex. Immatures of these species are known as *larvae.* Familiar examples are moth and butterfly caterpillars, fly maggots, and beetle grubs. Larvae usually have lifestyles radically different from those of their corresponding adults. This allows the species to take advantage of two or more habitats or food sources, and may prevent adults and immatures from interfering with each other. At the end of its final instar the larva changes into a *pupa,* often spinning a cocoon around itself first or encasing itself in its last larval skin. This is the familiar *chrysalis* of butterflies and moths. The form of the insect changes radically within the pupa by a dynamic process known as *metamorphosis* (fig. 6.4). A completely changed adult with wings finally emerges from the pupal case. In some parasites, such as blister beetles[†], the larva undergoes *hypermetamorphosis,* in which nearly every instar is different in form and behavior.

Larvae and nymphs spend most of their time feeding in preparation for their adult careers. The primary activities of adult insects center on seeking mates, mating, and egg-laying, and, in some species—such as termites, ants, social wasps, and bees—protecting the next generation or helping nestmates. At some point in the development of egg to adult insect, there may be an intervening resting phase, or *diapause,* during which most development

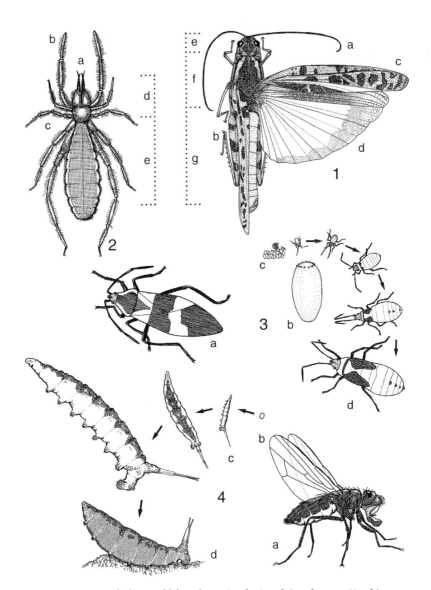

Figs. 6.1–6.4. Morphology and life cycles. 1. Coral-winged Grasshopper, *Xanthippus corallipes* (35 mm): (a) antenna; (b) legs; (c) forewing; (d) hind wing; (e) head; (f) thorax; (g) abdomen. 2. A wind scorpion, *Eremobates* species (25 mm): (a) jaws; (b) sensory legs; (c) walking legs; (d) cephalothorax; (e) abdomen. 3. Simple development in the Large Milkweed Bug, *Oncopeltus fasciatus:* (a) adult (19 mm); (b) single egg (1 mm); (c) egg mass; (d) progressive nymphal instars. 4. Complex development in the Alkali Fly, *Ephydra hians:* (a) adult (6 mm); (b) egg (0.5mm); (c) larva (12 mm); (d) pupa (10 mm). (Drawings by Evan A. Sugden)

and activity stops. Diapause is an important adaptation that allows some species to "sleep through" unfavorable conditions, such as winter cold. Each insect species diapauses at a particular stage, allowing for immediate resumption of the life cycle when conditions are likely to be favorable.

Most arthropods are "cold-blooded": They cannot generate their own metabolic heat and are therefore restricted in their behavior by ambient temperature. There are, however, some notable exceptions to this. Winter-active arthropods and bumble bees are among them and are discussed below.

CLASSIFICATIONS

Arthropods (from the Greek *arthro-,* "joint," and *pod-,* "foot") have jointed appendages and a hard, shell-like exoskeleton. They breathe by gills (aquatic insect larvae, crustaceans), air ducts (most terrestrial insects), or book lungs (spiders), or through diffusion (minute insects such as Collembola[†]). The creatures classified as arthropods include the arachnids (spiders, scorpions, and their kin), crustaceans (shrimp, crabs, copepods, lobsters, etc.), horseshoe crabs, millipedes, centipedes, insects, and several less familiar groups. People commonly call any small creature with many legs an insect or bug, but here we use these terms in a scientific sense, applying *insect* to specific subgroups of the phylum Arthropoda (most six-legged arthropods are insects) and *bug* to species of a group of juice-sucking insect families including the seed bugs[†] and assassin bugs[†].

Insects, then, are a class of arthropods having three pairs of legs, only two antennae, and three body parts (the sensory head, thorax, and abdomen). The abdomen houses most of the organs of digestion, elimination, respiration, circulation, and reproduction (fig. 6.1). Insects are the only invertebrates that fly, although not all have wings. Several groups of primitive insects (which may resemble the early ancestors of modern species) are wingless in all stages (for example, Collembola[†]). In addition, some species in every major insect category have lost their wings through evolution in adapting to particular conditions. In some species, such as termites[†] and ants[†], only one of several castes (the reproductive males and females) has wings. In other species, such as the velvet ants[†], only one sex (the males) has wings. But for most insects, both sexes possess two pairs of wings in the adult stage. One pair may be reduced to sensory knobs, as in the flies and some parasites, or it may have functions in addition to flight, as in the beetles.

The external anatomy of a noninsectan arthropod is illustrated in figure 6.2. The wind scorpion's head and middle body region are combined as the *cephalothorax;* what appears to be the first pair of legs is actually a pair of sensory mouth parts.

EXPLANATION OF TERMS, STYLE, AND SYMBOLS

Species mentioned in the text are representative of their families unless otherwise noted; closely related species not mentioned here are likely to have similar morphology, behavior, and ecology. A species and its scientific name are printed in boldface, followed by the family and sometimes the order to which the arthropod belongs, when it is first mentioned in the text. Interrelationships of arthropods are both interesting and important in nature. The junction of three major biological regions in the Eastern Sierra is reflected in the diversity of its habitats for arthropods and of the arthropods themselves. Most arthropods discussed here range beyond the Eastern Sierra into one or more of these regions; some are common throughout western North America. The most distinguishing geographic feature of Eastern Sierra arthropods is whether they inhabit only the eastern slope of the Sierra Nevada or whether they can be found on the western slope as well. In this chapter the names of species found *mostly* east of the Sierra Crest are followed by the letter *E* when they are first mentioned. A few species are limited to high mountains or certain aquatic situations or may be rare. If a species is likely to be found *only* in the Eastern Sierra region or only in the Sierra at high altitudes, its name is followed by the letters *ES*.

Illustrations and plates show the adult form unless otherwise indicated. Arthropods are mostly small or even minute compared to other animals. Most are less than an inch in length and are best measured in millimeters (mm), which is the unit used in this chapter. One millimeter is equivalent to about 0.04 inch; 25 mm is about 1 inch. Such measurements refer to *body length* of a typical specimen from head to tip of abdomen, not including long antennae or tail filaments. In the case of butterflies and moths, measurements are preceded by the letter *W* and refer to the span of the outstretched forewings. Common names follow the usage in Powell and Hogue, *California Insects* (1980), and the Entomological Society of America's *Common Names of Insects and Related Organisms* (edited by M. B. Stoetzel, 1989). We have composed simple, descriptive common names for a few species to which none have been previously assigned.

AQUATIC ARTHROPODS

Aquatic habitats of the Eastern Sierra vary almost as much as terrestrial ones. From the cold snowmelt streams of the high peaks to the hot springs and alkali lakes of the valleys, arthropods inhabit virtually all water habitats. Mountain streams are the home of larvae of the three most common orders of running-water aquatic arthropods. These are caddisflies (Trichoptera), whose larvae bear different types of tubular cases and whose adults resemble small moths; stoneflies (Plecoptera[†]) (fig. 6.69); and mayflies (Ephemeroptera) (fig. 6.70), whose larvae possess three tail filaments. Together, the various species of these insects are a vital part of a stream's food chain, which may terminate in the fisherman's frying pan. The presence of these aquatic nymphs is a rough gauge of a stream's purity. Turn over a medium-sized rock in a well-aerated riffle, and you will find them clinging to the rock's bottom. Many of these species are also winter-active.

In still water insects are sometimes eclipsed in numbers by their cousins, the crustaceans. The most common include brine shrimp, primitive species from the fairy shrimp order Anostrica. These inhabit saline lakes throughout the Eastern Sierra, although the Mono Lake population is the best known. The **Mono Lake Brine Shrimp**[ES], *Artemia monica* (fig. 6.7), is specifically adapted to the unique chemistry of Mono Lake. In spring larval stages hatch out of protective, winter-dormant cysts (fig. 6.7b and c), then mature as the lake warms in summer. *Plumes* (swimming clouds of adults) are associated with freshwater springs bubbling up from the lake bottom near shore. Smaller, local plumes occur in shallows where solar-heated water causes convection currents that trap the shrimp. Summer-produced eggs hatch in the female's body.

The **Alkali Fly**, *Ephydra hians* (Ephydridae) (fig. 6.4), inhabits the shores of many Eastern Sierra brine shrimp lakes. Dense masses of adults congregate in summer, mating and feeding on debris near the water's edge. They can walk underwater aided by a thin bubble of air held in place by microscopic bristles on their body. Pupal cases and larvae are sometimes knocked from their underwater rock holds by severe storms and wash ashore in piles. The Kuzedika Paiutes of the Mono Lake region harvested and prepared the pupae as a staple food, which they called *kutsavi*.

At Mono Lake both the brine shrimp and Alkali Fly are primary consumers in a simple food web beginning with bacteria and the photosynthetic algae upon which the flies feed (fig. 6.6). Each year close to a million migratory birds depend on these tiny arthropods for food while they rest and build fat reserves, before continuing their winter migration. The absence of fish in Mono Lake provides the shrimp and larval flies with an environment relatively free of predators and food competitors. Consequently, their numbers become astronomical during the summer, giving Mono its reputation as one of the world's most biologically productive bodies of water.

Honey Lake, near Susanville, harbors tadpole shrimp (Notostrica) (fig. 6.9). They are branchiopods, branched-foot crustaceans like brine shrimp, with a shell-like dorsal plate that makes them resemble small horseshoe crabs.

Volcanic activity along the Eastern Sierra has left a legacy of thermal springs that provide other important habitats for arthropods. Shore fly

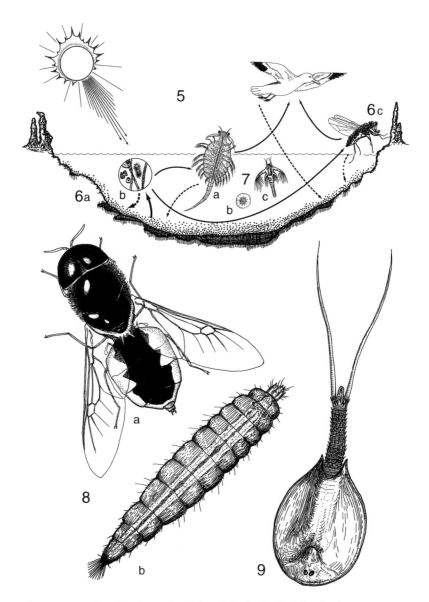

Figs. 6.5–6.9. Aquatic arthropods. 5. Mono Lake food web. 6. Food web components: (a) detritus; (b) algae; (c) brine flies. 7. Mono Lake Brine Shrimp, *Artemia monica:* (a) adult (8 mm); (b) cyst (0.5 mm); (c) nauplius (1 mm). 8. Rabbitbrush Soldier Fly, *Hedriodiscus truquii:* (a) adult (15 mm); (b) larva (24 mm). 9. A tadpole shrimp, *Notostrica* species (25 mm). (Drawings by Evan A. Sugden; fig. 6.5 redrawn with permission of Mono Lake Committee)

relatives of the Alkali Fly are common along the edges of thermal ponds. Warm-spring bathers sometimes find the predaceous larvae of the **Rabbit-brush Soldier Fly, *Hedriodiscus truquii*** (Stratiomyidae) (fig. 6.8), floating in the water. These larvae inhabit pond muck and may share their domain with larvae of the Big Black Horse Fly[†] and punkies[†]. In most water habitats the primary producer organisms are microscopic and specialized, such as the heat-tolerant blue-green algae and bacteria of hot springs or the salt- and alkali-tolerant ones of Mono Lake. Such microbes are the food of many aquatic arthropods, such as the brine and tadpole shrimps. An extensive system of temporary ponds, hot springs of various temperatures, and fresh water occurs near Whitmore Hot Springs, southeast of Mammoth Lakes.

DECOMPOSERS

Dead organic matter, whether animal or plant, provides abundant food for many decomposer species. They recycle dead matter by converting it into energy and into their own living tissue. Larvae of some of our largest and most beautiful beetles consume wood. They may mine the inner bark layers of weakened, dying, or dead trees. Early stages of the **Golden Buprestid, *Buprestis aurulenta*** (Buprestidae) (plate 7a), are often found in Douglas Fir, especially in fire-damaged trees. These larvae and others in the family are known as flatheaded borers (fig. 6.10). Roundheaded borers (fig. 6.11b) are minor forest pests. They are the larvae of longhorn beetles, such as the **Spined Woodborer, *Ergates spiculatus*** (Cerambycidae) (fig. 6.11a), the largest insect in the Eastern Sierra. The larvae bore into the heartwood of valuable Douglas Fir trees and sometimes telephone poles. The borer is parasitized by the tachinid fly[†] *Helicobia rapax*. Roundheaded borers may incidentally devour larval bark beetles[†] in their galleries.

Termites and ants are the most conspicuous decomposers of wood. The **Sierran Dampwood Termite, *Zootermopsis nevadensis*** (Hodotermitidae), a common species, tunnels in old logs. Workers (fig. 6.12a) are relatively small and numerous. Soldiers (fig. 6.12b) have formidable jaws. Sexually mature, winged reproductives (fig. 6.12c) emerge from large colonies in swarms of hundreds or thousands in midsummer. A single king and queen occupy the core of the colony.

Termites often abandon their galleries to the **Modoc Carpenter Ant, *Camponotus modoc*** (Formicidae). Colonies of this species have no distinct soldier caste; however, the workers (fig. 6.13a) specialize in various tasks, such as defense, foraging, and brood care. The colonies send off reproductive swarms when they become large, filling the air with sexually mature males and females (fig. 6.13b), many of which are eaten by birds and other

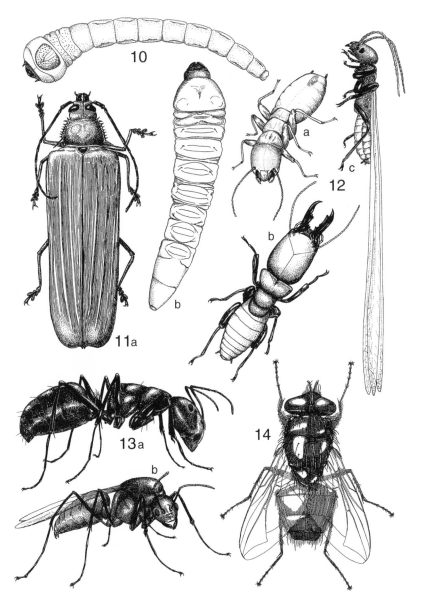

Figs. 6.10–6.14.   Decomposers. 10. A flatheaded borer, Buprestidae species: larva
(50 mm). 11. Spined Woodborer, *Ergates spiculatus:* (a) adult (64 mm); (b) larva (76 mm).
12. Sierran Dampwood Termite, *Zootermopsis nevadensis:* (a) worker (6 mm); (b) soldier
(9 mm); (c) winged adult (8 mm). 13. Modoc Carpenter Ant, *Camponotus modoc:*
(a) worker (8 mm); (b) winged adult (14 mm). 14. Greenbottle Fly, *Phaenicia sericata*
(9 mm). (Drawings by Evan A. Sugden)

predators. A few survivors mate; a fertilized female will become queen of a new colony that she founds with her first egg clutch.

Blow flies rapidly appear at the carcass of a dead animal. They lay eggs in or near the flesh, which soon hatch into larvae that infest the decaying meat. The carcass may become a seething mass of maggots, which in their vigorous feeding transform the carcass into a withered bag of skin and bones. Near the remains of their banquet the larvae burrow into the soil, encase themselves in a cocoon (pupate), and emerge later as adults. A common blow fly is the familiar metallic-green **Greenbottle Fly, *Phaenicia sericata*** (Calliphoridae) (fig. 6.14). Sometimes it lays eggs near a wound on a live animal. If the larvae begin to feed on healthy tissue, the wound may worsen; this often happens in sheep. In other animals, wound-feeding larvae generally restrict their feeding to infected tissue and may actually stimulate the healing process. The Greenbottle Fly has been used successfully to treat open wounds in humans.

Also attracted to dead animals are flesh flies, members of the genus *Sarcophaga* (Sarcophagidae) (fig. 6.15), and carrion beetles, including the **Garden Silphid, *Silpha ramosa*** (Silphidae) (fig. 6.16). As the animal decomposes further, other insects appear, such as the **Common Carrion Dermestid, *Dermestes marmoratus*** (Dermestidae) (fig. 6.17), which feeds on dried skin and hair, and the **Hairy Rove Beetle, *Staphylinus maxillosus*** (Staphylinidae) (fig. 6.18), a predator of maggots and other larvae.

The **European Dung Beetle, *Aphodius fimetarius*** (Scarabaeidae) (fig. 6.19), is a specialist in dung recycling. It may be joined in cow pats and other moist animal droppings by larvae of the **Golden-haired Dung Fly, *Scatophaga stercoraria*** (Scatophagidae) (fig. 6.20).

FOREST PESTS

Only a few arthropod species have reputations as pests, especially in the Eastern Sierra. But among those that do, the most serious compete with humans for forest resources. The **Pandora Moth**[E], *Coloradia pandora* (Saturniidae) (plate 7b), is an intermittent forest pest that feeds on the needles of the Jeffrey Pine. The moth's two-year cycle begins in June, when it lays eggs. The larvae (fig. 6.21) soon emerge and feed until September, then overwinter as mid-instar caterpillars. They cluster at branch tips during the cold months. The following year the caterpillars continue to feed and then pupate under the soil surface in midsummer. The pupae enter diapause in their cocoons and remain in a resting state until the third spring. They then emerge as adult moths that mate and lay eggs, beginning the cycle again. Local populations are synchronized in development. In alternating years either adults plus young larvae or older larvae plus pupae predominate. Needle feeding is

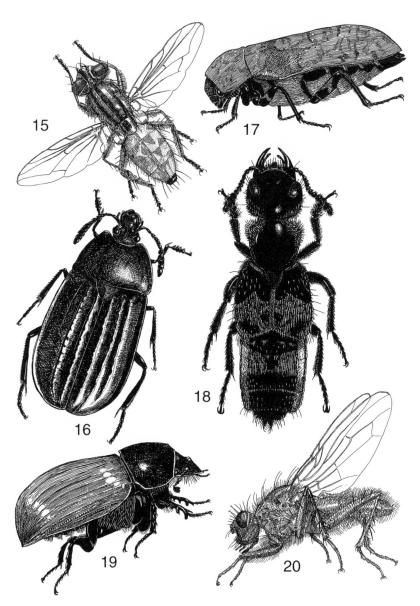

Figs. 6.15–6.20.   Decomposers. 15. Flesh fly, *Sarcophaga* species (11 mm). 16. Garden Silphid, *Silpha ramosa* (12 mm). 17. Common Carrion Dermestid, *Dermestes marmoratus* (11 mm). 18. Hairy Rove Beetle, *Staphylinus maxillosus* (15 mm). 19. European Dung Beetle, *Aphodius fimetarius* (8 mm). 20. Golden-haired Dung Fly, *Scatophaga stercoraria* (10 mm). (Drawings by Evan A. Sugden)

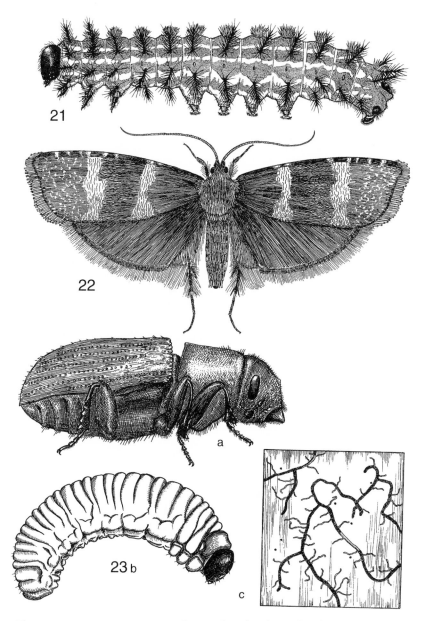

Figs. 6.21–6.23. Forest pests. 21. Pandora Moth, *Coloradia pandora:* larva (60 mm).
22. Jeffrey Pine Seed Moth, *Cydia injectiva* (10 mm). 23. Western Pine Beetle, *Dendroctonus brevicomis:* (a) adult (4.5 mm); (b) larva (5 mm); (c) gallery. (Drawings by Evan A. Sugden)

often tolerated by the same host trees for several years. New shoots usually emerge from the defoliated branch tips after the caterpillars stop feeding, giving the trees a tufted appearance. Populations of the Pandora Moth are normally small, and the damage they cause is insignificant. But occasionally, certain population characteristics and ecological factors combine to allow a sudden explosion in numbers, which can cause considerable damage. After several years of continuous defoliation, weakened trees may die or become susceptible to attack by bark beetles or tree diseases. Such an infestation began along Highway 395 between Crestview and Mammoth Lakes in 1979, when the presence of large caterpillars and massive losses of needles were first noticed. As in most such infestations, this one naturally subsided to normal levels within two years, and most of the trees survived. Pandora Moth larvae are parasitized by several insects, including a tachinid fly[†].

The Pandora Moth was once important in the diet of the local Paiute Indians. They dug shallow trenches around infested trees and drove the caterpillars, known as *piagee* (pronounced pee-ah-gee, the Paiute name), down from their feeding perches and into the trenches with smudge fires. They were baked in hot earth, air-dried, and stored or eaten.

Another needle-feeding caterpillar is the **Pine White,** *Neophasia menapia* (Pieridae) (plate 7c). It feeds on several pine and fir species although it is seldom economically important. Seeds of the Jeffrey Pine are often infested with larvae of the **Jeffrey Pine Seed Moth,** *Cydia injectiva* (Olethreutidae) (fig. 6.22). It may compete with birds, such as the Clark's Nutcracker, and with rodents for these seed resources.

The **Western Pine Beetle,** *Dendroctonus brevicomis* (Scolytidae) (fig. 6.23a), is the most serious pest of western pine forests, for it infests Ponderosa Pine, a major timber source. Adult beetles are attracted to a diseased or weakened tree by the volatile aromas it gives off and by *pheromones* (airborne chemical sex signals) given off by other bark beetles. They burrow into the bark, excavating a complex tunnel system, or *gallery,* in which they lay eggs. The initial adult excavation is soon extended by the larvae (fig. 6.23b), which hatch in summer and feed on the inner layers of the tree bark. The pattern of the gallery (fig. 6.23c) is characteristic of the species. Burrowing bark beetles transmit blue stain fungus, a tree disease that blocks the water-conductive tissue. The fungus weakens the tree, creating ideal conditions for the establishment of new beetle galleries. Adult beetles (fig. 6.23a) accumulate and carry fungal spores on their bodies as they emerge from the larval galleries and eventually carry the fungus to new host trees. Both fungus and beetle benefit by such an association, called *symbiotic* (from the Greek for "living together"). Severe Western Pine Beetle infestations can kill a tree quickly, particularly if it is already damaged or weakened. However, healthy trees have a way of protecting themselves. A Western Pine Beetle attacking a healthy tree may be ejected from its initial excavation by a flow of tree resin under pressure. The resulting resin plug projecting from the tree trunk is known as a *pitch tube.*

HUMAN-BITING ARTHROPODS

Mosquitoes are generally less abundant in the Eastern Sierra than in warmer, wetter climates, but they may still be a nuisance to humans. A mosquito's early life is aquatic; larvae (fig. 6.24b) and pupae—known respectively as *wrigglers* and *tumblers*—may infest ponds, stream or lake edges, and other water-filled cavities. They feed on debris in the water and breathe at the surface through a tiny siphon tube on the abdomen. They wriggle rapidly to the bottom if disturbed. Adult female mosquitoes must feed on blood for their eggs to mature. Males can be distinguished from females by their ornate, feathery antennae. The bodies of most adult mosquitoes are clad in a beautiful mosaic of microscopic scales; their wings are also characteristically scaly and hairy. The **Common Side-spotted Mosquito, *Aedes dorsalis*** (Culicidae) (fig. 6.24a), which breeds in a wide variety of habitats, is common in the Eastern Sierra. Another common species is the **Giant Spring Mosquito, *Culiseta inornata*** (fig. 6.25). Females of these species feed mainly on large animals. Both are capable of transmitting an encephalitis-causing virus; naturally infected *Aedes dorsalis* have been found in Owens Valley. However, there is little danger of mosquitoes transmitting the virus to humans in the Eastern Sierra, because mosquito populations are typically small.

Punkies, or "no-see-ums," are the most annoying insects along the Eastern Sierra. The **Bodega Gnat, *Leptoconops kertesi*** (Ceratopogonidae) (fig. 6.26a), is a common species. It emerges in early summer from mud flats and wet meadows where the detritus-feeding larvae congregate (fig. 6.26b). Swarms of adults may be blown far from the larval habitat. They will relentlessly attack exposed ankles, wrists, and heads of campers. The bite, initially painless, leaves a small pink welt with a red central dot. Itching may last many days. A similar fly, the **Blue Tongue Punkie, *Culicoides variipennis,*** transmits the virus that causes blue tongue disease in sheep, deer, and goats. The possibility of the punkie transmitting this disease from domestic sheep to wild Bighorn Sheep was a major consideration in reintroducing Bighorn to the central Sierra in the 1980s. *C. variipennis* also bites humans, although humans do not contract the disease.

The **Big Black Horse Fly**[†], ***Tabanus punctifer*** (Tabanidae) (fig. 6.27a), the largest of its family in North America, is widespread throughout the region. Its loud buzz may momentarily be mistaken for the sound of a distant airplane. The thorax is clothed in a velvety, silver-gray pile, and the forelegs are

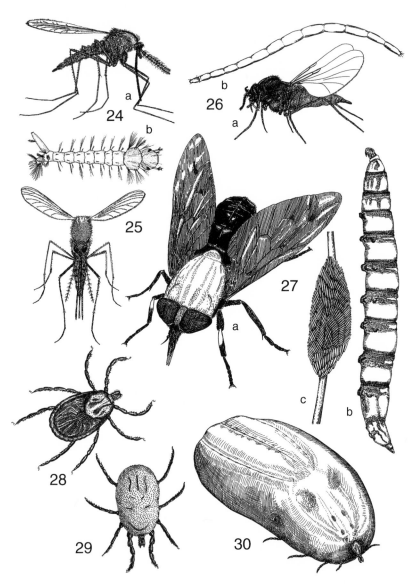

Figs. 6.24–6.30. Human-biting arthropods. 24. Common Side-spotted Mosquito, *Aedes dorsalis:* (a) adult (7 mm); (b) larva (9 mm). 25. Giant Spring Mosquito, *Culiseta inornata* (8 mm). 26. A biting gnat, *Leptoconops* species: (a) adult (1.5 mm); (b) larva (3 mm). 27. Big Black Horse Fly, *Tabanus punctifer:* (a) adult (20 mm); (b) larva (50 mm); (c) egg mass (8 mm). 28. Rocky Mountain Wood Tick, *Dermacentor andersoni* (4 mm). 29. A soft tick, *Ornithodoros* species (4 mm). 30. Herms' Tick, *Ornithodoros hermsi,* engorged with blood (17 mm). (Drawings by Evan A. Sugden)

banded with white. Wary and fast, this big-eyed fly is attracted to large moving objects, such as horses, cattle, deer, and even automobiles. It occasionally mistakes a human for one of its normal animal hosts. Like other horse and deer flies, the Big Black Horse Fly inflicts its painful bite with bladelike mouth parts, which simultaneously stab and slice the victim. The Big Black Horse Fly lays its eggs in masses of 200–800 on grass stems or twigs overhanging fresh water (fig. 6.27c). Many eggs are parasitized by a tiny wasp, *Telenomus emersoni* (Scelionidae) (fig. 6.54). The horse fly larva (fig. 6.27b) is predaceous and has been observed to capture and feed on tadpoles of a spadefoot toad in Arizona—the only known example of a fly as a predator of a vertebrate. The relationship is usually reversed! The larva inhabits pond muck and other moist habitats for two years before pupating and later emerging as an adult.

Hikers in the Eastern Sierra should check themselves for ticks after a day in the field. Ticks belong to two categories, hard and soft. Hard ticks feed mostly on mammals and feed only once between molts. After the final blood meal, the female produces a single clutch of several thousand eggs, which she releases to the ground just before her death. On hatching, immature ticks climb up into the vegetation and await a passing host. The **Rocky Mountain Wood Tick,** *Dermacentor andersoni* (Ixodidae) (fig. 6.28), is a common species. The **Pacific Rodent Tick,** *Ixodes pacificus* (Ixodidae), a similar tick species, carries Lyme disease, a serious bacterial disease of the nervous system that infects humans. Immature ticks feed on rodents, which harbor the bacteria but do not suffer any symptoms. Adult ticks normally feed on deer, but may bite humans and inject the bacteria. *Ixodes pacificus* is common elsewhere in California, but in the Eastern Sierra is found only in the warmer, southern part of Owens Valley. The chances of contracting Lyme disease here are very low compared to the level of risk elsewhere in California or in the eastern states.

Soft ticks, which typically inhabit desert regions, tend to choose birds and reptiles as hosts, although we will mention two that have mammalian hosts. They feed rapidly and may take many blood meals between molts. The female produces several separate egg batches. The **Pajaroello,** *Ornithodoros coriaceus* (Argasidae) (fig. 6.29), is a soft tick associated with cattle and deer; it often bites people as well. **Herms' Tick,** *Ornithodoros hermsi* (fig. 6.30), normally a rodent parasite, occasionally transmits relapsing fever germs to people when it bites a human by mistake. Relapsing fever can be fatal to humans. These germs are usually localized in just a few rodent populations, usually between 5000 and 10,000 feet (1524 and 3048 m) elevation, often at a specific site, such as a cave or cabin.

Another serious and potentially fatal insect-transmitted human disease in the Eastern Sierra is plague. The disease-causing organism is a bacterium, and its natural reservoir in this area is populations of rodents, such as the Golden-mantled Ground Squirrel. Rodent fleas, which naturally infest ground squirrels, may carry the bacteria from an infected animal to an accidental human host. Plague outbreaks occur in localized areas in the Sierra where native rodent populations are high and their contact with humans is frequent. Signs may be posted warning people to avoid contact with living or dead rodents.

MIMICS AND LOOK-ALIKES

An organism that benefits from its resemblance to another species is a *mimic.* Mimicry is usually an adaptation for deceiving predators, in which the otherwise vulnerable mimic resembles a dangerous or distasteful creature, the *model.* If obnoxious itself, the mimic may resemble other obnoxious species in a *mimicry complex,* in which all species benefit from reinforcing in their mutual predators the common warning that they are distasteful or dangerous. It is often difficult to establish whether one species mimics another or whether it has merely assumed similar form or behavior due to a coincidence of evolution. There are many such coincidental look-alikes among arthropods. In cases of true mimicry, the mimic and the model overlap in time and space, and the mimic derives some true benefit to its survival from the ruse. The examples below include both true mimics and look-alikes.

Bees and wasps are often mimicked, probably because of their bright color patterns and their ability to sting. Many share a black-and-yellow striped pattern, which is the basis for one of the most common insect mimicry complexes. Some examples are shown in plate 8e, including the **Zebra Hover Fly**[E], *Chrysotoxum ypsilon* (Syrphidae); the **Graceful Carder Bee,** *Anthidium formosum* (Megachilidae); the **Black-mouthed Soldier Fly,** *Stratiomys melastoma* (Stratiomyidae); and the **Aerial Yellowjacket,** *Dolichovespula arenaria* (Vespidae). **The Locust Clearwing,** *Paranthrene robiniae* (Sesiidae), is a moth remarkably similar to the **Golden Paper Wasp,** *Polistes fuscatus* (Vespidae), in shape, color, and even behavior. The moth is easily mistaken for the wasp as it hovers above flowers for nectar. Clearwing larvae tunnel in the wood of aspen, cottonwood, poplar, and willow.

The **Sierran Bumble Bee Moth,** *Hemaris senta* (Sphingidae) (plate 8c), mimics bumble bees. It forages by day, often in the company of the bees themselves. With scaleless wings and dense black-and-yellow body hair, it is easily mistaken for the **Yellow-faced Bumble Bee,** *Bombus vosnesenskii* (Apidae) (plate 8c). Robber flies, such as **Sacken's Robber Fly,** *Laphria sackeni* (Asilidae) (plate 8c), and **Fernald's Robber Fly,** *Laphria fernaldi* (Asilidae) (plate 8d), are excellent bumble bee mimics and sometimes even prey on bumble bees. Another bumble bee mimic, the **Fire-tipped Hover Fly,** *Arctophila flagrans* (Syrphidae) (plate 8d), feeds on flower nectar and pollen as an adult and probably plays a minor role in pollinating some plants, such as rabbitbrush. The Fernald's Robber Fly and the Fire-tipped Hover Fly closely resemble the **Red-skirted Bumble Bee,** *Bombus huntii* (Apidae) (plate 8d).

Many bee flies resemble bees, as their name suggests. The **Bomber Fly**[†], *Heterostylum robustum* (Bombyliidae) (fig. 6.39a), looks and flies like the **Elegant Anthophora**[†], *Anthophora urbana* (Anthophoridae) (fig. 6.40), and visits many of the same flowers for nectar.

PLANT-FEEDERS

Plant-feeding arthropods are primary consumers in the food chain, which begins with green plants, the primary producers. This group includes some crustaceans (algae-eating brine shrimp), certain mites, and millipedes, but mostly consists of insects. Plant-feeders are generally more numerous than their predators, the next link in the chain. They often occur in large masses or infestations. Examples include aphids, tent caterpillars, and blister beetles. Animals higher in the food chain consume plant-feeders in large numbers. Consequently the plant-feeders have evolved protective forms and behaviors that help them survive in a world of predators and parasites, and these are expressed in a wide variety of interesting adaptations. They include protective coloration or camouflage (many species); stinging hairs (some caterpillars); toxic body fluids (blister beetles); "warning" coloration or patterns (some beetles, bugs, and caterpillars); special escape behaviors, such as hopping or flying away (many species); and the production of protective shelters (spittle bugs, gall aphids, tent caterpillars).

Some plant-feeding insects consume only a small part of their host. In this respect, they are similar to parasites of animals, for the host continues to live. Cicadas are an example. They belong to the order Homoptera, which also contains the aphids, known as *plant lice.* Sucking, soda-straw-like mouth parts of the **Bloody Cicada**[E], *Okanagana cruentifera* (Cicadidae) (fig. 6.31), enclose a piercing double needle, which enables it to suck liquid from the inner, sap-containing layers of a host shrub or tree. Cicada eggs are inserted under the bark of twigs. Hatchling nymphs fall to the ground, burrow, and begin to feed on the juice of plant roots. They may live a subterranean life for several years before emerging as adults. You may find papery skins of last-stage nymphs clinging to shrubs where the adults have recently emerged. Adult cicadas are known for the male's mating call, a loud, metallic buzz that differs in pitch and rhythm among species.

Another large group of juice-sucking insects is the closely allied order

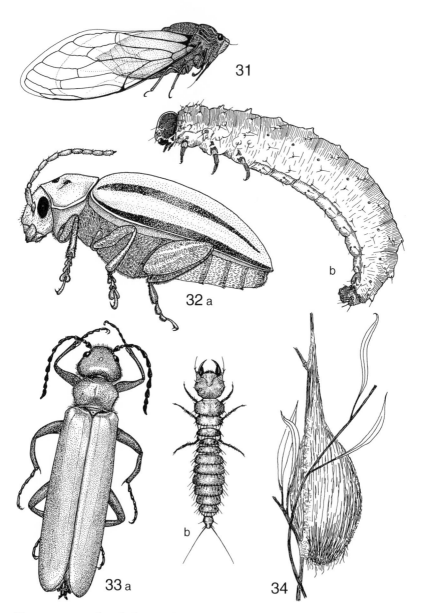

Figs. 6.31–6.34.   Plant-feeders. 31. Bloody Cicada, *Okanagana cruentifera* (32 mm).
32. Five-striped Willow Beetle, *Disonycha alternata:* (a) adult (22 mm); (b) larva,
Disonycha species (12 mm). 33. Nuttall's Blister Beetle, *Lytta nuttalli:* (a) adult (15 mm);
(b) triungulin larva (2 mm). 34. Glover's Silk Moth, *Hyalophora gloveri:* pupal cocoon
(50 mm). (Drawings by Evan A. Sugden)

Hemiptera, or *true bugs*, many of which are plant-feeders. The **Large Milk-weed Bug**, *Oncopeltus fasciatus* (Lygaeidae or seed bugs) (fig. 6.3), for example, feeds on milkweed seeds.

In contrast to these plant parasites are insects that have chewing mouth parts and may consume or fatally injure an entire plant in a short time. The **Coral-winged Grasshopper**, *Xanthippus corallipes* (Acrididae) (fig. 6.1), which occurs in open, grassy areas, is one of these. Although the body is mottled in a camouflage pattern, the hind wings flash a brilliant coral red when the grasshopper leaps and flutters skyward. Males use this display to attract females. Other related species in the Eastern Sierra have orange-red, transparent, yellow, or blue hind wing patterns. *X. corallipes* is among the largest of the group.

The leaf beetles are a large group of plant feeders. Adults typically have mouth parts that point downward and threadlike or beadlike antennae. Many species, when viewed from above, have their head hidden by the thorax. Their larvae (fig. 6.32b) are grubs that feed on roots or leaves. A brilliantly colored example is the **Five-striped Willow Beetle**, *Disonycha alternata* (Chrysomelidae) (fig. 6.32a). Its bottom is red, and the wing covers are cream-colored with black stripes. It belongs to the special "flea beetle" subfamily Alticinae, which is characterized by swollen basal segments of the hind legs, a modification for jumping.

Adults of **Nuttall's Blister Beetle**, *Lytta nuttalli* (Cantharidae) (fig. 6.33a), feed on members of the pea family. These striking beetles are iridescent purple and green. As is typical of the family, the neck is narrower than the head, and the wing covers are relatively thin and papery. Larvae[†] are parasites. Blister beetles' bodies contain cantharadin, a substance highly irritating to human skin; the family takes its name from this chemical. Many species are brilliantly colored and may occur in localized infestations on their host plants.

Moths and butterflies are the most conspicuous insects of the Eastern Sierra during summer. Their larvae—caterpillars—are as varied and often as beautiful as the adults. Their powerful vegetation-grinding jaws enable them to rapidly munch their way through many times their own body weight of their favorite plant foods. After metamorphosis, the mouth parts form a sucking tube that the moth or butterfly uses to suck up flower nectar and other juices for nourishment. Some butterflies feed as caterpillars only on rare plants and are therefore rare themselves. The **Apache Fritillary**, *Speyeria nokomis apacheana* (Nymphalidae) (plate 7d), is one of these. As a larva, it feeds on the Kidney-leaf Violet, *Viola sororia affinis*, which grows only in undisturbed, perennially wet meadows. This habitat is rapidly disappearing from the Eastern Sierra due to housing developments, grazing, groundwater pumping, and water diversion. Another rare species, now officially "federally endangered" and therefore fully protected, is the **Mono Lake Checkerspot,** *Euphydryas editha monoensis* ES (Nymphalidae). Its larva feeds on plantago.

A relatively common butterfly is **Becker's White,** *Pieris beckeri* E (Pieridae) (plate 7e). Its larvae feed on Princes' Plume and other members of the mustard family. Unlike the Apache Fritillary, its diverse and broadly distributed food-plants help assure its survival. Note its similarity to the Pine White† (Pieridae) (plate 7c).

Sheep moths are fuzzy, brightly colored moths of the family Saturniidae. Three closely related ones frequent the Eastern Sierra (plate 7f). Variation in seasonal and daily flight times and species-specific pheromones prevent hybridization among them. The species are also separated by larval food-plant differences. Larvae of the **Common Sheep Moth,** *Hemileuca eglanterina,* feed on more than ten chaparral plant species. But larvae of **Nuttall's Sheep Moth,** *H. nuttalli,* eat bitterbrush, and those of the **Hera Moth** E, *H. hera* (fig. 6.38), feed exclusively on Great Basin Sagebrush.

Another common saturniid is **Glover's Silk Moth** E, *Hyalophora gloveri* (plate 8a). Late-instar larvae are large, green caterpillars adorned with white, yellow, and black spines. They prefer willow, Antelope Bitterbrush (*Purshia tridentata*), and Desert Peach (*Prunus andersonii*), but may also feed on Cherry Plum (*Prunus virginiana*) and manzanita, and less successfully on Wild Rose (*Rosa woodsii*). The gray, fibrous cocoon (fig. 6.34) is attached to a branch of a shrub or tree. It is not easily found, as its color and texture blend with bark or tangled undergrowth. *H. gloveri* is a resident of the Great Basin and occurs in the Eastern Sierra from Monitor Pass south. It is separated from its western cousin *H. euryalis* by the peaks of the Sierra. The two interbreed in a small zone near Monitor Pass, and the latter species takes over farther north in the Eastern Sierra.

Hornworms are larvae of hawk moths (Sphingidae), a group that includes the familiar **Tomato Hornworm,** *Manduca sexta,* an infamous defoliator. The larva and pupa of the **Snowberry Clearwing Sphinx,** *Hemaris diffinis,* appear in figure 6.35a and b.

Sawflies are nonstinging primitive wasps with plant-feeding larvae. The abdomen is broadly joined to the thorax in the adult, unlike that of most Hymenoptera (ants, bees, and wasps). Larvae closely resemble butterfly and moth caterpillars, except that they possess legs on all segments (compare figs. 6.21 and 6.37b). The yellow and black adult of the **Tiger Sawfly,** *Tenthredo anomocera* (Tenthredinidae) (fig. 6.36), often feeds at flowers. This

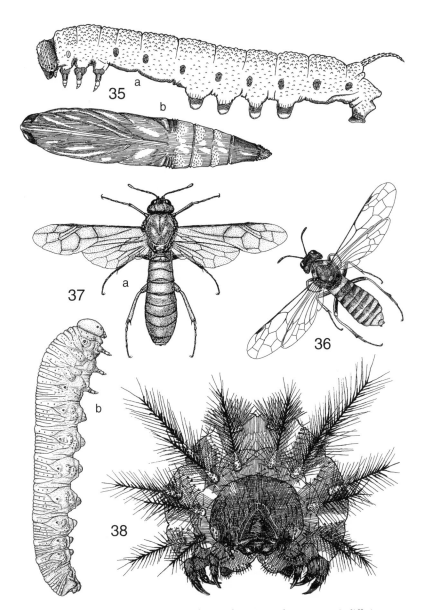

Figs. 6.35–6.38.   Plant-feeders. 35. Snowberry Clearwing Sphinx, *Hemaris diffinis:*
(a) larva (34 mm); (b) pupa (28 mm). 36. Tiger Sawfly, *Tenthredo anomocera* (12 mm).
37. Red Cimbex, *Cimbex rubida:* (a) adult (20 mm); (b) larva (50 mm). 38. Hera Moth,
Hemileuca hera: larval head and thorax (4 mm). (Drawings by Evan A. Sugden)

and some related species are unusual in that the adults are partly predatory, occasionally feeding on other insects.

The **Red Cimbex,** *Cimbex rubida* (fig. 6.37a), belongs to the Cimbicidae, a small and distinct family of sawflies with heavily clubbed antennae and robust bodies. Although they are uncommon, you may find adults clinging to streamside vegetation on cool summer mornings. The larvae (fig. 6.37b), which feed nocturnally on willow, are rarely seen.

POLLINATORS

Pollination is one of the most important ecological roles played by arthropods. Many plants provide nectar and/or pollen that entices food-gathering insects. The transfer of pollen from male to female flower organs usually occurs incidentally as the insects travel from flower to flower while they feed. Flowers also provide shelter for some insects, such as syrphid flies and some male bees.

Bee flies (Bombyliidae[†]) visit flowers regularly. The **Silver-aproned Bee Fly,** *Exoprosopa calyptera* (plate 8b), seeks its nourishment from exposed blossoms, including those of the sunflower family. Newly emerged adults display an apron of silvery scales on the abdomen. Larvae parasitize immature wasps, such as those of the Western Sand Wasp[†]. Some long-tongued species, such as the Bomber Fly[†] (fig. 6.39), are able to feed on nectar or pollen from deep-corolla or tubular blossoms, including the Scarlet Monkey Flower (*Mimulus coccineus*).

Bees are the most specialized group of pollinators. The Elegant Anthophora[†] (fig. 6.40), although common, often goes unnoticed owing to its rapid flight. Adult females burrow in cliffsides and hollow out solitary earthen nest cells at the end of long tunnels. They provision the cells with pollen and then deposit an egg in each one, from which hatches a hungry larva. This species commonly feeds on bell-shaped blossoms, such as those of the Timberline Phacelia (*Phacelia hastata* ssp. *compacta*), although it visits a wide variety of flowers. The **Alkali Bee,** *Nomia melanderi* (Halictidae) (fig. 6.41), often visits flowers of the pea family, as well as others. Its brilliant abdominal bands are pearly white. The female prefers alkaline soil in which to dig its nest, hence its name. It is important to commercial alfalfa seed farmers in the Pacific Northwest, for it readily pollinates alfalfa flowers. In some areas farmers construct artificial *bee beds* near the fields to encourage growth of the local Alkali Bee population. Bee beds are specially prepared layers of soil that imitate the natural alkaline soil conditions in which the bees nest. Other solitary bees, such as the **Mountain Osmia,** *Osmia montana* (Megachilidae) (fig. 6.42), build their nests in hollow twigs. This species restricts its foraging to members of the sunflower family.

Figs. 6.39–6.43.   Pollinators. 39. Bomber Fly, *Heterostylum robustum:* (a) adult female (12 mm) flinging egg; (b) egg (0.5 mm); to left, head. 40. Elegant Anthophora, *Anthophora urbana* (13 mm). 41. Alkali Bee, *Nomia melanderi* (10 mm): adult female feeding on an alfalfa flower. 42. Mountain Osmia, *Osmia montana* (12 mm). 43. Honey Bee, *Apis mellifera:* worker (11 mm). (Drawings by Evan A. Sugden)

Bumble bees (plate 8c and 8d) are the most conspicuous and familiar native pollinators. In spring fertilized queens emerge from winter hibernation and search for nest sites, typically abandoned rodent burrows. From the labors of an individual queen, a summer colony of several hundred workers may be produced. The size of newborn worker bees increases through the season owing to increasing amounts of food provided to the larvae. By the end of summer only males and large, fertile females are produced. After they have finished mating in the fall, the males die and the new queens burrow into the soil until the following spring. Bumble bees may be active at temperatures close to freezing, too cold for most other insects. You may see them busily working the earliest of spring flowers, often with snow still on the ground. They have the ability, unusual among insects, to raise their body temperature by shivering their flight muscles as a warm-up exercise. A thick coat of hair, which keeps them warm, also serves as a brightly colored warning flag. Predators that have previously experienced a bumble bee sting will heed the warning (see Mimics).

The familiar hive **Honey Bee, *Apis mellifera*** (Apidae) (fig. 6.43), was brought to California by ship in the 1850s, after Europeans had introduced it to the eastern states and to Mexico two or three hundred years earlier. An extremely adaptable species, Honey Bees escaped into the wild and within a few decades became ubiquitous and permanent residents except at very high elevations. Honey Bees are closely related to bumble bees, but their colonies last throughout the year, unlike those of bumble bees. Worker Honey Bees cluster around the queen to conserve heat in winter. Both Honey Bees and bumble bees must be able to forage on a wide variety of flowers in order to maintain a constant food supply for their populous colonies throughout the warm season. In addition, Honey Bees must store enough extra food to last the colony through the winter. Thus, these social bees are flower *generalists* and are important pollinators of many plants. Honey Bee populations have declined in recent years due to two parasitic mite species introduced accidentally from Europe or Asia. Throughout North America (including the Eastern Sierra) domestic bee hives and wild colonies alike have been decimated by these blood-feeding mites. In some areas, bumble bees are once again more common than Honey Bees.

PREDATORS

Solpugidae, or wind scorpions, are nocturnal predators that hide during the day under rocks, boards, and debris. They often appear in tents or under sleeping bags. Adult solpugids are harmless to humans, yet they possess what are possibly the largest jaws of any animal relative to total body size. They are ravenous eaters. A hungry adult kept in a jar will eat continuously for several hours if live insects are provided. The appendages that appear to be the first pair of legs are actually sensory mouth parts. There are only four pairs of true legs; the foremost are considerably reduced. Their legs are extremely sensitive to ground vibration, and this aids wind scorpions in detecting prey movement. Local species are of the genus *Eremobates* (Eremobatidae) (fig. 6.2).

The liquid-feeding true bugs, order Hemiptera, are discussed under Plant-feeders. Among the predatory representatives are assassin bugs (Reduviidae). They have powerful grasping forelegs and are often cryptically colored. Their sharp, daggerlike mouth parts are adapted for stabbing their insect prey. They are defensive if handled and can inflict a painful wound in humans, which is aggravated by their venomous saliva. A common species is the **Spiny-collared Assassin Bug,** *Pseliopus spinicollis* (fig. 6.44).

The **Oregon Tiger Beetle,** *Cicindela oregona* (Cicindelidae) (fig. 6.45a and b), and other members of the same genus occur along sandy lakeshores, on dirt roads near streams, and at the edges of melting snowfields. Adults move in a series of rapid, darting runs and low-level flights. They are ferocious hunters, as evidenced by their large, saw-toothed jaws. The unusual larvae (fig. 6.45c) inhabit burrows in moist earth or mud flats, usually near the active adults. They also are predatory. The head of the larva blocks the burrow entrance and blends into the ground coloration. When the larva senses an insect victim, it flips its head and gaping jaws backward and grasps the prey in an upside-down position. The larva is very sensitive to vibration, however, and will quickly drop into its hole to avoid danger, suddenly revealing the burrow entrance.

The most prominent insect hunters of ponds and lakes are dragonflies. A representative family are the skimmers (Libellulidae) (fig. 6.46a). The immatures are gill-breathing, aquatic nymphs. They have a double-hinged jaw (fig. 6.46b), which extends their reach. Nymphs may spend many months underwater before crawling out of their last molt to become sleek, winged adults. Adult male dragonflies maintain favorite perches and patrol their territories regularly. Their diet and feeding behavior are similar to those of swallows and some bats, for while in flight they scoop up small insects, such as biting gnats[†] and mosquitoes[†]. In mating, the male grasps the female by the neck with special tail-end claspers. She responds by pressing her genital openings to the male's sperm-transferring organ, located on the second abdominal segment. This position is maintained as the adults fly in a characteristic loop configuration (fig. 6.46c).

Robber flies[†] (Asilidae) are common insect-eating predators of the Eastern Sierra. They pounce on prey in mid-flight and inject a digestive saliva with their sharp mouth parts, then withdraw the victim's blood. You may see

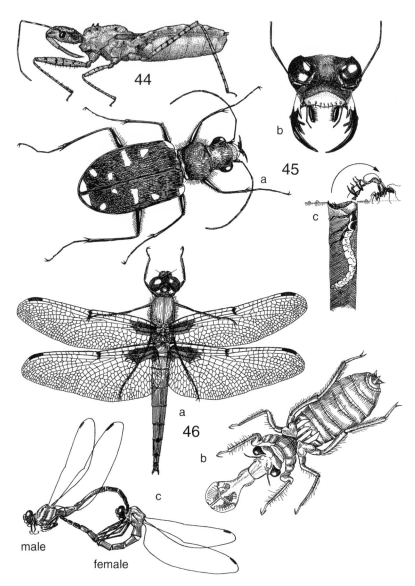

Figs. 6.44–6.46. Predators. 44. Spiny-collared Assassin Bug, *Pseliopus spinicollis* (10 mm). 45. Oregon Tiger Beetle, *Cicindela oregona:* (a) adult (12 mm); (b) adult face (3 mm); (c) larva (15 mm) in burrow, showing prey capture. 46. Skimmer dragonflies, Libellulidae species: (a) adult (50 mm); (b) nymph (25 mm); (c) adults, mating pair. (Drawings by Evan A. Sugden)

them carrying their booty from roost to roost. The Honey Bee[†] is a favorite prey for the moderate-size **Martin's Robber Fly**[E], *Stenopogon martini* (fig. 6.47). Larger robber flies, such as *Laphria* species[†] (plate 8c and 8d), may take the Hera Sheep Moth[†] and cicadas[†] as prey. Asilids are sometimes cannibalistic. The larvae (fig. 6.48a) are wormlike predators that, along with their pupae (fig. 6.48b), are found in soil, debris, or rotten stumps.

Spider wasps feed their larvae with spiders that they paralyze with their potent stings. They are usually found hunting near the ground in open areas. The females' legs are specialized for carrying prey back to the burrow. Adults, which feed primarily on flower nectar, are important pollinators of milkweeds. The **Giant Tarantula Hawk, *Pepsis thisbe*** (Pompilidae) (fig. 6.49), is the largest species of the group. This large, blue-black wasp with red wings engages in dramatic duels with its heavyweight prey, sometimes dragging the victim home along the ground. The tiny **Darkling Spider Wasp, *Anoplius tenebrosus*** (fig. 6.50), attacks small wolf spiders.

Perhaps the most diverse family of predators is the thread-waisted or digger wasps (Sphecidae), of which the ***Golden Digger Wasp, Sphex ichneumonia*** (fig. 6.51), is a good example. It provisions its subterranean larval galleries with katydids or crickets, sometimes stocking several burrows simultaneously. Members of the genus *Ammophila* (fig. 6.52) are truly thread-waisted. They capture and paralyze caterpillars, carrying them home in their sickle-shaped jaws or between their forelegs. Females of some species may block their burrows with a small stone while hunting; upon returning, they remove the stone and carefully set it aside to be reused later.

The sand wasps are an atypical, beelike subgroup of the sphecids. The **Western Sand Wasp, *Bembix occidentalis*** (fig. 6.53), attacks the Big Black Horse Fly[†] and other large insects. This wasp is marked with bright yellow-and-black stripes and nests in sandy ground. It is difficult to see in flight, owing to its speed and its broken color pattern. As with other sphecids, larvae consume insect prey captured by their mothers; adults of both sexes take flower nectar. Wasps that live socially in a single nest are in the family Vespidae. The Golden Paper Wasp[†] (plate 8e) is a caterpillar hunter. Its nest is the familiar inverted paper umbrella of hexagonal cells found under the eaves of buildings or in attics. The wasp builds its delicate home of chewed plant fiber cemented with saliva, hence the name "paper wasp," applied to this and closely related genera. A relative, the **Aerial Yellowjacket, *Dolichovespula arenaria*** (plate 8e), builds a similar nest but makes several layers. Each layer is connected to the next by a fine central strand. When completed, the tiered galleries are surrounded by a multi-layered paper covering and may attain football size. Adults are predators of small, soft-bodied insects. Ground-nesting yellowjackets may be a nuisance at picnic areas. They are very fond

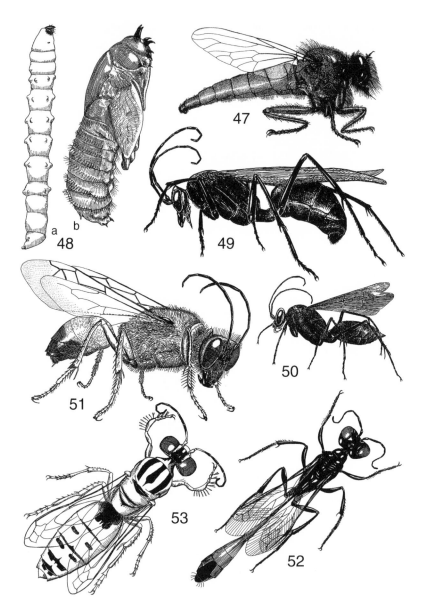

Figs. 6.47–6.53. Predators. 47. Martin's Robber Fly, *Stenopogon martini* (30 mm).
48. Sacken's Robber Fly, *Laphria sackeni:* (a) larva (34 mm); (b) pupa (16 mm). 49. Giant
Tarantula Hawk, *Pepsis thisbe* (35 mm). 50. Darkling Spider Wasp, *Anoplius tenebrosus*
(7 mm). 51. Golden Digger Wasp, *Sphex ichneumonia* (22 mm). 52. A thread-waisted
wasp, *Ammophila* species (25 mm). 53. Western Sand Wasp, *Bembix occidentalis* (20 mm).
(Drawings by Evan A. Sugden)

of high-protein food, especially meat, such as raw hamburger. In some areas where they are numerous, this has earned them and other yellowjackets the improper nickname "meat bees."

PARASITES

A *parasite* is an organism that lives at the partial expense of its host—another, usually larger organism. A well-adapted parasite allows its host to survive because it depends on a living host to complete a stage of development or an entire life cycle. However, it may interfere with the host's reproductive ability or otherwise debilitate it. Most of the species mentioned below eventually kill their host, but only after feeding on its tissues or eating its food for some time. Predators, by comparison, are generally larger than their prey. (Carnivorous ants are an exception: They may subdue relatively large prey with their overwhelming numbers and their stings.) Predators usually kill their prey quickly and consume it entirely. Most parasites have only one, or a few closely related, host species, while predators have a greater range of acceptable food items. Many flies and wasps, and some beetles, follow parasitic lifestyles; this is especially true in the ant, bee, and wasp order Hymenoptera. The larvae of some bees eat the pollen provisions of other bees before the host larvae can mature, earning themselves the label *cleptoparasite* (food-stealing parasite). Parasites play vital roles in controlling populations of their hosts, which include some of our most destructive pests.

Larval blister beetles[†] are parasites of immature solitary bees, wasps, and grasshoppers. First-instar larvae are tiny, host-searching creatures known as *triungulins* (meaning "three-clawed") (fig. 6.33b). Some hatch from eggs on the ground and crawl up onto flowers and other vegetation, where they wait for a flying host to land. When a female bee or wasp comes by, the triungulin grasps it and is carried to the host's nest. Here it molts and begins to consume the larval occupants. The triungulins of some species seek out buried grasshopper egg masses.

Bee fly[†] larvae (Bombyliidae) are also parasites. As in the blister beetles, first-instar larvae are free-living, host-searching larvae. They hatch from eggs that were deposited in or near their hosts or host burrows. In some species, the adult female fly flings the eggs from mid-air (fig. 6.39). Bee and wasp larvae are common hosts. Some bee flies deposit their eggs directly into the ground; then, when the larvae emerge, they actively search for caterpillars and bore into them.

A fly called the **Texas Conopid, *Physocephala texana*** (Conopidae) (fig. 6.56a), is a parasite of sand wasps[†], the Alkali Bee[†], and bumble bees[†]. The female attacks an adult host in mid-flight and deposits an egg on it. The larva

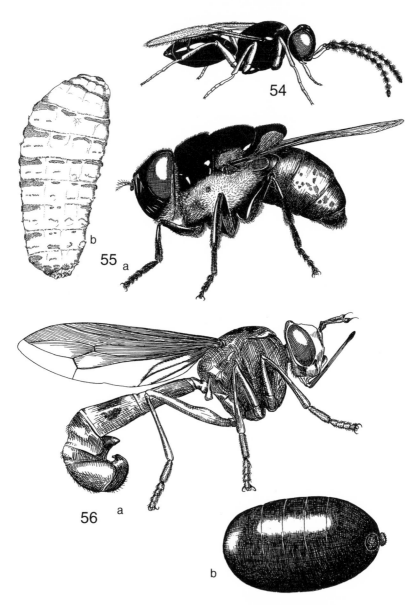

Figs. 6.54–6.56. Parasites. 54. *Telenomus* species (1 mm). 55. Jellison's Bot Fly, *Cuterebra jellisoni:* (a) adult (20 mm); (b) bot fly larva, *Hypoderma* species (20 mm). 56. Texas Conopid, *Physocephala texana:* (a) adult (12 mm); (b) pupa (6 mm). (Drawings by Evan A. Sugden)

soon hatches and burrows into the host's abdomen. The parasite feeds initially on blood, and then, as it matures, on other internal tissue. The bee or wasp may lead a normal life for some time but usually dies near the end of the larva's development. The larval fly pupates (fig. 6.56b) and emerges as an adult from the dead host.

Tachinid flies are a large and varied family. Stiff bristles cover the bodies of most species. As larvae, most tachinids are specialized caterpillar or grub parasites. After eggs that have been laid on the host hatch, the fly larvae eat their way into the body cavity. The caterpillar may continue to feed until it becomes essentially a bag of tachinid maggots ready to exit and pupate. The host dies at a late instar. Pandora Moth[†] larvae are often parasitized in such a manner (fig. 6.58). You may see large, strong-flying representatives, such as the **Cold-loving Tachinid,** *Nowickia algens* (Tachinidae) (fig. 6.57), in flight in open country even during high summer winds.

Stingless wasps are the most diverse parasitic arthropod group, although they are frequently overlooked owing to their small size. They include, among many others, ichneumon wasps and some of the smallest known insects, members of the fairy wasp family, Mymaridae, parasites of insect eggs. A braconid wasp, **Brunner's Bark Beetle Destroyer,** *Coeloides brunneri* (Braconidae) (fig. 6.59), parasitizes the Douglas Fir Beetle, which is in the same genus as the Western Pine Beetle[†]. This wasp locates its host with special sensors on the antennae that detect infrared radiation given off by the bark beetle larvae in their galleries. The wasp drills a hole through the bark with its specialized *ovipositor* (literally, "egg-placer") and deposits an egg directly on target.

Cuckoo wasps are so named because their behavior is similar to that of the cuckoo bird. The female locates the ground nest of a bee or wasp host. Then, while the adult host is away, she lays an egg on the food supply (pollen or insect cadavers) cached in the nest. The larval parasite develops rapidly, consuming the food intended for the host larva. The eggs of some cuckoo wasps hatch only after the host larva has begun to pupate; the larval parasite then consumes the host prepupa directly. **Edwards' Cuckoo Wasp,** *Parnopes edwardsii* (Chrysididae) (fig. 6.60), follows this scheme. One of its hosts is the Western Sand Wasp[†]. Adults are relatively large among Eastern Sierra chrysidids, but size may vary depending on the larval host. They are brilliant metallic-green or blue-green and appear jeweled due to their finely sculptured cuticle. Chrysidids are also known as jewel wasps. Another parasite of the Western Sand Wasp is **Sacken's Velvet Ant,** *Dasymutilla sackenii* (Mutillidae) (fig. 6.61). Untrue to its common name, this velvet ant is actually a solitary wasp. The wingless females possess a long, scimitarlike sting. This

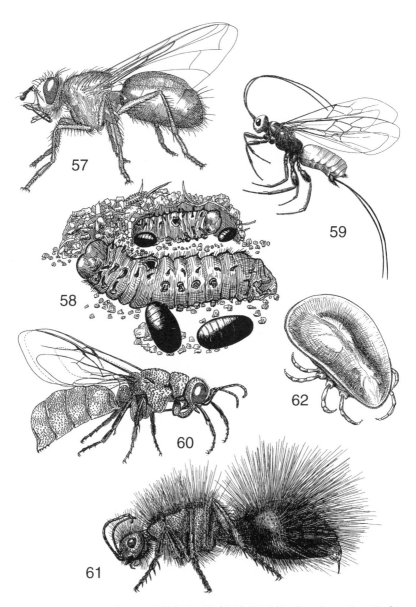

Figs. 6.57–6.62. Parasites. 57. Cold-loving Tachinid, *Nowickia algens* (15 mm). 58. Tachinidae species: pupae (10 mm) after exiting Pandora Moth host larvae. 59. Brunner's Bark Beetle Destroyer, *Coeloides brunneri* (4 mm). 60. Edwards' Cuckoo Wasp, *Parnopes edwardsii* (10 mm). 61. Sacken's Velvet Ant, *Dasymutilla sackenii* (15 mm). 62. A bird tick, *Argas* species (7 mm). (Drawings by Evan A. Sugden)

species sometimes feeds on Edwards' Cuckoo Wasp larvae, making it a parasite of another parasite.

Most arthropod parasites are parasites of other arthropods, but some may infest warm-blooded animals (see Human-biting Arthropods). A bird tick, **Argas sp.** (Argasidae) (fig. 6.62), infests the California gull rookery on the islets of Mono Lake. These ticks cluster in masses around the legs of a hatchling, taking blood and weakening their host. Intense infestation may kill a chick.

Jellison's Bot Fly, *Cuterebra jellisoni* (Cuterebridae) (fig. 6.55a), is representative of the rodent bot flies, which parasitize small mammals. From a distance this fast-flying species resembles the Big Black Horse Fly[†] and is sometimes similarly attracted to automobiles. Mouth parts of the adult cuterebrids are nonfunctional. Females lay their eggs on rabbits, woodrats, or mice. The parasitic larva (fig. 6.55b) burrows under the skin, leaving a breathing hole to the surface. At the end of the last larval instar, when it has reached the size of a grape, the larva exits from the hole, burrows into the ground, and pupates. A deer mouse supporting a final-instar cuterebrid larva could be compared to a person suffering an internal parasite the size of a basketball. Some cuterebrids deposit their eggs on vegetation, which may then be consumed by a future host.

WINTER-ACTIVE ARTHROPODS

How do arthropods survive winter? A few species, such as the Monarch Butterfly, migrate long distances to milder climates, but most remain in diapause at some stage, temporarily suspending most active life processes. The onset and termination of diapause depend on the hormonal secretions of special brain cells, the activity of which may be controlled by day length, temperature, or accumulation of metabolic products. Thus a biological clock regulates the beginning and end of diapause, regardless of the direct effects of winter. This allows insects to begin diapause before their environment becomes completely frozen or devoid of food and allows them to emerge early enough to take advantage of the first available food or other resources in spring. Through evolution the clock has become well synchronized with the seasons. Diapausing arthropods are usually secretive; they hide in soil, dead logs, leaf litter, and other refuges insulated from the bitter cold. Those that do not hide, such as young Pandora Moth[†] caterpillars, produce blood "antifreeze" agents, such as glycerol, which prevent their tissues from freezing.

Yet many species are active in winter. Thousands of tiny snowfleas often swarm over snowbanks. An example is the **Snow-loving Springtail, *Achorutes nivicola*** (Poduridae) (fig. 6.63). They are primitive insects in the order

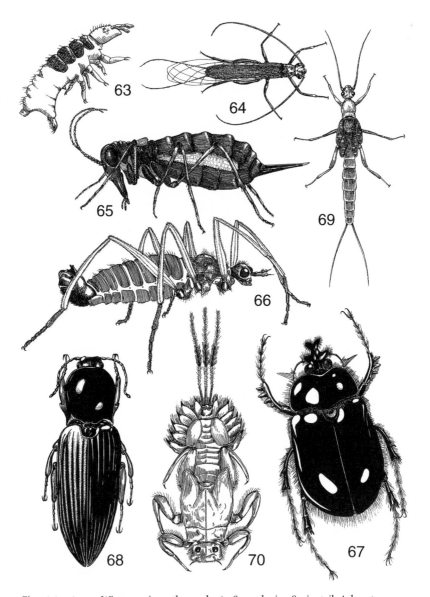

Figs. 6.63–6.70. Winter-active arthropods. 63. Snow-loving Springtail, *Achorutes nivicola* (2.5 mm). 64. A winter stonefly, *Capnia* species (8 mm). 65. California Snow Scorpionfly, *Boreus californicus* (4.5 mm). 66. A thick-legged snow gnat, *Chionea* species (6 mm). 67. Sierran Rain Beetle, *Pleocoma rubiginosa* (25 mm). 68. Shadowy Click Beetle, *Cardiophorus tenebrosus* (10 mm). 69. A stonefly, Nemouridae species: nymph (8 mm). 70. A mayfly nymph, *Ephemerella* species (12 mm). (Drawings by Evan A. Sugden)

Collembola, also known as springtails. A powerful springing device hinged underneath the abdomen allows them to jump hundreds of times their own body length. Snowfleas feed primarily on decayed organic matter and come to the snowbanks to mate.

Stoneflies in the genus *Capnia* (Nemouridae) (fig. 6.64) are often seen flying or crawling on snow near partially frozen streams. In the northern part of the Eastern Sierra, you may find the flightless **California Snow Scorpionfly**[E], *Boreus californicus* (Boreidae) (fig. 6.65), ambling across the snow in search of a mate. *B. californicus* larvae occur in clumps of thick moss. The snow scorpionfly should not be confused with thick-legged snow gnats, *Chionea* **species** (Tipulidae) (fig. 6.66), wingless flies that frequent snow-covered glades near partially frozen streams in midwinter.

An unusually large insect active in cold weather is the **Trans-Sierran Rain Beetle**[ES], *Pleocoma rubiginosa transierrae* (Scarabaeidae) (fig. 6.67); it occurs at elevations up to 10,000 feet (3048 m) in the southern Sierra. The male's mating flight takes place during the typical winter thaw in February. Flightless, burrowing females emit a pungent pheromone. Males follow the aerial pheromone "trail" to the burrow. Rain beetle grubs are thought to feed on sagebrush roots. Coastal species of *Pleocoma* fly only at dusk after the first autumn rains, hence the generic name "rain beetle." The **Shadowy Click Beetle**, *Cardiophorus tenebrosus* (Elateridae) (fig. 6.68), which also flies during the winter thaw, has been observed above the barren pumice slopes of the Mono Craters.

Cold-weather arthropods also lurk in ponds or streams where the temperature remains above freezing. Immature forms of many aquatic insects persist there in a nondiapausing state. Some feed actively, while others await the flush of nutrients from spring runoff and the warm-weather bloom of aquatic microorganisms, which they or their prey consume. Stonefly and mayfly nymphs are common in swiftly moving water (figs. 6.69, 6.70), while those of dragonflies (fig. 6.46b) occur in still or slow-moving water.

OBSERVING ARTHROPODS

Arthropods are everywhere; they often find you before you are ready for them! But often—unlike with most rocks, plants, or large animals—it is difficult to predict where and when a certain species can be found. Even when close at hand, many arthropods are inconspicuous: They hide or move too quickly to be seen. Many are small and short-lived; populations may be plentiful one year and nearly nonexistent the next. Some depend on host plants, whose abundance may also vary. Few of the thousands of arthropods in the Eastern Sierra are both conspicuous and predictable. Those that are typically occur nearly everywhere (mosquitoes and Honey Bees) or in large numbers (Mono Lake Brine Shrimp). The best strategy for observing them is to find types of places and conditions in which certain arthropods are likely to occur. Often, such places are represented by patches of *microhabitat*.

Most insects, particularly flying ones, prefer warmth, sunlight, and a source of food. The greatest abundance and diversity of insects generally occur where vegetation is dense and varied. They also congregate around moist places, such as streams, ponds, and thermal springs. In general, natural, pristine sites support more species than disturbed or polluted locations, although numbers of certain species may be greater in unnatural situations—for example, greenbottle flies[†] near a landfill or **Alfalfa Butterflies,** *Colias eurytheme,* near a hay field. Sand dunes provide unique habitats that support many interesting species. Butterflies and some other flying insects are often abundant at the tops of isolated hills or mountain summits on warm days—a phenomenon known as *hilltopping*. Flowers are an attractive source of food for many arthropods. Blossoming willows are good places to look in the spring. Rabbitbrush blooming in late summer attracts swarms of bees, flies, butterflies, beetles, and many others. Their predators, including spiders, can also be found there.

A Coleman lantern or a street lamp can attract hundreds or thousands of flying insects on a warm summer evening. Lay a white sheet under a bright light and await a fascinating display of activity, forms, and color. Fluorescent lights, and especially ultraviolet (UV) or "black lights," are particularly good for this. Ground-creeping arthropods, such as solpugids and some predaceous beetles, tend to be more active at night when it is warm. An easy way to catch night-active arthropods for close observation is to trap them in pitfalls with steep-sided plastic cups or slippery cereal bowls. Sink the cups into the ground up to their rims, smooth the edges of the excavation, leave several traps overnight, and check them in the morning. Pitfall traps are most successful in sandy areas.

Turning stones at the edge of a stream sometimes yields an interesting array of aquatic insects. Since game fish depend on aquatic insects, a healthy trout stream is a good place to start. Turning over logs or stones on dry land is also a good way to find secretive, land-dwelling species. Be careful of your fingers, and gently roll heavy objects toward you. Be sure to replace stones and logs as you find them; such microhabitats are vital to the survival of many creatures and are easily disturbed.

Another interesting way to find arthropods to examine—in this case precaught—is to dissect the discarded stomach of trout bound for the frying pan. Fishermen use this technique to find out what the fish are eating, and choose their fishing flies accordingly. Dead but still interesting insect specimens can be plucked from automobile radiators or from glass light fixtures.

Often simply watching arthropods in their natural habitat is the most gratifying way to study them. A hand lens (10× or more) and a pair of close-focusing binoculars (up to 10 ft, or 3 m) can help you explore the fascinating world that this chapter has introduced and can take you on an amazing journey into the lives of our most abundant planet-mates.

RECOMMENDED READING

Borror, D. J., and R. E. White. 1974. *A Field Guide to the Insects of America North of Mexico.* Peterson Field Guide Series. New York: Houghton Mifflin.

Milne, L., and M. Milne. 1980. *The Audubon Society Field Guide to North American Insects and Spiders.* New York: Alfred A. Knopf.

Powell, J. A., and C. L. Hogue. 1980. *California Insects.* Berkeley: University of California Press.

CHAPTER SEVEN

NATIVE FISHES

DONALD SADA

Eastern Sierra lakes and streams are widely known for their scenic beauty and angling opportunities. Yet many people are surprised to learn that rainbows, goldens, browns, and the other popular sport fishes are not native to the region and that the region's native fishes are rarely seen and now occupy only a small portion of their historical ranges. While many articles and books have described the beauty and the fishing opportunities east of the Sierra, the ecology, distribution, and abundance of native fishes have received little attention. Here, then, we focus on the native fishes and discuss other, better known fishes only in terms of their effect on native fishes.

ORIGIN AND AFFINITIES OF REGIONAL FISHES

Eastern Sierra native fishes occupy a variety of habitats in the drainages flowing east of the Sierra Crest: thermal springs, small streams, rivers, and large lakes. Although it is difficult to determine which areas the sixteen types of native fish (table 1) once occupied, these probably included most of the lower-elevation waters, but very few waters higher than 8000 feet (2438 m) elevation. Mountain glaciers during the past 2 million years prevented fish from inhabiting most high mountain lakes and streams. Following retreat of the glaciers, these waters remained fishless because waterfalls and other obstacles prevented fish from migrating upstream to populate them. Also, most Eastern Sierra native fish prefer the lower-gradient, warmer waters found at lower elevations. However, the current distribution of native fish bears little resemblance to their distribution before settlement. Prospectors dredged creek bottoms, and ranchers and farmers diverted streams and encouraged the introduction of nonnative fish (table 2)—altogether drastically altering aquatic habitats. Most fish communities today are dominated by introduced species.

TABLE 1 NATIVE FISHES OF THE EASTERN SIERRA.

COMMON NAME	SCIENTIFIC NAME
Salmon and Trout Family	Family Salmonidae
Lahontan Cutthroat Trout[1]	*Oncorhynchus clarki henshawi*
Paiute Cutthroat Trout[1]	*Oncorhynchus clarki seleniris*
Eagle Lake Trout[1]	*Oncorhynchus mykiss iridens*
Mountain Whitefish[1]	*Prosopium williamsoni*
Minnow Family	Family Cyprinidae
Lahontan Speckled Dace[1]	*Rhinichthys osculus robustus*
Owens Speckled Dace[2]	*Rhinichthys osculus* ssp.
Lahontan Redside[1]	*Richardsonius egregius*
Lahontan Creek Tui Chub[1]	*Gila bicolor obesa*
Lahontan Lake Tui Chub[1]	*Gila bicolor pectinifer*
Owens Tui Chub[2]	*Gila bicolor snyderi*
Sucker Family	Family Catostomidae
Tahoe Sucker[1]	*Catostomus tahoensis*
Owens Sucker[2]	*Catostomus fumeiventris*
Lahontan Mountain Sucker[1]	*Catostomus platyrhynchus lahontan*
Cui-ui[1]	*Chasmistes cujus*
Pupfish Family	Family Cyprinodontidae
Owens Pupfish[2]	*Cyprinodon radiosus*
Sculpin Family	Family Cottidae
Paiute Sculpin[1]	*Cottus beldingi*

[1] Species in the Lahontan Basin. [2] Species in the Owens Basin.

As a result, the prehistoric distribution of native fishes and the composition of their communities can only be determined from records compiled during scientific studies made in the early 1900s.

Two drainages flow from the Eastern Sierra slope, the Death Valley system and the Lahontan Basin. They are isolated from all other drainages in California and Nevada (fig. 7.1), and they lie along the western boundary of the Great Basin. Both basins are closed, with no outlets to the ocean. Although similar fish species occupy these basins, the fish communities are distinct from one another and from communities in other drainages.

COMMON NAME	SCIENTIFIC NAME
Salmon and Trout Family	Family Salmonidae
Rainbow Trout[1,2]	*Oncorhynchus mykiss gairdneri*
Golden Trout[2]	*Oncorhynchus mykiss aguabonita*
Colorado Cutthroat Trout[2]	*Oncorhynchus clarki pleuriticus*
Sockeye Salmon[1,2]	*Oncorhynchus nerka*
Brown Trout[1,2]	*Salmo trutta*
Brook Trout[1,2]	*Salvelinus fontinalis*
Lake Trout[1]	*Salvelinus namaycush*
Arctic Grayling[1]	*Thymallus arcticus*
Minnow Family	Family Cyprinidae
Carp[1,2]	*Cyprinus carpio*
Golden Shiner[1]	*Notemigonus crysoleucas*
Sacramento Blackfish[1]	*Orthodon microlepidotus*
Catfish Family	Family Ictaluridae
Channel Catfish[1,2]	*Ictalurus punctatus*
Brown Bullhead Catfish[1,2]	*Ictalurus nebulosus*
Black Bullhead Catfish[1]	*Ictalurus melas*
White Catfish[1]	*Ictalurus catus*
Livebearer Family	Family Poeciliidae
Mosquitofish[1,2]	*Gambusia affinis*
Temperate Bass Family	Family Perichthyidae
White Bass[1]	*Morone chrysops*
Sunfish Family	Family Centrarchidae
Sacramento Perch[1,2]	*Archoplites interruptus*
Largemouth Bass[1,2]	*Micropterus salmoides*
Smallmouth Bass[1,2]	*Micropterus dolomieui*
Bluegill Sunfish[1,2]	*Lepomis macrochirus*
Green Sunfish[1]	*Lepomis cyanellus*
Redear Sunfish[2]	*Lepomis microlophus*
Pumpkinseed[2]	*Lepomis gibbosus*

TABLE 2 (*continued*)

COMMON NAME	SCIENTIFIC NAME
Perch Family	Family Percidae
Yellow Perch [1]	*Perca flavescens*
Walleye [1]	*Stizostedion vitreum*
Stickleback Family	Family Gasterosteidae
Three-spined Stickleback [2]	*Gasterosteus aculeatus*

[1] Species in the Lahontan Basin. [2] Species in the Owens Basin.

Fig. 7.1.   Location of the Lahontan Basin and Death Valley system. (Map by Donald Sada)

The diversity and distribution of native fishes in these basins are the products of genetic divergence caused by geological processes and changing climates during the past several million years. Higher precipitation and lower Sierra Nevada elevations during the Pleistocene Epoch (the time when glaciers were widespread, beginning about 2 million years ago and ending about 10,000 years ago) permitted the most recent connections between adjacent drainages that allowed fish to move between basins. Ancestral fishes gained access to the Eastern Sierra when its waters were connected to adjacent drainages. Recent mountain building and drier climates isolated these drainage basins from one another, prevented fish from moving between basins, and allowed fish populations in isolated waters to evolve special characteristics that made them distinct from one another and sometimes enabled them to survive. Evidence of these prehistoric connections comes from examining morphologic and genetic similarities and differences between fishes in the Lahontan Basin and Death Valley system, and fish in surrounding drainages.

DEATH VALLEY SYSTEM

The Death Valley system is a prehistoric drainage consisting of the Owens, Amargosa, and Mojave drainages, which were tributary to Death Valley during the Pleistocene Epoch (fig. 7.2). Although drainages in this system are no longer connected because the present-day climate is drier, perennial water persists in isolated springs and rivers. The Owens Basin is the only portion of this prehistoric drainage that receives water from the Sierra, and the Owens River drainage is the only drainage in this southern portion of the Eastern Sierra occupied by native fish. There is no evidence that the Mono Basin or Indian Wells Valley historically supported native fish.

There are only four native fishes—two minnows, one sucker, and one pupfish—in the Owens Basin (table 1), from a total of ten native species in the entire Death Valley system. Taxonomic analysis (study of the relationships among species) shows that the Owens Pupfish, Owens Tui Chub, and Owens Sucker are endemic to the Owens Basin. Further study is needed to determine whether the Owens Speckled Dace is a distinct species. Early fisheries studies concluded that native fish inhabited most valley-floor waters and some springs in the basin. Higher-elevation lakes and streams contained no fish at all before the introduction of sport fish in the mid-1800s by individuals and by fish and wildlife agencies.

Morphological and genetic similarities between fishes in the Owens and adjacent basins indicate that the Owens Basin was at one time connected with the Lahontan Basin and also with the lower Colorado River. The Owens Tui Chub, Owens Sucker, and Owens Speckled Dace are most closely related to

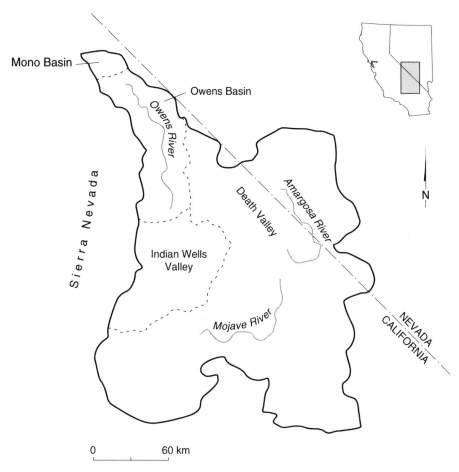

Fig. 7.2. Schematic map of drainages in the Death Valley system. Only the Mono Basin, Owens Basin, and Indian Wells Valley lie in the Eastern Sierra. (Map by Donald Sada)

species in the Lahontan Basin; the Owens Pupfish is most closely related to the Desert Pupfish *(Cyprinodon macularius)* in the lower Colorado River drainage.

LAHONTAN BASIN

The Lahontan Basin includes most of northern Nevada, northeastern California, and a small portion of southern Oregon (fig. 7.3). The Eastern Sierra portion of the basin is drained by the Walker, Carson, Truckee, and Susan rivers. The Eagle Lake Basin is an isolated drainage at the northern edge of the Sierra. Although these drainages are now isolated from each other, during the Pleistocene they were all tributary to ancient Lake Lahontan—a gigantic

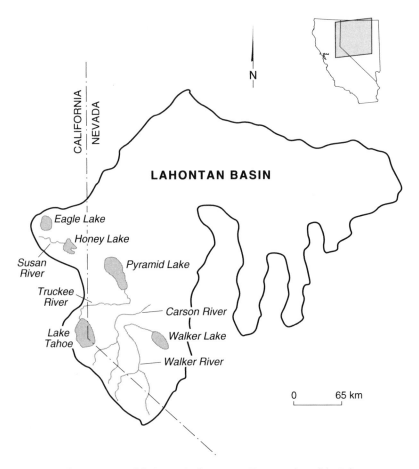

Fig. 7.3.   Schematic map of drainages in the Eastern Sierra portion of the Lahontan Basin. (Map by Donald Sada)

lake that covered 8665 square miles (22,442 sq km) at its highest level and whose arms extended into most of the basin's valleys. As a result, there are few differences between the fish fauna in each drainage.

There are twelve types of fish native to the Eastern Sierra portion of the Lahontan Basin. Most of them are widespread throughout the basin. However, the Paiute Cutthroat Trout and Eagle Lake Trout are restricted to Eastern Sierra waters. Only the Paiute Sculpin and Mountain Whitefish occupy waters outside of the Lahontan Basin. Ancestral fishes probably entered the Lahontan Basin during the Pleistocene from the Klamath River of northern California and the Columbia River in southern Oregon. These connections probably did not last long, because only a small portion of the Klamath and Columbia fauna is represented in the Lahontan Basin. (A long connection would have permitted more species to enter the Lahontan Basin.) Volcanic eruptions, mountain building, and drier climates altered these ancient drainage patterns, isolating these waters from one another as they are today.

FISH ECOLOGY AND AQUATIC HABITATS

Life is difficult for Eastern Sierra fish. Rivers and streams are small compared to those in wetter climates, and they are affected by wide variations in temperature and discharge. To survive in these conditions species must be able to withstand water temperatures varying from 32°F to 90°F (0–32°C), dissolved oxygen concentrations that are very low, frequently less than 3 milligrams per liter (mg/l), and streamflows that may range from a torrent in spring and winter to barely a trickle in summer. For fish to persist in these environments, they must have reproductive strategies that allow populations to recover quickly from high mortality caused by periodic floods and drought. Most lakes in the region provide additional challenges to fishes. All terminal lakes in the region, such as Eagle, are highly alkaline, presenting a physiologically stressful environment; several of them, such as Honey, are periodically dry. Fish in subalpine lakes and high-gradient streams must survive not only short growing seasons—early autumn freezes and late spring thaws—but also limited food supplies, because the nutrients needed for aquatic life are scarce.

Life history strategies and physiological characteristics reflect the adaptations these species require to survive such harsh environments. For example, Lahontan Cutthroat Trout tolerate lower dissolved oxygen concentrations and higher water temperatures than almost any other trout in western North America. They also persist in the highly alkaline waters of Pyramid and Walker lakes, which neither Rainbow nor Brown Trout can tolerate. Most native fish compensate for high mortality caused by irregular, unforeseen events by producing many eggs each year and/or becoming sexually mature early in life. Also, most of these fish are relatively small, which allows them to live in the region's small streams and springs.

Five general types of aquatic habitat in the region contain native fishes: high-gradient streams, low-gradient streams, subalpine lakes, *terminal* lakes (having no outlet), and springs and spring brooks. Although native fish occupy all of them, not all species occupy every habitat. The Mountain Sucker and fishes in the salmon, trout, and sculpin families generally prefer high-altitude, cold streams and lakes. Other fishes, such as the Lahontan Redside, may also inhabit these waters, but they are generally more abundant in lower-altitude, warmer, placid waters.

Compared to those of most other North American drainages, the number of native fish species in the Eastern Sierra is low. This probably results from infrequent and short-term connections with other drainages and from the region's harsh environment.

WILL THEY SURVIVE?

Eastern Sierra native fishes are well adapted to withstand their harsh environments, but they usually cannot adapt to highly altered habitats and introduced sport fish. As a result, the U.S. Fish and Wildlife Service currently lists the Paiute and Lahontan cutthroat trouts as threatened, and the Owens Pupfish, Cui-ui, and Owens Tui Chub as endangered. Populations of these species have declined to such low levels that they require special protection to prevent extinction, but they are not the only native species to have suffered: The population size and distribution of all native fishes have declined drastically in modern times.

Populations are low because almost all habitats have been physically and/ or biologically altered in many ways. Diversions of water for hydroelectric power generation, municipal consumption, and agriculture have decreased flows in most streams. Pollutants and nutrients from mining, excessive livestock grazing, agriculture, timber harvest, and municipal runoff have adversely altered the chemical characteristics of lakes and streams. Grazing and timber harvest have altered stream courses and removed riparian and upland vegetation, destabilizing watersheds and causing more frequent and more violent flooding.

These activities harm fish by reducing population size. Low flows decrease the habitat needed for food production, living space, and reproduction. Low flows may also limit streamside vegetation and reduce the habitat diversity that is necessary to support the varied lifestyles of the native fishes. Nutrients such as nitrogen and phosphorus may be lethal to fish; they also tend to stimulate excessive algae growth, which depletes oxygen in the water. Low dissolved oxygen concentrations are fatal to fish or their foods.

Modifications to native fishes' physical habitat are conspicuous in the Truckee River system, where they have caused the decline of Lahontan Cutthroat Trout and Cui-ui. Cities, agriculture, and hydroelectric power plants all use Truckee River water. Since the early 1900s agricultural diversions have reduced by more than 50 percent the amount of water reaching Pyramid Lake, causing the lake level to drop more than 70 feet (21.5 m) and causing a delta to form at the mouth of the Truckee, with water flowing through a shallow, braided channel. The delta blocked Lahontan cutthroat from migrating to upstream spawning grounds, and by the late 1930s it had eliminated them from Pyramid Lake. In addition, lowered streamflows and increased water temperatures in the lower Truckee usually cause cutthroat eggs incubating there to die. Diversions for hydroelectric power generation frequently prevent fish from moving upstream as well as downstream. Small dams that have not been fitted with adequate fish ladders prevent upstream movement. Diversions may shunt downstream-moving fish into canals, where they may die in agricultural fields or in electric turbines.

Pollution has also affected the lower Truckee River, where phosphorus and nitrogen from sewage-treatment facilities cause excessive growth of algae and other vegetation. This growth may suffocate fish or kill incubating embryos by lowering the amount of oxygen in the water. In the upper Carson River, mercury discharging from old mines is toxic to all aquatic life. The movement of this mercury into the lower Carson is believed to have elevated the levels of mercury in some sport fish, which has caused health authorities to suggest that those types of fish should not be eaten.

Most Eastern Sierra waters have also been polluted by the introduction of nonnative fishes, which displace native fishes through predation, competitive interactions, and hybridization. Species such as Largemouth Bass, most species of sunfish, Brown Trout, and Mosquitofish prey on native fish and eliminate populations. Rainbow Trout readily hybridize with both Paiute and Lahontan cutthroat trout; as a result, there are almost no genetically pure Cutthroat Trout populations here. Hybridization between Owens Tui Chub and Lahontan Lake Tui Chub nearly eliminated the Owens Tui Chub and provided the basis for listing it as endangered.

Reestablishing native fishes requires restoring habitat and controlling introduced species. Since many factors have adversely affected Eastern Sierra fishes, many changes in land and wildlife management are needed to help protect native fishes from extinction and rebuild their populations. In the Truckee River, proper diversion management is needed to provide the water quality and quantity necessary for maintaining Lahontan Cutthroat Trout and Cui-ui. In the Owens Basin, the marshes and oxbows that bordered the Owens River must be restored to provide habitat for Owens Pupfish. Throughout the Eastern Sierra livestock-grazing practices must be changed to decrease the amount of silt entering streams. This silt covers the gravel and stones that provide habitat for foods eaten by fish and habitat required by spawning fish. Grazing livestock also decrease habitat quality by reducing the amount of vegetation covering streambanks, by polluting water with fecal material, and by trampling streambanks, which eliminates the undercuts providing cover for fish.

All of the native fishes are important to the Eastern Sierra. Lahontan Cutthroat and Eagle Lake trout are valued sport fish. The Cui-ui and Owens Pupfish were important to native peoples who used them as staple foods. Today, the native fishes are important indicators of the quality and health of Eastern Sierra waters. Their decline is a warning signal that the activities that have damaged their habitats also adversely affect all life in the region that depends on water. Their disappearance reduces the Eastern Sierra's characteristic grandeur and uniqueness.

Many programs are attempting to prevent native fishes' extinction, but few are rebuilding populations so that rare species will no longer need special protection. Lahontan Cutthroat and Eagle Lake trout populations are maintained by hatchery production, in which brood stocks of adult fish massively produce fish that are returned to waters as sport fish. Without hatcheries, populations in Pyramid, Walker, and Eagle lakes could not survive. Other programs maintain Owens Pupfish and Owens Tui Chub in small, isolated habitats called *refugia,* where they are protected from introduced species and habitat alteration.

Although it is doubtful that these species could have survived without special programs, it is clear that captive breeding programs by themselves cannot rebuild populations in the natural environment. It is now clear that hatchery populations will be maintained only as long as funds are provided and that they generally produce fish better adapted to survive in a hatchery environment than in natural lakes and streams. Refugia populations are generally small and these artificial habitats poorly resemble the fishes' natural habitat. Although fish may survive well in refugia, they are extremely susceptible to unusual natural events and the introduction of predatory fishes. For example, since the mid-1980s the largest Owens Pupfish population, in the Owens Valley Native Fish Sanctuary, has been destroyed by events as varied as earthquakes and vandalism. Recovery can occur only when habitats are reconstructed to provide the food, cover, and native aquatic communities where these species can persist without the assistance of special programs. Protecting and enhancing native fish habitats also ensures the protection of Eastern Sierra streams and lakes.

SPECIES ACCOUNTS

In the following species accounts, *substrate* refers to a lake or stream bottom. An asterisk (*) indicates species that occur only east of the Sierra Crest.

SALMON AND TROUT: FAMILY SALMONIDAE

*LAHONTAN CUTTHROAT TROUT, *Oncorhynchus clarki henshawi*   A fish of silver color highlighted by large, circular black spots covering the body. Fish in streams usually weigh less than 1 lb (0.5 kg) and are not longer than 12 in (30 cm), while fish in lakes weigh up to 40 lbs (18 kg) or more. Adults live no longer than 10 years.

Cutthroat Trout occupy all major drainages in North America west of the Rocky Mountains. There are at least 22 different subspecies, inhabiting isolated drainages. In streams Cutthroat Trout feed on insects drifting downstream; in lakes adults are predaceous, feeding on Tui Chub, Tahoe Sucker, and Lahontan Redsides. In streams they prefer the slower-moving waters typical of low gradients found in meadows; in lakes they occupy the shallows near shore where prey is most abundant. Lahontan cutthroat spawn only in streams and only during the spring. Lake populations may migrate long distances to spawn. For example, before their demise in the Truckee River system fish migrated from Pyramid Lake more than 60 miles (96 km) to spawn in tributaries of Lake Tahoe. Females deposit eggs in nests they dig into stream gravels where there is sufficient current to adequately aerate the incubating embryos. Eggs usually do not survive water temperatures above 57°F (14°C).

The Lahontan Cutthroat Trout is endemic to the Lahontan Basin, where it once occupied 5 lakes and at least 3500 miles (5635 km) of stream. All of its lake habitat was on the western edge of the Lahontan Basin, where it also occupied approximately 1170 miles (1883 km) of stream in the Carson, Walker, Truckee, and Susan rivers. It was sufficiently abundant in the early 1900s to support commercial fisheries in Lake Tahoe, and Pyramid and Walker lakes. Competition with Lake Trout contributed to its elimination from Lake Tahoe. Today neither Pyramid nor Walker has a self-sustaining population. In Pyramid Lake it grew to the largest size known for a landlocked trout in western North America. It has disappeared entirely from the higher elevation habitats in the Eastern Sierra owing to hybridization with Rainbow Trout, overexploitation by commercial fishing, and habitat deterioration from placer mining and overgrazing. Agricultural water diversions from the Walker River have affected Walker Lake just as diversions from the Truckee affected Pyramid Lake. The deltas at the river mouths are so shallow that they block fish from migrating to upstream spawning grounds. Today Lahontan cutthroat populations in Pyramid and Walker lakes are maintained by a hatchery stock of Lahontan cutthroat from Summit Lake in northwestern Nevada. The native Pyramid and Tahoe cutthroat are extinct. *Distribution* In the Eastern Sierra today, self-sustaining populations persist only in 17 miles (28 km) of small tributaries in the upper Walker and Carson rivers and in Independence Lake, less than 1 percent of its historic range. There has been a sharp decline of Lahontan Cutthroat Trout throughout the remainder of its range. Within the Lahontan Basin it currently occupies only 240 miles (386 km) of stream, 6.5 percent of its historic range. Today the largest Lahontan cutthroat populations are supported by hatchery production.

Early transplants established Lahontan Cutthroat Trout in a few Owens Basin lakes and streams. Some of these populations persist today.

*PAIUTE CUTTHROAT TROUT, *Oncorhynchus clarki seleniris* Closely related to the Lahontan Cutthroat Trout, the Paiute cutthroat is distinguished from it only by the absence of black spots and a more rosy purple color. Its life history is similar to that of stream populations of Lahontan cutthroat. Its historic range was limited to approximately 6 miles (10 km) of isolated stream in the Silver King Creek drainage of the upper East Fork Carson River. Populations have declined because it has hybridized with Lahontan Cutthroat and Rainbow trout. Fishing for Paiute cutthroat is currently not allowed because populations are so small. Increasing the size and number of populations will probably make fishing possible in the future. *Distribution* Paiute cutthroat persists in only a few streams within its native range and in several lakes and streams where it was introduced, including some in the Owens Basin.

EAGLE LAKE TROUT, *Oncorhynchus mykiss irideus* The only native Rainbow Trout population in the Great Basin, established in prehistoric times when water from Eagle Lake flowed west into the upper Sacramento River Basin. Recent genetic and morphological analyses indicate this is an isolated and distinctive population of Coastal Rainbow Trout, which occupies streams along the Pacific coast from Alaska to Mexico. Adults rarely weigh more than 11 lbs (5 kg). They feed on Tui Chubs, which are abundant in the shallow waters of the lake. In spring they spawn over gravel substrates in Pine Creek, the only stream that flows into Eagle Lake year-round. However, the population is maintained by hatchery plantings. Natural spawning is usually unsuccessful because irrigation diversions from Pine Creek often lower the streamflow and eliminate suitable spawning habitat. *Distribution* Endemic to Eagle Lake.

*MOUNTAIN WHITEFISH, *Prosopium williamsoni* Bears little resemblance to its close relative, the trout; has larger scales and a moderately small mouth. However, the presence of an *adipose fin* (a small, fleshy fin behind the main, bony fin) identifies it as a member of the salmon and trout family (Salmonidae). It may reach 20 in (50 cm) in length and weigh up to 7 lbs (3kg). Although these fish may spawn in lakes, they prefer to spawn in streams, where groups congregate during autumn over gravel substrates. They reach sexual maturity in 3 years and may live for almost 20. A sexually mature female may carry up to 14,000 eggs. Whitefish feed on insects, collecting most of their food from the substrate. Lake populations also consume *zooplankton* (microscopic aquatic animals). Their diet changes seasonally, generally consisting of whatever food is most abundant. *Distribution* The bottoms of colder streams and lakes in the Lahontan Basin and tributaries to the Snake River in northeastern Nevada.

MINNOWS: FAMILY CYPRINIDAE

*SPECKLED DACE, *Rhinichthys osculus* The widest-ranging native fish in western North America. Body coloration varies between populations. Bellies are typically tan, backs usually olive to dark green. Fish in some populations may have 2 black lateral stripes along the body. Other populations may have a single stripe, some have no stripe. Varying amounts of black speckles may cover the body, but some populations have no speckles. Fins and lips may become bright orange during spawning season. Usually no longer than 4 in (10 cm).

The Speckled Dace occupies a wide variety of habitats, including thermal springs, cool- and warm-water rivers and streams, swift riffles flowing over cobble-size substrate, quiet backwaters, or shallow, muddy-bottom streams. Spawning occurs primarily during springtime, but may occur throughout the year in spring habitats where water temperatures are constant. Fish may reach sexual maturity during their first year. They rarely live longer than 4 years. Speckled Dace feed primarily on insects. They are moderately gregarious and continuously move through their habitat in groups of 2–15 individuals.

Speckled Dace have not survived well in the Owens Basin, where they occupy approximately 17 percent of their historic range. Their populations, isolated in springs and streams, are poorly understood; recent morphological studies suggest there may be 4 distinct subspecies. Of the 9 known populations, 5 have been totally destroyed by habitat alteration and/or introduction of nonnative species. *Distribution* Speckled Dace occur from British Columbia south into Mexico and from the Pacific coast to the Continental Divide. Several subspecies occupy isolated drainages and habitats. A single subspecies, *R. o. robustus,* occupies the Lahontan Basin portion of the Eastern Sierra. Although currently less abundant than in historic times, *R. o. robustus* continues to occupy most mid- and low-elevation streams in the northern portion of the Eastern Sierra.

*LAHONTAN REDSIDE, *Richardsonius egregius* Bright red-and-black lateral stripes during the spawning period distinguish the Lahontan Redside as one of the most colorful native fish in the western United States. It is rarely longer than 5 in (12 cm). Spawning occurs during late spring to early summer, when fish congregate over gravel bottoms in lakes and streams. Males fertilize eggs as females broadcast them over the gravel. Fish reach sexual maturity during the third or fourth year, but rarely live longer than 4 years. Although primarily insectivorous, they may eat whatever foods are readily available. Studies in Lake Tahoe found that they consume terrestrial and aquatic insects, plankton, and fish eggs, and that they usually school over rocky bottoms in water less than 30 ft (9 m) deep. In streams, Lahontan Redsides hold at mid-water to capture insects drifting downstream; they prefer pools with protective cover, such as overhanging banks, woody debris, or rough surface water. *Distribution* Throughout the Lahontan Basin. Large schools of fish are common in mid- and low-elevation lakes and streams throughout the Eastern Sierra.

*TUI CHUB, *Gila bicolor* Widespread throughout the western United States. In lakes Tui Chubs may grow to be 16 in (40 cm) long; in streams they are rarely longer than 8 in (20 cm). Age studies in Eagle Lake determined that a fish 14 in (36 cm) long was 32 years old.

Tui Chubs spawn during the spring. Although spawning has never been observed, it is probably similar to that of other minnows, which spawn in groups by broadcasting eggs and milt over vegetation or bare substrate. Sexually mature at age 2, they are prolific: Lake Tahoe females carry as many as 25,000 eggs. They are omnivorous but feed primarily on invertebrates. *G. b. obesa* and *G. b. snyderi* prefer foods found near the bottom, such as insects, snails, crayfish, and aquatic vegetation; *G. b. pectinifer* feeds on plankton. In large, open habitats Tui Chubs move in schools, but they do so less in shallow habitats with more abundant cover.

Tui Chubs occupy lakes, streams, rivers, and springs; they prefer calm, deep areas with dense protective cover. Lake populations of *G. b. obesa* mainly occupy shallow habitats near the shore with dense cover. *G. b. pectinifer* rarely occupies streams; in lakes it inhabits mid-water areas not frequented by *G. b. obesa*. Even though lake populations of the 2 Lahontan Basin subspecies occupy different habitats much of the year, they frequently hybridize.

The total number of Owens Tui Chubs, *G. b. snyderi,* at any one time probably never exceeds 500. Owens Tui Chub populations declined because of predation from introduced sport fish and hybridization with *G. b. pectinifer,* which was introduced as a bait fish. Most Tui Chubs in the Owens Basin are these hybrids. *Distribution* Abundant in many locations. Most drainages support a distinct subspecies; however, their differences are not yet understood, and many Great Basin spring-dwelling populations have not been named or described. The Lahontan Creek Tui Chub *(G. b. obesa)* and Lahontan Lake Tui Chub *(G. b. pectinifer)* are endemic to the Lahontan Basin and widespread throughout its larger waters. The Owens Tui Chub *(G. b.*

snyderi) is endemic to the Owens Basin, where only 4 small populations persist. Of these populations, 2 are in natural habitat, and 2 occur in artificial ponds established for their conservation. The Owens Tui Chub has been introduced into the Mono Basin and the Upper San Joaquin River drainage.

SUCKERS: FAMILY CATOSTOMIDAE

*TAHOE SUCKER, *Catostomus tahoensis*   Endemic to the Lahontan Basin, and at present the commonest sucker in the region. Tahoe Suckers in lakes are larger than those in streams. Tahoe Suckers in Pyramid Lake are among the largest known, some measuring almost 24 in (60 cm) long. In streams they are usually no longer than 12 in (30 cm).

They feed most actively at night—on aquatic plants, detritus, and invertebrates gleaned from the substrate. Their most distinctive feature is a fleshy, *subterminal* mouth (located below and behind the end of the nose) that is well adapted to efficiently remove food from the substrate. They have a high reproductive capacity and are comparatively long-lived. Suckers in Pyramid Lake live up to 27 years; large females may carry 35,000 eggs. In Lake Tahoe they reach sexual maturity at age 4 or 5, when they are between 8 and 12 in (20 and 30 cm) long. During the spring they migrate into streams and congregate over spawning sites. Females broadcast eggs that drift onto gravel substrates; the drifting eggs are fertilized by attending males. *Distribution* Most Eastern Sierra waters in the Lahontan Basin, including high-altitude lakes and small streams, but most common in warmer, lower-elevation waters.

*OWENS SUCKER, *Catostomus fumeiventris*   Inhabits streams and lakes below 7500 ft (2286 m). It is the most abundant native fish in the Owens Basin, yet even so its populations have been depleted. No studies have examined its life history. Its closest relative is the Tahoe Sucker, which suggests that most aspects of their ecology are similar. *Distribution* Endemic to the Owens Basin but now limited to northern Owens Valley and the upper Owens River drainage. Introduced into the Mono Basin and the Santa Clara River Basin in southern California.

*LAHONTAN MOUNTAIN SUCKER, *Catostomus platyrhynchus lahontan* This subspecies is endemic to the Lahontan Basin. It is distinguished from the Tahoe Sucker by its smaller size and rounded snout. It has a deep notch between its upper and lower lips, and is rarely longer than 10 in (25 cm).

The Lahontan Mountain Sucker is herbivorous. It reaches sexual maturity somewhere between its second and fifth year when it is at least 4 in (9 cm) long. Females may produce up to 4000 eggs. Spawning occurs from late spring through midsummer in streams over gravel substrate. Spawning behavior is believed to be similar to that of the Tahoe Sucker. Distributions of these 2 suckers broadly overlap in the Lahontan Basin, and they frequently hybridize. However, the Lahontan Mountain Sucker does not inhabit lakes and seems to prefer swifter waters and colder streams; the Tahoe Sucker is most abundant in lakes and warmer streams at lower elevations. Early studies found that Mountain Suckers were more abundant than Tahoe Suckers in cool waters of the Eastern Sierra, but today Mountain Suckers are rare and the Tahoe Sucker common to abundant. Perhaps the dams and reservoirs constructed during the past 30 years have eliminated the stream habitats preferred by Lahontan Mountain Suckers and replaced them with lakes that Tahoe Suckers prefer. *Distribution* Broadly distributed throughout the West in parts of the Columbia, Colorado, and Missouri drainages, and in the Bonneville and Lahontan basins.

CUI-UI, Chasmistes cujus Endemic to Pyramid Lake, where it grows to almost 36 in (90 cm) and may live longer than 40 years. The Cui-ui spawns during the spring over gravel substrate only in the lower Truckee River downstream from Reno. Large females carry up to 100,000 eggs. After the eggs are broadcast, males fertilize them as they drift to the bottom. The Cui-ui population declined in response to the same effects of water diversions that destroyed a self-sustaining Lahontan cutthroat population in Pyramid Lake. Heavy precipitation during the mid-1980s and recent programs to conserve the species have improved its status. However, many years of successful reproduction are required to stabilize the population, which managed to reproduce itself only twice between 1950 and the mid-1970s. *Distribution* Restricted to Pyramid Lake.

PUPFISH: FAMILY CYPRINODONTIDAE

OWENS PUPFISH, Cyprinodon radiosus Its life history and efforts to prevent its extinction make the Owens Pupfish the best known native fish in the Owens Basin. Male pupfish are rarely longer than 2.5 in (6 cm); females are somewhat smaller. Males are bright blue during the spring and summer spawning period; females are olive-gray year-round.

Pupfish feed on invertebrates, consuming whatever creatures are most abundant. They rarely live longer than 2 years, but reproduce during their first year. During the spawning season males are extremely active and highly territorial in defense of the microhabitat each one occupies for feeding and spawning. When spawning occurs, a female enters a male's territory and begins a ritualized dance of nipping at the substrate and lying motionless and horizontal, awaiting the male's approach. As the male settles alongside, she quivers and releases one to several eggs, which he fertilizes as they attach to the substrate. A female may deposit several hundred eggs as spawning continues at intervals for several months.

Observations made in the early 1900s documented that the Owens Pupfish was abundant in all shallow and slow-moving waters that bordered the Owens River on the Owens Valley floor. It was also sufficiently plentiful in the past to be a staple food of native peoples. Populations declined with the introduction of sport fish and with water diversions for local farms and the Los Angeles Aqueduct. As diversions dried up the pupfish's shallow habitats, the only habitats remaining were the deep waters and springs inhabited by predatory nonnative fish, where the pupfish did not fare well. It was believed extinct for many years until isolated populations were discovered in 1956. Again in 1964 approximately 300 fish were found in a single, isolated pool in Fish Slough. *Distribution* Endemic to the Owens Basin. Currently persists only in 3 habitats isolated from introduced fishes and managed for its conservation.

SCULPIN: FAMILY COTTIDAE

*PAIUTE SCULPIN, *Cottus beldingi* The most abundant bottom-dwelling fish in the Eastern Sierra, the Paiute Sculpin is small and drably colored, enabling it to hide between rocks on the stream bottom, where it feeds on insects and smaller fish hiding between and underneath the rocks. It is up to 5 in (12.7 cm) long in Lake Tahoe, but is usually less than 4 in (10 cm) long.

The Paiute Sculpin is most active during the night when it preys on aquatic insects and snails, which it captures by ambush. Studies conducted in Lake Tahoe and Eastern Sierra streams found little evidence that it preys on fish. It can reproduce during its second year, spawning during the spring, and rarely lives longer than 5 years. In streams Paiute Sculpin usually spawn in riffles; in lakes, usually in wave-swept areas along the shoreline or close to a stream mouth. Males select and defend mating territories located over rocky bottoms where females lay eggs in rock crevices. Females deposit a single cluster of 100–200 eggs under rocks; males protect the cluster until the eggs hatch. The Paiute Sculpin prefers riffles in cold streams, in habitats typically occupied by trout. In Lake Tahoe it is common over rocky substrates. It also occurs over gravel stream bottoms and in dense aquatic vegetation. *Distribution* Occurs throughout the Lahontan Basin, portions of the Columbia and Colorado drainages, and the Bear River in the Bonneville Basin of Utah and Idaho.

UNCERTAIN FUTURE

Here at the edge of the Great Basin, life is difficult for native fishes. In the lowland valleys, where they spend most of their lives, streams may dwindle to a trickle in fall, water temperatures may vary from near-freezing in winter to over 90°F (32°C) in summer, and desert lakes may be naturally alkaline. At all elevations, food is scarce in winter. Native fishes have evolved to cope with these natural hardships in several ways. Most are relatively small; they produce many eggs each year, and some become sexually mature early in life. The unnatural hardships they have been subjected to by human activities are another matter. We have altered their habitats so greatly that Pyramid and Tahoe cutthroat trout are already extinct. By diverting water to farms and cities, we have dried up streams and lowered lake levels so that some lakes have become highly alkaline. We have introduced species such as Rainbow Trout and Largemouth Bass that have displaced native fishes, preyed on them, or hybridized with them. Dams that block upstream spawning grounds, pollution from sewage plants and old mines, and overgrazing: All of these disturbances make it questionable whether native fishes will survive unless solutions are found in the near future.

RECOMMENDED READING

La Rivers, I. 1994. *Fishes and Fisheries of Nevada*. Reno: University of Nevada Press.
Moyle, P. B. 1993. *Fish: An Enthusiast's Guide*. Berkeley: University of California Press.
Sigler, W. F., and J. W. Sigler. 1987. *Fishes of the Great Basin*. Reno: University of Nevada Press.

CHAPTER EIGHT

AMPHIBIANS

DIANA F. TOMBACK

Amphibians are the only vertebrates that can breathe both underwater and on land. How is this possible? Many species lay their eggs in water; the eggs hatch into larvae with gills that enable them to take oxygen from the water. When the larvae metamorphose into land-dwelling adults, most species develop small lungs; they also have an amazing thin, moist skin through which they "breathe" oxygen. Although most adult amphibians live on land, they return to water at crucial stages in their lives. Some species are fully aquatic, others live most of their lives on land, but nearly all must return to water to reproduce. To regulate their body temperature, they move among microhabitats to suitable temperatures. Many species are immobilized by cold and cannot stand prolonged exposure to heat. The first terrestrial vertebrates, amphibians appeared in the fossil record of the mid-Devonian period, about 375 million years ago. Until the reptiles appeared about 60 million years later, the amphibians were the *only* land vertebrates.

Two groups of amphibians—the salamanders, and frogs and toads—are well represented in the Eastern Sierra Nevada. In fact, several newly found species of lungless salamander are currently being studied. Wetlands and moist areas at all elevations provide breeding sites and feeding places for amphibians. In moist forest understory, Long-toed Salamanders forage silently in the shelter of logs or under leaves, and groups of male Western Toads court females at night.

In recent years scientists have noticed that many amphibian species around the world, including those of the western mountains of the United States, are declining in numbers. Many species in the Sierra Nevada are currently found in only a fraction of their previous ranges. It is not yet known whether these declines represent natural fluctuations in population sizes or whether they are the consequence of environmental problems. The decline of some species may well be caused by wetland destruction and disturbance. For others, factors such as acid rain or increasing intensities of ultraviolet radiation may be harming eggs and larvae in breeding waters. Because amphibians depend on both aquatic and terrestrial habitats, they are a highly vulnerable group whose decline may be warning us of important environmental changes.

EXPLANATION OF TERMS

Amphibians' *body length* is measured from snout to vent. *Breeding season* begins with courtship and mating, and extends to the end of parental duties (which for amphibians means the end of egg laying, since they do not care for their young). If a *subspecies* (ssp.) is listed, the appearance and behavior described refer to that subspecies in the Eastern Sierra. *Distribution,* however, includes the range for the *entire* species, not just the subspecies listed. An asterisk (*) indicates species that occur only east of the Sierra Crest.

SALAMANDERS: ORDER CAUDATA

Of all terrestrial vertebrates, salamanders are the least tolerant of high temperature. They need cool, humid conditions and are able to maintain activity at very low temperatures. To move, they undulate their body, in concert with leg movements. For the Mount Lyell Salamander and its relatives, the skin is the main respiratory organ, whereas the Long-toed Salamander and most others also have small lungs. In order for them to "breathe," the thin, delicate skin of salamanders is moistened by many small mucous glands. Also in the skin are poison glands, which secrete a substance irritating to predators. Salamander species vary in their dependence on water. Some species lay their eggs in moist places on land, and the young hatch into small replicas of adults. Other species lay their eggs in water, and the young hatch into gilled larvae (usually a form without legs) that metamorphose into adult, terrestrial forms. Still other species, both young and adult, are completely aquatic; the adult form is that of a large larva, often with external gills.

Most salamanders are voiceless. Scent glands, used in courtship, provide identity and sex cues. Courtship and mating rituals vary. The Long-toed Salamander male grasps the female from behind and shakes her. Most salamanders have a special kind of internal fertilization, which begins when the male deposits several packets of sperm near the female. In some species the male is brightly colored around the vent. The receptive female focuses her attention on the vent and then nudges the male, which responds by producing sperm packets. The female positions herself over a sperm packet and takes it into her vent, where the sperm fertilizes her eggs before she lays them in jelly-covered packages.

MOLE SALAMANDERS: FAMILY AMBYSTOMATIDAE

LONG-TOED SALAMANDER, *Ambystoma macrodactylum;* Southern ssp., *A. m. sigillatum* Length: 2.1–3.2 in (5.3–8.1 cm). Dark brown or black above, irregular yellow markings down center of the back from head to tail. Requires rotting logs and stumps near water for shelter. Eats insects (including crickets, grasshoppers, and caterpillars), snails, slugs, worms, centipedes, and spiders. Breeds from May to July in ponds, lakes, and streams after snow and ice melt; female lays 85–350 eggs. Eggs hatch into free-swimming larvae that become adults in first or second year. The Southern subspecies occurs in the northern portion of both Sierra slopes; on the east slope from Alpine Co. north, near water in Sagebrush Scrub, Montane and Subalpine forest, and Alpine Rock and Meadow Communities. *Distribution* Ranges from northern British Columbia south throughout the Pacific Northwest and northeastern California, and east through western Montana and northern Idaho.

*TIGER SALAMANDER, *Ambystoma tigrinum* Occurs in the Eastern Sierra only in Owens Valley. The origin of the population is unknown; it may be from escape or release of salamanders intended for fish bait.

LUNGLESS SALAMANDERS: FAMILY PLETHODONTIDAE

MOUNT LYELL SALAMANDER, *Hydromantes platycephalus* Length: 1.8–2.8 in (4.6–7.1 cm). Fig. 8.1c. Brown-gray above with dark speckling, resembling granite. Flattened body and partly webbed toes. Occurs on granite slabs in fissures, on cliffs, and in cave openings, in areas moistened by snowmelt or spray from creeks or waterfalls. Can squeeze into crevices; climbs well, aided by webbed toes and tail. Diet includes a variety of insects and spiders. Female lays 6–11 eggs in moist places in fall; young hatch in adult form the following spring or summer. *Distribution* Limited to scattered locations in the central Sierra; on the east slope from central Inyo Co. through Eldorado Co. at the higher elevations.

OWENS VALLEY WEB-TOED SALAMANDER, *Hydromantes* species A little-known species that occurs at high elevations on the eastern Sierra slope adjacent to Owens Valley, Inyo Co.

KERN PLATEAU SLENDER SALAMANDER, *Batrachoseps* species The range of the recently discovered Kern Plateau Slender Salamander is restricted to the higher elevations of southern Inyo and northern Kern counties on the east slope and adjacent areas of Tulare Co. on the west slope. Like other *Batrachoseps* salamanders, it has a thin body, long tail, and small limbs.

TOADS AND FROGS: ORDER ANURA

Although toads and frogs have the same general shape, frogs are slimmer and have long hindlegs, webbed feet, and fairly smooth skin. Poison glands that secrete substances repellent to predators are located along the sides of the body. In contrast, toads are more robust, with warty skin and a large parotid gland behind each eye. The parotid glands and warts secrete a milky toxin that, if ingested by predators, causes muscular pain and paralysis. The eyes of the bufonid or "true" toads have horizontal pupils, which are evident in strong light, whereas the pupils of frogs are more oval or round. Toads eat insects, sowbugs, spiders, snails, and crayfish. Frogs have a similar diet, although the larger species also eat snails, small fish, other amphibians, small snakes, and even birds and mice.

The skin of toads and frogs, like that of salamanders, serves as a supplementary breathing device. Gases are exchanged across the thin, moist skin, which contains many blood vessels. Mucous glands lubricate the skin, aiding breathing as well as helping these amphibians slip out of predators' grasp. Most male toads and frogs have vocal sacs that serve as resonating chambers. For some toads and frogs, a single vocal sac swells the throat to huge proportions. The Northern Leopard Frog has a pair of vocal sacs, one swelling on each side of the head, while the entire throat of the Bullfrog enlarges. In contrast, the Western Toad has no vocal sac at all. Each species has a different courtship call, which enables females to find males of their own species.

During the breeding season, at aquatic breeding places, females move toward groups of calling males. A male, responding automatically to the presence of a female swollen with eggs, clasps her from behind with his front legs. Clasping may continue for several hours or days. Fertilization of the eggs is external, occurring as the mating pair shed eggs and sperm simultaneously into the water. Frogs lay eggs in jelly-covered masses; a single female may lay as many as 20,000 eggs. Female toads usually lay eggs in long, jelly-covered strings containing more than 16,000 eggs, although some species lay eggs in short strings or in small masses.

Related species of frogs and toads that inhabit the same area occasionally hybridize. This should not be surprising, since the mating behavior of many species is automatic and similar. But for such mis-pairings, the breeding effort is wasted: Eggs and young rarely develop because their genes are not compatible. However, differences in breeding behavior among close relatives—

Fig. 8.1.   Amphibians. (a) and (b) Yosemite Toad, *Bufo canorus:* (a) female; (b) male.
(c) Mount Lyell Salamander, *Hydromantes platycephalus.* (d) Pacific Treefrog, *Hyla regilla.* (e) Mountain Yellow-legged Frog, *Rana muscosa*

such as timing of breeding, preferred breeding sites, and male courtship call—minimize the chances of hybridizing and reduce breeding errors.

Treefrogs are small climbing frogs with long legs and flattened toepads that secrete sticky mucus. They rapidly undergo major color changes that help regulate their body temperature and also help camouflage them from predators. Changes in color, which may take only ten minutes, are under both nervous and hormonal control. In cold, humid, or dark environments treefrogs darken and absorb more sunlight; in warm, dry environments they lighten their color and absorb less sunlight. These color changes occur as pigment is rearranged within skin cells. The concentration of dark pigment in the center of these cells produces a green color; the dispersion of dark pigment in the cells produces gray-brown.

TOADS: FAMILY BUFONIDAE

WESTERN TOAD, *Bufo boreas;* California ssp., *B. b. halophilus* Length: 2.5–5 in (6.4–12.7 cm). A common toad of the western United States. Gray, brown, or green above with narrow light stripe down back. Warts surrounded by large, black blotches. Breeds in ponds, lakes, and streams from February to July; breeding peak varies with elevation. Males may court in groups of 12–15, calling in chorus with low, chirping notes. The California subspecies is common on both Sierra slopes (except in the range of the Yosemite Toad) in moist habitats, including Eastern Sierra Riparian Woodlands; Alkali, Montane, and Subalpine meadows; and Montane and Subalpine forest up to about 10,000 ft (3048 m). *Distribution* Ranges from southwestern Alaska south through western Canada and the Pacific states, except for southeastern California; east to western Montana and south to western Colorado.

YOSEMITE TOAD, *Bufo canorus* Length: 1.8–3 in (4.6–7.6 cm). Fig. 8.1a and b. A toad of the High Sierra. Male is yellow-green to dark olive above with sparse, dark speckling. Female is larger, gray to light olive above with many large, black blotches. Emerges from hibernation after snow melts, about April; active both day and night until October. Breeds from mid-May until mid-July; lays eggs in shallow ponds, slow streams, and lakeshores. Male courtship call is a rapid trill of 10–20 notes. Occurs on east slope from northern Inyo Co. north through Alpine Co. Common in wet Alpine Rock and Meadow Communities, wet Subalpine Meadows, Subalpine Forest borders, and high-elevation Montane Riparian Woodland, up to more than 11,300 ft (3450 m) elevation. *Distribution* Limited to the Sierra Nevada.

*RED-SPOTTED OR DESERT TOAD, *Bufo punctatus* Length: 1.5–3 in (3.8–7.6 cm). A small toad of desert wetlands. Light gray, brownish, or olive above with orange-tipped warts and small, round parotid glands; white or buff below. Flat body and spade-shaped head, possible adaptations to crevice-dwelling. Primarily nocturnal, except during breeding. Breeds opportunistically from April to September, using temporary pools, streamflows, and springs. Courtship call is a high-pitched trill lasting 4–10 seconds. In the Sierra, only on the east slope from northern Inyo Co. south. Occurs in desert canyons and rocky habitats, High Desert Riparian Woodland, and to some extent in desert shrub communities and grasslands. *Distribution* Ranges from northern Inyo Co. east to Oklahoma and Texas, south through Baja California and Mexico to Hidalgo.

TREEFROGS: FAMILY HYLIDAE

PACIFIC TREEFROG, *Hyla regilla*   Length: 0.8–2 in (2–5.1 cm). Fig. 8.1d. A small, climbing frog often found on the ground or in low vegetation. Assumes different colors, including green, gray, tan, or gray-brown, with short irregular black stripes. Black stripe through eye runs down to shoulder; underparts are light, but male has dark throat. Flat body, long legs, toepads, and only slight toe webbing. Both nocturnal and diurnal; breeds from February to July, using marshes, ponds, springs, streams, and lakes. Courtship call is a loud, continuous *kreck-ek*. Females lay eggs on vegetation in water. On both slopes of the Sierra Nevada, inhabits diverse, moist habitats including Eastern Sierra Riparian Woodlands; Desert Marshlands; Montane Chaparral; Alkali, Montane and Subalpine meadows; and Alpine Rock and Meadow Communities, as high as 13,000 ft (3960 m) in the central Sierra. *Distribution* Ranges from southern British Columbia south through the Pacific Northwest to California and Baja California; east to western Montana and eastern Nevada.

FROGS: FAMILY RANIDAE

MOUNTAIN YELLOW-LEGGED FROG, *Rana muscosa*   Length: 2–3.2 in (5.1–8.1 cm). Fig. 8.1e. The only true frog of the High Sierra. Brown above with dark spots; toe tips dark; belly and undersides of hindlegs yellow or orange. Smells like garlic. Both diurnal and nocturnal. Breeds at higher elevations from May to August after snow and ice melt. Tadpoles may overwinter in water under the ice and change into adults the following spring. Found on both Sierra slopes; on the east slope from about southern Inyo Co. north. Occurs from about 5000 to 13,000 ft (1525–3960 m) elevation in Alpine Rock and Meadow Communities, Subalpine and Montane meadows, Subalpine and Montane forest, and Montane Riparian Woodland—near streambanks, ponds, and lakeshores. *Distribution* Range includes the Sierra Nevada and southern California mountains.

FOOTHILL YELLOW-LEGGED FROG, *Rana boylii*   Length: 1.8–2.8 in (4.6–7.1 cm). Olive, gray, or brown above with dark flecks; pale triangular patch on snout. Belly and undersides of hindlegs yellow. Active year-round during the day. Requires rapid streamflows for breeding; breeds mid-March through May. Courtship call is a series of deep croaks. Occurs on the west slope; on the east slope only in Plumas Co. Prefers sunny streambanks up to about 6000 ft (1830 m) elevation in Montane Riparian Woodland and Montane Forest. *Distribution* Ranges from southwestern Oregon south through northwestern California, and the Coast Range to the Transverse Mountains.

*NORTHERN LEOPARD FROG, *Rana pipiens*   The range of the Northern Leopard Frog is now restricted to small populations in northwestern Inyo Co., and Sierra, Alpine, and Eldorado counties. Inhabits Eastern Sierra Riparian Woodlands and Montane Meadows.

BULLFROG, *Rana catesbeiana*   Length: 3.5–8 in (8.9–20.3 cm). Native to the eastern United States. Large frog introduced to California in 1915, and to other locations west of the Rocky Mountains and around the world. Green, olive, or brown above, often with light green head; male has yellow throat. Dark bands or streaks on legs. Frequents communities with permanent water and cover, including Desert Marshlands and Eastern Sierra Riparian Woodlands in Inyo, southern Mono, and Eldorado counties. Widely distributed on the west slope.

OASES

Even in the Eastern Sierra's arid foothills and valleys, in this land of low humidity and high evaporation rates, you are almost sure to find salamanders or frogs, but you may have to hunt for the tiny oases that sustain them. Although they live part of their lives on land, amphibians must have moist places or ponds to lay their eggs. At every elevation, from Alpine Rock and Meadow Communities to the desert scrub communities, hidden springs, temporary ponds, or slow-moving streams provide the breeding sites and feeding places that amphibians need.

RECOMMENDED READING

Smyth, H. R. 1962. *Amphibians and Their Ways.* New York: Macmillan.
Stebbins, R. C. 1972. *California Amphibians and Reptiles.* Berkeley: University of California Press.
———. 1985. *A Field Guide to Western Reptiles and Amphibians.* 2d ed. New York: Houghton Mifflin.

CHAPTER NINE

REPTILES

DIANA F. TOMBACK

Like amphibians, reptiles cannot generate their own body heat but must depend on their environment to regulate body temperature. Consequently, reptiles and amphibians are limited to the warmer regions of the world; relatively few occur at high elevations or at high latitudes. With the exception of turtles, reptiles of desert and semiarid regions, such as those found in the Eastern Sierra, are adapted to a nearly water-free life.

In contrast to the amphibians, most reptiles have evolved certain behaviors and physical characteristics that free them from moist or aquatic environments for reproduction. When reptiles breed, males inseminate females internally rather than spawn in streams or ponds. Females retain the fertilized eggs in their reproductive tracts for a few days to a few weeks (or, for a few lizards and snakes, until the eggs hatch), and then bury the eggs under vegetation or rocks, or in burrows, sand, or pits. Surrounded by an outer shell that may be hard or leathery in texture, quite unlike a bird's eggshell, the eggs resist dehydration—an important adaptation for a fully terrestrial lifestyle. Because the eggs are hidden, they are less vulnerable to predators or accidents than are the egg masses of amphibians; they also have a higher hatching success. Consequently, reptiles lay fewer eggs per breeding season than amphibians do. Hatchlings resemble small adults; they do not go through an aquatic larval stage, as do many amphibians.

EXPLANATION OF TERMS

Body length of lizards is measured from snout to vent. Turtles' body length is the length of the shell. Snakes' body length is measured from the snout to the tip of the tail. *Breeding season* begins with courtship and mating, and extends to the end of egg laying. Hatchlings resemble small adults and fend for themselves. *Estivation* is a dormant condition (metabolism slows) that some animals enter in summer to escape heat and drought; usually they go underground. If a *subspecies* (ssp.) is listed, the appearance and behavior described refer to that subspecies in the Eastern Sierra. *Distribution,* however, includes the range for the *entire* species, not just the subspecies listed. An asterisk (*) indicates species that occur only east of the Sierra Crest.

TURTLES: ORDER TESTUDINES

Turtles' most notable feature is their shell, which is made of bony plates formed in the skin. If threatened, turtles pull their head and limbs inside the shell. To support the shell, the turtle skeleton has ribs that flare out to the sides. Modern turtles do not have teeth; instead, they have sharp, bony ridges on their jaws. Turtles have evolved varying body and leg shapes that reflect their lifestyles. The completely terrestrial tortoises have thick, sturdy legs and strong claws for digging; their shells tend to be domelike. A very flat African tortoise, known as the Pancake Tortoise, squeezes between rocks to escape predators. Aquatic turtles tend to have flatter, more streamlined forms and webbed feet. Sea turtles' legs have evolved into flippers. All aquatic turtles lay their eggs on land.

TURTLES: FAMILY EMYDIDAE

WESTERN POND TURTLE, *Clemmys marmorata;* Northwestern ssp., *C. m. marmorata* Length: 3.5–7 in (8.9–17.8 cm). The only turtle in the Sierra Nevada. Somewhat flattened upper shell, black, brown, or olive, often with markings radiating from the center of each shell plate. Inhabits vegetated ponds, marshes, streams, rivers, and irrigation ditches. Basks on rocks or logs. Preys on invertebrates, fish, aquatic plants, and carrion. Breeds from March to May; lays 5–11 eggs in moist soil or sand. The Northwestern subspecies ranges widely on the west Sierra slope; on the east slope, in Plumas Co. and east of Lake Tahoe in the Truckee and Carson River drainages. *Distribution* Southwestern British Columbia south in the western portions of the Pacific states to northern Baja California. Also local populations east of the Sierra-Cascade Crest.

LIZARDS AND SNAKES: ORDER SQUAMATA

Reptiles of the order Squamata (meaning "the scaly ones") are common to the dry desert, scrub, and lower montane communities of the Eastern Sierra. These species reduce water loss in such arid environments by having a covering of dry body scales, by excreting uric acid (a solid) instead of liquid urea, and by estivating in their burrows during extremely dry periods. Usually these animals drink little or no water, obtaining most from their food. However, after a rain, they drink large volumes.

Mistakenly called "cold-blooded" animals, reptiles are able to use external heat sources, such as solar radiation and warm surfaces, to achieve and maintain high body temperatures. In contrast, birds and mammals generate their own body heat. Reptiles regulate their body temperatures with changes in behavior. For example, in the morning when lizards emerge from their burrows with low body temperatures, they assume sunbathing postures. To hasten heating, they orient their bodies at right angles to the sun. As the air temperature warms and they begin to overheat, they face toward or away from the sun, minimizing exposed skin surface, and assume an erect posture with legs extended and tail raised to reduce contact with warm soil or rock. At uncomfortably high temperatures lizards pant (thus unloading heat by evaporative cooling) and retreat into the shade or into their burrows. Snakes similarly bask in the sun to warm and later in the day avoid high temperatures. During the colder season, most lizards and snakes hibernate underground, although the Side-blotched Lizard remains active year-round. In the Eastern Sierra, most reptiles are active from April through July at the lower elevations and from May through September at the higher elevations.

As in amphibians, the body color of reptiles serves both to blend the animals with their background and to regulate body temperature. Although the color of a reptile is probably a compromise between these two needs, lizards and some rattlesnakes can lighten or darken their background color. In the morning when body temperature is low, lizards are darker and, consequently, absorb more heat from the sun. As body temperature increases, body color lightens. Lizards may thermoregulate in another way: Different soils or rocks may heat to different temperatures and provide lizards with heating or cooling surfaces. For example, at high elevations in the White Mountains in Inyo County, the Sagebrush Lizard prefers to bask on black-brown sandstone, which heats to higher temperatures than the white dolomite found in the same area.

The social behavior of reptiles is often more complex than that of amphibians. Lizards tend to be territorial—that is, males defend an area to ensure access to food, females, and basking sites. Interactions among males include displaying, chasing, and fighting. Males usually perform territorial displays from prominent places, such as boulders, logs, stumps, rock outcrops, and fence posts. These displays include postural changes to expose brightly colored patches on the throat or sides, "push-ups," and head bobs; they provide clues to species identification, because each species has its own display patterns. This is particularly important when several different lizard species occur together in one community. Male lizard displays also provide cues for species identification by females. A courting male lowers his head and approaches a female. If she is receptive, he grips the skin of her neck in his jaw, straddles her, and mates.

Male snakes engage in ritualized fighting, involving contests of strength. The outcome determines which males are dominant and have access to females. Males detect receptive females by their odor and behavior; in most species, the male approaches from behind the female, crawls along her back, and attempts to wrap his body around hers before copulation. In some species of snakes and lizards, females may defend their eggs from predators; that, typically, is the extent of parental care.

Many of the morphological differences between lizards and snakes, such as elongated body form, rearrangement of organs, and loss of legs (although vestigial rear legs may occur in some species), are adaptations to a burrowing lifestyle. The snake skull, too, is modified. Bones of the skull are loosely jointed, the upper jaw may move independently of the skull, and each half of the lower jaw may move independently. Snakes use their sharp teeth for gripping prey, not for chewing. All of these features permit snakes to maneuver and to swallow whole animals larger than their head.

The senses of hearing, color vision, and smell are well developed in lizards and snakes. However, in snakes the eardrum and eustachian tube are missing. One of the ear bones connects loosely with the lower jaw, so snakes best detect ground vibrations. Both lizards and snakes have *Jacobson's organs* (paired pouches opening into the roof of the mouth, which detect tastes and smells). Reptiles with forked tongues will rapidly flick the tongue in and out of the mouth, inserting the tips into the Jacobson's organs in order to sense odors in the air.

Rattlesnakes have several special characteristics, including the remarkable ability to detect warm objects, such as birds and mammals, from a distance of several feet. The thermo-sensitive organs of rattlesnakes are the facial pits, one on each side of the head between the nostril and the eyes. (Thus, rattlesnakes are also known as pit vipers.) The rattle is formed from a series of jointed, horny rings and makes a hissing sound when the tail is vibrated. The fangs of the rattlesnake are also unusual. They fold back along the upper jaw when it is closed but spring out nearly perpendicular to the jaw when the mouth is opened. Highly toxic venom travels by ducts into a canal in each fang when the rattlesnake bites. When threatened, rattlesnakes coil, assume an S-shape with the front of the body erect, vibrate the tail, and strike, injecting venom with their fangs. Caution: If provoked or surprised, rattlesnakes will readily bite humans. Anyone who is bitten must immediately seek medical attention.

EYELIDDED GECKOS: FAMILY EUBLEPHARIDAE

*WESTERN BANDED GECKO, *Coleonyx variegatus;* Desert Banded ssp., *C. v. variegatus* Length: 2.5–3 in (6.4–7.6 cm). Slender nocturnal lizard with brown cross-stripes and irregular markings above on cream, pale yellow, or pink background; pale underparts. Prominent eyelids and soft skin with small, fine scales. Active April to June. Feeds on insects and spiders. Breeds from April to May; lays only 2 eggs. A species absent from the west slope. The Desert Banded subspecies occurs in the Sierra only on the east slope, from Inyo Co. south. Inhabits rocky areas, particularly canyon bottoms, in Alkali Sink Scrub and Shadscale Scrub. *Distribution* Range includes southern California, southern Utah, Arizona, Baja California, and northwestern Mexico.

COLLARED AND LEOPARD LIZARDS: FAMILY CROTAPHYTIDAE

*GREAT BASIN COLLARED LIZARD, *Crotaphytus bicinctores* Length: 3–4.5 in (7.6–11.4 cm). Fig. 9.1b. Heavy-bodied lizard with wide head and long tail flattened along the sides; 2 broad, black collar bands edged with white. Above, light cross-stripes on torso and dark spots on tail, legs, and face. Male has green, blue, or orange on throat and underparts. Breeding female has orange bars and spots on sides of neck and body. Uses rocks for shelter, sunbathing, and displaying. Active only 4 or 5 months of the year, depending on elevation. Hunts insects, spiders, small snakes, lizards, and small mammals, and eats some vegetable matter. Actively pursues prey, leaping from rock to rock; attains high speeds by running on its hind legs with tail raised. Breeds from June to July, laying 4–9 eggs in sandy soil. Occurs only east of the Sierra Crest, in easternmost Plumas and Sierra counties and from northeastern Mono Co. south. Inhabits rocky areas in Shadscale, Sagebrush, and Alkali Sink scrub, and open Pinyon-Juniper Woodland. *Distribution* Ranges through the Southwest and the Great Basin, east to Illinois, and south to Baja California and northern Mexico.

*LONG-NOSED LEOPARD LIZARD, *Gambelia wislizenii;* Large-spotted ssp., *G. w. wislizenii* Length: 3.5–5 in (8.9–12.7 cm). Large, sturdy lizard with round body and long tail. Dark spots and light crossbars above. Body color can change from light to dark gray. During breeding season, male underparts are pink or rust, and female has orange on each side and under tail. Active 4 or 5 months of the year, depending on elevation. Feeds on large insects such as grasshoppers and beetles, lizards, spiders, flowers, and seeds. Catches prey by ambush or stalking; at high speeds, runs on hind legs. Breeds

Fig. 9.1   Reptiles. (a) Western Fence Lizard, *Sceloporus occidentalis*. (b) Great Basin Collared Lizard, *Crotaphytus bicinctores*

from May to June; lays about 6 eggs. Species occurs only east of the Sierra, in eastern Plumas and Sierra counties and from eastern Mono Co. south. Lives in open, flat areas in Creosote Bush, Shadscale, Alkali Sink, and Sagebrush scrub, and Pinyon-Juniper Woodland.   *Distribution*   Ranges from southeastern Oregon and southern Idaho south through the Great Basin, the Southwest, Baja California, and parts of northern Mexico.

SAND AND SPINY LIZARDS: FAMILY PHRYNOSOMATIDAE

*ZEBRA-TAILED LIZARD, *Callisaurus draconoides;* Common ssp., *C. d. draconoides*   Length: 2.5–3.5 in (6.4–8.9 cm). A slender, gray lizard with dark stripes across the upper tail surface and across the white undersurface. Small black stripes in blue patch on each side; dark throat, sometimes with orange or pink central spot. High-speed runner, aided by long legs and tail; fastest on hind legs (clocked at 18 mph or 29 kph). Diet includes insects, spiders, lizards, and some plant material. Breeds from June to July, laying 2–6 eggs. Species occurs only east of the Sierra Crest, from Mono Co. south. Prefers flat, open areas in Alkali Sink, Sagebrush, Shadscale, and Creosote Bush scrub and in Pinyon-Juniper Woodland.   *Distribution*   Ranges from central Nevada south to southeastern California, Arizona, Baja California, and northwestern Mexico.

*DESERT HORNED LIZARD, *Phrynosoma platyrhinos;* Southern ssp., *P. p. calidiarum*   Length: 2.8–3.8 in (7.1–9.6 cm). Squat lizard with short, pointed tail, rounded body, fringed scales on sides and legs, blunt snout. Frequently called "horned toad" because of the large, pointed spines projecting from back of head. Body color matches soil, ranging from light to dark; wavy, dark stripes on back, blotches on sides of neck, a light stripe down the back. Skin color can change in a few minutes, providing camouflage. Other anti-predator defenses include squirting blood from the eyes or distending the body by inflating the lungs. A poor runner because of body shape; when frightened, retreats into burrows and other shelters or buries itself in sand. Large stomach for diet primarily of ants; sometimes eats other insects. Usually found near ant nest. Breeds in May and June; lays about 8 eggs. Absent from the west slope. The Southern subspecies occurs in the Eastern Sierra from northeastern Mono Co. south. Inhabits sandy and gravelly areas in all High Desert scrub communities and in open Pinyon-Juniper Woodland. *Distribution* Ranges from southeastern Oregon and southern Idaho south through the Great Basin, southeastern California, western Arizona, northeastern Baja California, and northwestern Sonora.

WESTERN FENCE LIZARD, *Sceloporus occidentalis;* Great Basin ssp., *S. o. longipes* Length: 2.5–3.5 in (6.4–8.9 cm). Fig. 9.1a. A common lizard of many habitats. Brown, dusky, or gray above with dark blotches down the back; belly blue with yellow or orange on back of limbs. Male has blue throat patch and sometimes blue on upper parts; female's blue markings faded or absent. *Sceloporus* lizards appear "spiny" because of long, pointed scales. Diet includes insects, spiders, scorpions, and centipedes, and leaves and buds. Breeds from April to June, laying 5–15 eggs. Occurs on both Sierra slopes. The Great Basin subspecies inhabits the eastern slope in rocky and wooded areas in montane and High Desert communities, including chaparral, riparian woodlands, forests, and desert scrub communities, up to 10,000 ft (3048 m) elevation. Often around old buildings, in woodpiles, on fence posts, rocks, and trees. *Distribution* Range extends from northern Washington south (except southeastern California) to northern Baja California, and from southern Idaho south through the Great Basin to southern Nevada.

SAGEBRUSH LIZARD, *Sceloporus graciosus;* Northern ssp., *S. g. graciosus* Length: 2–2.5 in (5.1–6.4 cm). Similar to Western Fence Lizard but less spiny; has black bar on each shoulder, rusty "armpits" and sides, and blue belly patches. Male also has blue throat, and female has orange or red on neck and sides. Prefers sites near brush, rocks, or logs for shelter. Eats insects and other arthropods. Breeds from May to July, laying 2–7 eggs. The Northern subspecies occurs on the eastern slope; a different subspecies occurs on the west slope. Inhabits Montane Chaparral, Montane and Subalpine forest, Pinyon-Juniper Woodland, Sagebrush and other desert scrub communities, usually above 3000 ft (915 m). *Distribution* Ranges through montane regions from north-central Washington east to Montana, south through California and Baja California, and east to New Mexico and western Texas.

*DESERT SPINY LIZARD, *Sceloporus magister;* Barred ssp., *S. m. transversus* Length: 3.5–5.5 in (8.9–14 cm). Robust lizard with large, prickly scales. Pale yellow or brownish above with dark crossbars or spots. Prominent dark patch on each side of neck, and rusty or yellow sides. Male has blue patch on throat and sides of belly; breeding female has orange head. Forages on the ground or in trees for insects, spiders, other arthropods, lizards, and vegetable matter. Breeds in May and June; lays 3–12 eggs. The Barred subspecies occurs east of the Sierra from northeastern Mono Co. south. Prefers rocky areas in Pinyon-Juniper, High Desert Riparian, and Joshua Tree woodlands, and in all desert scrub communities. *Distribution* Ranges from southern Nevada east across southern Utah; south through southeastern California to western Texas; Baja California and northwestern Mexico.

SIDE-BLOTCHED LIZARD, *Uta stansburiana* Length: 1.5–2.3 in (3.8–5.8 cm). Small, slim, abundant lizard. Brown above with dark blotches; a blue or black blotch behind each armpit. Light-phase males speckled above with blue. Female has brown and white blotches and no blue above. Basks on rocks. Hunts insects, spiders, scorpions, and other arthropods. Breeds from April to June; lays 2–9 eggs. Common prey for birds and other lizards. Found on the west slope; east of the Sierra, from northeastern Mono Co. south. Inhabits Pinyon-Juniper Woodland and all desert scrub communities, usually in sandy, gravelly, or rocky places. *Distribution* Ranges from north-central Washington south through the Great Basin; from southern California east to western Texas and south through Baja California and northwestern Mexico.

SKINKS: FAMILY SCINCIDAE

WESTERN SKINK, *Eumeces skiltonianus;* Skilton ssp. *E. s. skiltonianus* Length: 2.5–3.2 in (6.4–8.1 cm). Slim, smooth, shiny lizard with small head, thick neck, and small legs; vaguely snakelike. Brown stripe edged with black down center of back, with white and dark stripes along the sides. Tail tipped with blue or gray; breeding adult has orange head markings. Found under logs, stones, and leaf litter; runs for cover with snakelike undulations. If the skink is seized by a predator, its tail breaks off and thrashes about, distracting the predator while the skink escapes; a new tail regenerates. Eats insects and spiders. Breeds from June to July, laying 5–10 eggs. Found in the northern Sierra; on the east slope, from Placer Co. north. Prefers rocky areas, ground litter, or herbaceous vegetation in Montane Forest, Montane Chaparral, Pinyon-Juniper and High Desert Riparian woodlands. *Distribution* Ranges through the Northwest east through Idaho and south to Utah and Nevada; northern California and along the coast through Baja California.

WHIPTAILS: FAMILY TEIIDAE

WESTERN WHIPTAIL, *Cnemidophorus tigris;* Great Basin ssp., *C. t. tigris* Length: 3–3.8 in (7.6–9.6 cm). Fig. 9.2b. An active, slim lizard with tail twice the length of body. Gray, yellow, or brownish color above with 4 light stripes down back; dark barring on sides, black markings on hind legs. Throat pink or orange. Forages for insects, spiders, scorpions, or other lizards. Walks with short, halting steps, looking from side to side. Agile runner, raises tail off ground and sprints on hind legs. Species occurs on both slopes. The Great Basin subspecies ranges east of the Sierra in eastern Sierra and Plumas counties, and from central Alpine Co. south. Inhabits dry, sandy, or rocky places in arid communities including Sagebrush, Creosote Bush, Alkali Sink, and Shadscale scrub, and Pinyon-Juniper Woodland. *Distribution* The Great Basin, most of California and Baja California, the Southwest, and northern Mexico.

ALLIGATOR LIZARDS: FAMILY ANGUIDAE

NORTHERN ALLIGATOR LIZARD, *Elgaria coerulea;* Sierra ssp., *E. c. palmeri* Length: 3.5–5.2 in (8.9–13.2 cm). Fig. 9.2a. Large, slim lizard with thick neck, narrow head, short legs, and tail nearly twice as long as the body and head. Olive or blue-tinged above with small, dark blotches or bars; usually dark stripes on belly. Good swimmer and climber; undulates body when moving quickly. When the lizard is seized by a predator, the tail breaks easily and thrashes about as distraction. Diet includes earthworms, insects, other arthropods, spiders, and snails. Breeds from May to June. Eggs are retained in female reproductive tract; females give birth to 2–15 live young during the fall. The Sierra subspecies inhabits both Sierra slopes; on the east slope in cool, moist areas in Montane and Subalpine forest and Montane Riparian Woodland, up to about 10,500 ft (3200 m) elevation. *Distribution* Range includes southern British Columbia, northern Idaho, western Montana, western Oregon; in California, the Coast Range, Sierra Nevada, and Cascades.

SOUTHERN ALLIGATOR LIZARD, *Elgaria multicarinata;* San Diego ssp., *E. m. webbii* Dark cross-stripes on body and tail. An agile climber, it uses its long, flexible tail to grasp branches. The San Diego subspecies occurs on

Fig. 9.2. Reptiles. (a) Northern Alligator Lizard, *Elgaria coerulea*. (b) Western Whiptail, *Cnemidophorus tigris*. (c) Rubber Boa, *Charina bottae*. (d) Common Kingsnake, *Lampropeltis getula*. (e) Common Garter Snake, *Thamnophis sirtalis*. (f) Western Rattlesnake, *Crotalus viridis*

the western slope; on the eastern slope, populations exist only near the towns of Olancha and Independence in Inyo Co., in open Montane Forest and Montane Riparian Woodland.

BOAS: FAMILY BOIDAE

RUBBER BOA, *Charina bottae;* Northern ssp., *C. b. bottae*   Length: 14–30 in (35.6–76.2 cm). Fig. 9.2c. Docile snake of coniferous forest. Small and thick with smooth, shiny scales; brown above, yellow below. No defined neck. Tail shaped like head and moves like a head, which distracts predators. Usually active at dusk or night. Climbs vegetation, swims, and burrows. Preys upon lizards and small mammals; kills by constriction (wrapping prey in coils). Spurs near male vent are remnants of hind limbs and are used to stroke females before mating. Breeds April to June. Young born alive in fall; 1–6 per litter. The Northern subspecies is found on both Sierra slopes. Usually under cover, such as loose soil, rotting logs, rocks, and tree bark in Montane Riparian Woodland or streamside in Subalpine and Montane forest and Montane Meadows. *Distribution*   From southern British Columbia through the Northwest and east to Wyoming; south through the Coast Range and Sierra, east though northern Nevada and the Utah mountains.

COLUBRID SNAKES: FAMILY COLUBRIDAE

RACER, *Coluber constrictor;* Western Yellow-bellied ssp., *C. c. mormon* Length: 22–78 in (55.9–198 cm). Slim snake with wide head, large eyes, and smooth scales; fast-moving. Brown or olive above, pale yellow below. Active by day; hunts insects, small mammals, amphibians, reptiles, and fledgling birds. Prey pinioned by body coils. Breeds from April to July; lays 16–28 eggs under stones, rotting logs, or in loose soil. The Western Yellow-bellied subspecies occurs on both Sierra slopes. Found near logs and rocks or near streams in open areas in Eastern Sierra Riparian Woodlands, Pinyon-Juniper Woodland, and Montane Forest and Meadows, up to about 6700 ft (2040 m), from central Mono Co. north. *Distribution*   Southern British Columbia through much of the United States and south to Guatemala.

COACHWHIP, *Masticophis flagellum;* Red ssp., *M. f. piceus*   Length: 36–102 in (91.4–259 cm). Long, slender snake with large eyes, wide head, and tapered tail. Reddish or pink above, with faint to distinct crossbands on neck. Climbs bushes and trees; hides in rodent burrows. Quick, feisty; defends itself by hissing and striking. Eats mammals, lizards, snakes, birds, eggs, insects, and carrion; pinions prey under body coils. Mates in April or May, laying 4–11 eggs. The Red subspecies inhabits both Sierra slopes; on the east slope, easternmost Sierra Co. and from Mono Co. south. Frequents open areas in Sagebrush, Creosote Bush, Alkali Sink, and Shadscale scrub communities. *Distribution* Across the southern half of the United States; Baja California, and south to central Mexico.

*STRIPED WHIPSNAKE, *Masticophis taeniatus;* Desert ssp., *M. t. taeniatus* Length: 30–72 in (76.2–183 cm). Quick-moving snake with large eyes. Dark brown, black, or gray above; on sides, light stripe divided by a solid or broken black line. Hunts for small mammals, snakes, and lizards. Breeds in June, laying 4–8 eggs. Species absent from west slope. The Desert subspecies occurs east of the Sierra around stream courses or among rock outcrops in Pinyon-Juniper Woodland and High Desert scrub communities, from eastern Nevada Co. north and from northeastern Mono Co. south. *Distribution* Ranges from south-central Washington through the Great Basin, the Southwest, and south to central Mexico.

*WESTERN PATCH-NOSED SNAKE, *Salvadora hexalepis;* Mojave ssp., *S. h. mojavensis* Length: 20–45 in (50.8–114 cm). Slender, fast-moving snake with large, rounded scale on tip of snout. Yellow or beige stripe down back; below, white or tinged with orange. Diet includes lizards, snakes, reptile eggs, and small mammals. Breeds in April and May; lays 4–10 eggs. The Mojave subspecies occurs east of the Sierra in rocky and sandy areas in High Desert scrub communities, from southeastern Mono Co. south. *Distribution* Range includes southern California, much of Nevada, the Southwest, Baja California, and northern Mexico.

GOPHER SNAKE, *Pituophis melanoleucus;* Great Basin ssp., *P. m. deserticola* Length: 36–100 in (91.4–254 cm). Robust snake with rough scales and tapering tail. Yellow or cream above with large, dark dorsal blotches; these are black and joined in the neck region. A dark line runs across the forehead and down to the jaw. When threatened, the Gopher Snake coils, hisses, flattens its head, and vibrates its tail, mimicking a rattlesnake. Hunts rabbits, rodents, birds, eggs, and lizards; kills prey by constriction. A good climber; raids bird nests high in trees. Breeds March through June; lays 3–18 eggs in damp soil. Species occurs on both slopes. The Great Basin subspecies is found on the east slope in diverse communities. All High Desert scrub communities, Eastern Sierra Riparian Woodlands, Pinyon-Juniper Woodland, and open Montane Forest up to about 9000 ft (2745 m). *Distribution* Range includes parts of the Southeast, most of the central and western states from southwestern Canada south through Baja California and northern Mexico.

*GLOSSY SNAKE, *Arizona elegans;* Mojave ssp., *A. e. candida* Length: 27–56 in (68.6–142 cm). Snake with shiny scales and "bleached" appearance. Light-colored; cream, pink, light brown, or gray above, with tan or gray blotches. Hunts at night for small mammals, snakes, and lizards; remains in burrow during day. Breeds from late spring to early summer; lays 4–23 eggs underground. The Mojave subspecies occurs on the east slope only, from Inyo Co. south. Prefers sandy soil and rocky areas in Sagebrush, Alkali Sink, and Shadscale scrub communities. *Distribution* Range includes west-central and southern California, the Southwest north to Nebraska; Baja California and northern Mexico.

CALIFORNIA MOUNTAIN KINGSNAKE, *Lampropeltis zonata;* Sierra ssp., *L. z. multicincta* Length: 20–40 in (50.8–102 cm). Uncommon, colorful snake with shiny scales. Banded with red, black, and white; black head. Typically, black bands border each red and white band, but red bands may be lacking in some individuals. Diet includes small snakes, lizards, small mammals, nestling birds, and eggs. Breeds in April and May; lays 4–8 eggs in loose soil or in rotten logs. The Sierra subspecies occurs on both Sierra slopes, on the east slope from Placer Co. north. Inhabits moist areas in Montane Forest and Montane Riparian Woodland up to 8000 ft (2440 m). *Distribution* Range extends from southwestern Oregon south through the mountains of California to northern Baja California. An isolated population in southern Washington.

COMMON KINGSNAKE, *Lampropeltis getula;* California ssp., *L. g. californiae* Length: 30–82 in (76.2–208 cm). Fig. 9.2d. Widely distributed snake with smooth, glossy scales and alternating, irregular dark and light bands. If threatened, vibrates tail and strikes like rattlesnake. Active in morning or late afternoon, but at night in desert areas. Primarily a snake-eater; also takes lizards, frogs, birds and eggs, and small mammals. Hunts rattlesnakes; immune to their venom. May swallow snakes longer than its own body. Breeds from March to July, laying 6–12 eggs in loose soil. The California subspecies inhabits both Sierra slopes; all High Desert and montane communities up to 7000 ft (2135 m), among rocks, vegetation, or rotting logs. *Distribution* Ranges across the southern United States; north to New Jersey in the east; north to southwestern Oregon in the west; south through northern Mexico.

LONG-NOSED SNAKE, *Rhinocheilus lecontei;* Western ssp., *R. l. lecontei* Length: 20–41 in (50.8–104 cm). Uncommon, slim snake with black saddles separated by pink or red bands along the back; pale underparts. Head barely wider than neck, with long, pointed snout. By day, in burrows or under boards, stones, debris. Active at dusk or at night. Eats lizards, small mammals, eggs, and insects. Breeds from March through May, laying 3–8 eggs in loose, moist soil. The Western subspecies occurs on both Sierra slopes; east of the Sierra from Mono Co. south. Inhabits all desert scrub communities, particularly Shadscale, Alkali Sink, and Sagebrush scrub. *Distribution* Range includes the Great Basin, central and southern California, the Southwest; western Kansas, Oklahoma, Texas, south to northern Mexico.

COMMON GARTER SNAKE, *Thamnophis sirtalis;* Valley ssp., *T. s. fitchi* Length: 18–51 in (45.7–130 cm). Fig. 9.2e. Common snake in moist areas. Slender with large eyes, rough scales, head slightly wider than body, and tapered tail. Brown or gray above with broad, light yellow stripe down back; top of head black. Light stripe low on sides; series of red spots between dorsal and side stripes. Retreats into water if frightened. If taken by a predator, defends itself vigorously, biting and smearing excrement and musk over its captor. Hunts fish, amphibians, small mammals, birds, slugs, leeches, earthworms, and insects. Breeds from mid-March to June; 3–85 young are born alive in summer. The Valley subspecies occurs on both Sierra slopes; on the east slope, from northern Alpine Co. north. Found near water, often in streamside vegetation, in Montane Riparian Woodland, Montane Forest and Meadows, and Pinyon-Juniper Woodland, up to 8000 ft (2440 m) elevation. *Distribution* Wide range includes southern Canada, most of the United States except the Great Basin and the Southwest, and small areas in New Mexico and Chihuahua, Mexico.

WESTERN TERRESTRIAL GARTER SNAKE, *Thamnophis elegans;* Mountain ssp., *T. e. elegans;* Wandering ssp., *T. e. vagrans* Length: 18–42 in (45.7–107 cm). Similar to Common Garter Snake. Mountain subspecies black or dark brown above with yellow or orange-yellow dorsal stripe and yellow or cream lateral stripe. Wandering subspecies has dull yellow or brown dorsal stripe fading near the tail and light ground color with dark spots between stripes. No red markings on either subspecies. Hunts fish, amphibians, reptiles, small mammals, birds, slugs, leeches, and earthworms. Breeds from April to June; 10–30 young born alive during the summer. The Mountain subspecies inhabits both Sierra slopes; the east slope from mid–Inyo Co. north. The Wandering subspecies occurs in the southern part of the east slope. Prefers moist habitats but found in most communities, particularly Eastern Sierra Riparian Woodlands and meadow communities, up to 10,000 ft (3048 m). *Distribution* Range extends from southwestern Canada south throughout most of the West and the Southwest and east to South Dakota; absent from southern California, southern Nevada, and western and southern Arizona.

SIERRA GARTER SNAKE, *Thamnophis couchii*   Length: 18–57 in (45.7– 145 cm). Narrow, faint dorsal stripe restricted to front part of body. Pale gray above with small dark blotches in checkered arrangement; rough scales above. Hunts frogs, tadpoles, fish, salamanders, earthworms, and mice. Breeds from mid-March to early June; bears 10–30 live young. Both Sierra slopes; on the east slope, from Mono Co. north. Highly aquatic snake; lives near ponds, lakes, streams, rivers, and marshes in most Eastern Sierra montane and woodland communities, particularly wet Montane Meadows and Montane Riparian Woodland, up to about 8000 ft (2440 m). *Distribution* Southwestern Oregon south (except for arid southeast California) to northern Baja California.

*WESTERN SHOVEL-NOSED SNAKE, *Chionactis occipitalis;* Nevada ssp., *C. o. talpina*   Length: 10–17 in (25.4–43.2 cm). Small, burrowing snake with flat, shovel-like snout. White or yellow above with large, dark brown crossbands; sometimes with dark markings in between bands. Burrows by tipping head down at 45° angle and moving head from side to side; smooth scales reduce resistance. Nocturnal. Feeds on insects, larvae, scorpions, and spiders. Breeds in spring, laying 2–4 eggs underground. In the Eastern Sierra the Nevada subspecies occurs in much of southern Inyo Co. Prefers open, sandy areas in Alkali Sink and Shadscale scrub. *Distribution* Range includes the deserts of southern California, western Nevada, western and southern Arizona, and northeastern Baja California.

SOUTHWESTERN BLACK-HEADED SNAKE, *Tantilla hobartsmithi*   Length: 5.5–15 in (14–38.1 cm). Small, rare black-capped snake with smooth scales and small head; sometimes with white collar behind cap. Pale brown to gray above, and wide orange or red stripe along belly. Nocturnal; eats insects, larvae, spiders, centipedes, and millipedes. Breeds from May to June, laying 1–2 eggs. Occurs on both Sierra slopes; east of the Sierra, throughout Inyo Co. Found under stones, in crevices, and in burrows, primarily in High Desert Riparian and Pinyon-Juniper woodland, and in Alkali Sink and Shadscale scrub communities. *Distribution* Range includes southern half of California to northern Baja California; scattered populations east to Colorado, southwestern Texas and south through northern Mexico.

NIGHT SNAKE, *Hypsiglena torquata*   Length: 12–26 in (30.5–66 cm). An uncommon snake that kills prey with venom. Beige or gray above with brown or dark gray spots along length; large spots on the back and smaller spots along the side. Dark stripe through eye and large, dark blotches on neck. Active at night. Preys upon salamanders, small toads, small rodents, and insects. Injects prey with venom using rear teeth. Both Sierra slopes;

on the east slope from southeastern Alpine Co. south. Frequents rocky and sandy areas in all High Desert scrub communities and in Pinyon-Juniper Woodland. *Distribution* Wide range extends from southern Washington south through the Great Basin, central California to Baja California; the Southwest to western Oklahoma and Texas, and south to Costa Rica.

*SPOTTED LEAF-NOSED SNAKE, *Phyllorhynchus decurtatus;* Western ssp., *P. d. perkinsi* A nocturnal, secretive snake restricted to Creosote Bush Scrub. Has an enlarged, patchlike scale on the tip of its snout. Occurs only east of the Sierra, from Owens Lake south.

RATTLESNAKES: FAMILY VIPERIDAE

WESTERN RATTLESNAKE, *Crotalus viridis;* Great Basin ssp., *C. v. lutosus* Length: 15–62 in (38.1–157 cm). Fig. 9.2f. Poisonous, heavy-bodied snake with broad, triangular head, rough scales, and a blunt tail with jointed rattle. Buff or gray-brown above with brown blotches down back and light stripe from eye to corner of jaw. Active day and night. Prey includes lizards, squirrels, rabbits, gophers, mice, and birds; kills with venom injections. At higher elevations groups of snakes may hibernate together in caves. Breeds in spring; 4–21 young born alive. Species occurs on both slopes. The Great Basin subspecies inhabits the east slope from central Mono Co. north. Frequents talus slopes, rock outcrops, ledges, and rock-strewn stream courses in all High Desert scrub communities, Pinyon-Juniper Woodland, and Montane Forest. *Distribution* Range includes southwestern Canada, most of the western United States east to Iowa and Texas, and south through northern Chihuahua.

*SIDEWINDER, *Crotalus cerastes;* Mojave Desert ssp., *C. c. cerastes* Length: 17–31 in (43.2–78.7 cm). Agile, poisonous snake with wide head, blunt tail with jointed rattle, and hornlike scale above each eye. Gray, cream, tan, or pink above. Can change body color to blend with background. Brown blotches on back and speckles on sides. Hornlike scales fold down over eye, like eyelid, for protection in rocky or brushy terrain. Moves by sidewinding, which reduces slippage on sand; crawls sideways with body shaped like an S, leaving trail of J-shaped marks. Nocturnal; hides by day in rodent burrows or lies coiled in shallow depression near shrub. Eats small rodents, lizards, and an occasional bird; kills prey with venom. Breeds May to June and September to October, bearing 6–16 live young. Absent from the west slope. In the Eastern Sierra the Mojave Desert subspecies occurs from southeastern Mono Co. south. Found in open, sandy areas in Shadscale and Alkali Sink scrub communities to about 6000 ft (1830 m). *Distribution* Range includes southeastern California, southern Nevada, southwestern Utah, western Arizona, northeastern Baja California, and western Sonora.

*SPECKLED RATTLESNAKE, *Crotalus mitchellii;* Panamint ssp., *C. m. stephensi*   Length: 24–52 in (61–132 cm). Alert, poisonous snake with characteristic wide head and rattle. Above, dark and light speckling like granite on brown, tan, gray, yellow, cream, or pink body color. Blends well with environment. Active day or night. Hunts mice, ground squirrels, kangaroo rats, lizards, and birds, killing prey with venom. Breeds in spring, giving birth to 6–10 young. In the Eastern Sierra, the Panamint subspecies occurs from southeastern Mono Co. south. Prefers areas of sparse vegetation and rocky sites in Sagebrush, Shadscale, and Alkali Sink scrub communities, Pinyon-Juniper Woodland, and Montane Chaparral.   *Distribution* Southern California, southern Nevada, western Arizona, and Baja California.

A WATER-FREE LIFE

The turtles, lizards, and snakes of desert and semiarid regions, unlike the amphibians, are adapted to a nearly water-free life. Those in the foothills and desert scrub communities obtain most water from their food, although after a rain they may drink large volumes. Females lay eggs with hard or leathery shells that resist dehydration. But like amphibians, reptiles cannot generate their own body heat; both must depend on solar heat and air temperature. Behavioral and physiological traits enable them to fine-tune their body temperature: They bask in the sun to warm; if they become too hot they retreat to the shade or to their burrows. During the colder months, most lizards and snakes hibernate underground. These remarkable traits enable reptiles to survive in the desert as well as in the high mountains at the edge of the Great Basin.

RECOMMENDED READING

Bellairs, A. 1970. *The Life of Reptiles.* New York: Universe Books.
Klauber, L. M. 1982. *Rattlesnakes: Their Habits, Life Histories, and Influence on Mankind.* Berkeley: University of California Press.
Stebbins, R. C. 1972. *California Amphibians and Reptiles.* Berkeley: University of California Press.
———. 1985. *A Field Guide to Western Reptiles and Amphibians.* 2d ed. New York: Houghton Mifflin.

CHAPTER TEN

BIRDS

DIANA F. TOMBACK

Birds manage to survive in almost all terrestrial and aquatic environments, from the Arctic to the Antarctic. In the Eastern Sierra some birds inhabit the highest, cold ridges while others inhabit the lowest, hot, arid valleys; some prefer dense mountain forests, and others prefer sparsely vegetated desert lowlands. So specialized are many of these birds that few could survive in another's habitat.

The various summer feeding behaviors of birds in Subalpine Forest (and adjacent meadows and streamside woodlands) illustrate their incredibly diverse lifestyles. Calliope Hummingbirds forage in flowers for nectar; Belted Kingfishers, perched on branches over creeks, suddenly dive for a fish. Dippers forage for invertebrates while swimming underwater in fast-running creeks; Spotted Sandpipers hunt for insects on lakeshores. Clark's Nutcrackers whack open ripening pinecones for their seeds; Red-breasted Sapsuckers drill shallow holes in trees to feed on sap. White-crowned Sparrows hunt for insects and seeds in willows near creeks. Red-tailed Hawks, while circling above the forest and meadows, watch for ground squirrels and small birds. Each of these species is successful in its particular lifestyle because of specialized traits. The Calliope Hummingbird, as it forages for nectar, rapidly dips its long, brush-tipped tongue in and out of a flower, sweeping up the nectar; special shoulder joints enable it to hover. However, hovering flight and a brushy tongue would not be useful traits for the Dipper, a songbird that feeds on invertebrates under the water of fast-flowing streams. Instead, Dippers have warm, waterproof plumage and a flap over the nostrils to keep water out during underwater forays. In turn, the Dipper's traits are not relevant to the survival of the Red-tailed Hawk, which relies on keen, binocular vision for aerial hunting and on strong, taloned feet for seizing prey.

Although there are about 9000 species of birds worldwide, some with very different evolutionary histories, the body plan of all birds is much the same. Whether it is an ostrich, a duck, or a hummingbird, the basic features of *birdness* are quite apparent: feathers, beak, wings, and similar profiles. Yet within the constraints of this basic "bird plan," evolution has filled numerous ecological niches with bird species. Over many thousands of years, each species has evolved a suite of adaptations for surviving in a unique ecological niche. Among related species, bill shape and size, body size, and leg length adapt each species to a slightly different lifestyle or niche. Other special adaptations that enable birds to survive include torpor (a nightly equivalent of hibernation), diving, elaborate nest construction, tool use, spatial memory, and complex song. Migration twice a year—sometimes just a few miles downslope, sometimes thousands of miles across continents and oceans, to areas where food is more plentiful—is another essential survival mechanism for many species. Adaptations such as torpor and migration underscore the importance of maintaining energy balance—that is, taking in enough calories to meet daily requirements. Because birds have a high metabolic rate and maintain high body temperatures, they must eat enough during the day to provide a reserve for the night. Maintaining energy balance during cold weather is extremely challenging; large numbers of small birds may die during unusually severe or long cold periods. Modified body form, plumage, behavior, physiology, diet, and sensory perception enable birds to pursue the remarkably varied lifestyles described in the species accounts that follow.

EXPLANATION OF TERMS

Measurements in the following species descriptions are based on museum specimens, as reported by others. *Length* (L) is the total body length from bill tip to tail tip; *wingspread* (Ws), reported for some groups, is the distance between the extended wings from tip to tip. Metric measurements for lengths are rounded to the nearest 0.5 cm; wingspread lengths are less precise. Birds that occur only on the eastern slope of the Sierra Nevada (rather than on both slopes) are noted by an asterisk (*). (*Slope,* by our definition, includes the valleys immediately adjacent to the mountain slope; *western slope,* then, includes the Sacramento and San Joaquin valleys.) To *fly-catch* is to dart from a perch, snatch an insect in mid-air, and immediately dart back to the same or a different perch. *Breeding season* begins with courtship and mating, and terminates with the end of parental care.

SPECIES AND DISTRIBUTIONS

By no means do the birds discussed in this chapter constitute an exhaustive list of the species that occur in the Eastern Sierra. We have included primarily those birds that are reliably observed year-round, are summer or winter residents, or are particularly interesting. The distributions of many species are not completely known in the Eastern Sierra and will require additional research; the information presented here is based on a combination of sources.

GREBES: ORDER PODICIPEDIFORMES

Specialized swimmers and divers, grebes are found in lakes and ponds over much of the world. Few birds are more completely adapted to an aquatic lifestyle. Grebes live most of their lives in water—feeding, sleeping, courting, and nest-building. They are superb swimmers, capable of diving to depths of 20 feet (6 m) and remaining underwater for thirty seconds or longer. Foot power—or "lobed" toes, to be exact—propels them through the water. Each toe has flaps that push water during a kick and collapse backward during the forward return. (Webbed feet are an alternative solution to the problem of propulsion through water.) However, grebes are weak fliers. In order to take off, they need a long stretch of water for a start. Despite this, some are long-distance migrants, usually traveling at night in the spring and fall.

Courtship is a strange affair of choreographed dancing through the water and of ritualized nest construction. Nests are built on small, floating islands of vegetation. Both sexes, which look alike, care for the young and ferry them about on their backs. Grebes make an assortment of odd sounds, including trills, squeals, wails, and chuckles. Feather-eating is a peculiar grebe habit with no known function. This order includes one family, the Podicipedidae. The **Common Loon** *(Gavia immer),* member of another highly aquatic order, the Gaviiformes, is an uncommon to rare winter resident on lower-elevation lakes.

GREBES: FAMILY PODICIPEDIDAE

EARED GREBE, *Podiceps nigricollis* L: 12.5–13.5 in (31.5–34 cm). Fig. 10.1g. This small waterbird has a brilliant red eye and a thin bill with a sharp, up-slanted tip. In winter, black above, gray or white below; in summer, a yellow tuft over each ear, brick-red sides and belly. Diet includes insects and aquatic organisms. Breeding may occur from late April to late September. Clutch size, 3–9 eggs; both sexes incubate about 20 days. Nest is a mat of cattails or rushes, floating in shallow water. In fall at Mono Lake up to 750,000 Eared Grebes at one time gather from their northern breeding sites to layer on fat and replace flight feathers before migrating south. Their principal foods are brine shrimp and alkali flies from the lake's saline waters. To prevent ingesting too much salt, they squeeze water out of their mouths with their thick tongues. Also, they excrete excess salt from nasal glands. Many of these grebes winter at the Salton Sea, along the Gulf of California, and on salt ponds in San Francisco Bay, but about 100 may remain at Mono Lake all winter. Occurs on both Sierra slopes; in the Eastern Sierra, most commonly in lakes and ponds of High Desert communities, occasionally at higher elevations. *Distribution* Breeds in the inland western United States and Canada. Winters along the west coast, in the Southwest and northern Mexico.

Other Eastern Sierra grebes include the small **Horned Grebe** *(Podiceps auritus)*, an uncommon winter resident from Mono Co. north; the **Western Grebe** *(Aechmophorus occidentalis)* and **Clark's Grebe** *(A. clarkii)*, two closely related, long-necked species that breed and occasionally winter in Mono and possibly adjacent counties; the **Pied-billed Grebe** *(Podilymbus podiceps)*, a small, heavy-billed year-round resident of the Eastern Sierra.

PELICANS AND CORMORANTS: ORDER PELECANIFORMES

Accomplished swimmers and divers, birds of this order have a unique aquatic adaptation: a foot with four webbed toes. (Ducks and geese have only three webbed toes.) All pelecaniform species also have an unfeathered pouch below the bill, called the gular pouch. The largest gular pouches are found in the pelicans. Avid fish eaters, pelicans scoop up fish, hold them in the bill, and squeeze out the water before swallowing. Contrary to popular notion, they do not use the pouch to carry fish about, like a shopping bag. Instead, they carry fish tucked back into the esophagus. Under heat stress, pelicans use their gular pouch for evaporative cooling by opening the bill and fluttering the enormous gular area. This action moves air across the tissue surface, evaporating moisture and dissipating body heat. Pelicans are gregarious, feeding in flocks and nesting in colonies of a few to hundreds of birds. To and from feeding areas, they fly in groups that typically assume V-formations or orderly lines. In flight, pelicans cannot be confused with any other bird; in addition to their heavy-billed, strangely proportioned profiles, their heads are pulled back with bills resting on necks. Pelicans are very sensitive to human disturbance during breeding and to high pesticide levels in their food, which disrupt egg production. Two families of this order, the Pelecanidae and Phalacrocoracidae, are represented in the Eastern Sierra.

PELICANS: FAMILY PELECANIDAE

AMERICAN WHITE PELICAN, *Pelecanus erythrorhynchos* L: 50–70 in (127–178 cm). Ws: 8–9.5 ft (2.4–2.9 m). Fig. 10.1c. Large, oddly proportioned bird. Big head, giant orange bill, small body, stumpy tail, and short, orange-red legs. All white but for black wingtips. Fishes cooperatively. Several pelicans will swim in an arc or line, driving fish to shallow water where they are easily taken. Eats mostly fish, occasionally amphibians and crayfish. Breeding season is late March to August. Nests vary from bare ground to mounds of earth or sand; 2 eggs per nest; both parents incubate about 30 days. Breeds on islands in saline and freshwater lakes in High Desert communities mostly north and east of the area covered in this book, including Pyramid, Eagle, Honey, and Tule lakes, and Clear Lake Reservoir. Flies between lakes and south through Mono Co. in search of food. Transient in the Central Valley. *Distribution* Breeds in the western United States and Canada. Winters along the Pacific coast from central California to Central America, the coasts of Florida and the Gulf of Mexico to Yucatán.

CORMORANTS: FAMILY PHALACROCORACIDAE

THE DOUBLE-CRESTED CORMORANT, *Phalacrocorax auritus* A dark bird with slender, hooked bill; frequents lakes in High Desert communities from Inyo Co. north and may be observed year-round, although uncommon in some seasons.

HERONS, EGRETS, AND VULTURES: ORDER CICONIIFORMES

Most of the Ciconiiformes live in marshy or shallow-water habitats. These species have long legs, long bills, rounded wings, short tails, and four long toes (three forward), sometimes slightly webbed. An introduced species, the **Cattle Egret (***Bubulcus ibis***),** has been sighted in Mono and Inyo counties. Recently the family Cathartidae, the vultures, was included in the Ciconiiformes.

HERONS AND EGRETS: FAMILY ARDEIDAE

The long-legged herons, egrets, and bitterns search for prey by walking slowly through shallow water or by stirring up the bottom with a foot to flush them out. They grasp prey, usually aquatic vertebrates, with their long, pointed bill and swallow them whole. Pellets of undigested matter are later regurgitated. Their flight—a slow, steady flapping—is distinctive: The legs dangle under the tail at takeoff and landing, the neck curves back to the shoulders in an S-pattern. They are monogamous breeders, and both sexes incubate the eggs and feed chicks regurgitated fish. Bitterns have shorter necks and legs than do the other ardeids. They are masters of camouflage, hidden in marsh vegetation and disguised by their plumage pattern and color. When the bittern freezes in position, clinging to the base of tall reeds with its feet and pointing its bill skyward, it practically disappears from view.

AMERICAN BITTERN, *Botaurus lentiginosus* L: 24–34 in (61–86.5 cm). Ws: 50 in (1.3 m). A solitary heron of marshes. Chunky body with short neck. Streaked brown and buffy above, brown and white below. White throat with

Fig. 10.1.   Birds. (a) Red-tailed Hawk, *Buteo jamaicensis*. (b) American Kestrel, *Falco sparverius*. (c) American White Pelican, *Pelecanus erythrorhynchos*. (d) Turkey Vulture, *Cathartes aura*. (e) Northern Pintail, *Anas acuta*. (f) Canada Goose, *Branta canadensis*. (g) Eared Grebe, *Podiceps nigricollis*

dark whisker stripe on each side of neck; dark-tipped wings. Pointed yellow bill and yellow eyes. Call is a croak. Feeds on aquatic organisms, amphibians, snakes, small mammals, and birds. Nests from April to late August in marshes, on a platform of marsh vegetation often among cattails and tules. Lays 2–7 eggs; female incubates 24–28 days. Populations have declined as a result of habitat destruction. In the Eastern Sierra the American Bittern is an uncommon breeder from Inyo Co. north; it occurs in Desert Marshlands, Alkali Meadows, and wet, lower Montane Meadows and ponds. Uncommon resident in western valleys. *Distribution* Breeds from southeastern Alaska, British Columbia, southern Mackenzie (a district in the western Northwest Territories of Canada), and east to Newfoundland; and south throughout two-thirds of the United States. Winters from southern British Columbia south through the Pacific states and east through the southern United States, north to New York and south through Central America.

GREAT BLUE HERON, *Ardea herodias* L: 42–52 in (106.5–132 cm). Ws: 88 in (2.2 m). Fig. 10.2f. Widespread, common heron, often observed in flight. Gray-blue over most of body, white head with black plume over each eye extending behind head. White and cinnamon on neck with white plumes on breast; yellow-orange bill. Feeds on fish, aquatic invertebrates, and small mammals such as mice, amphibians, and snakes. Herons usually nest in colonies from mid-February to early September in a grove of tall trees near water. Nest consists of a platform of sticks, built higher each year, lined with vegetation. The 2–6 eggs are incubated for 25–29 days. Frequents shallow water in streams, lakes, marshes, and wet meadows in lower Montane and High Desert communities. Winters along the east slope; usually breeds from Nevada Co. north but is sighted south through Inyo Co. Resident along the west slope. *Distribution* Breeds in southern Alaska, the Pacific coast of Canada, southern Canada through the United States and much of Mexico. Winters in southern Alaska, the Pacific coast of Canada and the United States, the southern two-thirds of the United States south through northern South America.

The **Snowy Egret** *(Egretta thula)* is a graceful white bird with head and tail plumes, and bright yellow feet. Breeds in the Eastern Sierra from Sierra Co. north; occurs south through Inyo Co. Resident in western valleys.

BLACK-CROWNED NIGHT-HERON, *Nycticorax nycticorax* L: 23–28 in (58.5–71 cm). Ws: 46 in (1.2 m). A nocturnal heron, with stout body, short neck and legs. Black crown, back, and bill; red eye; white otherwise with 2–3 white hindneck plumes. Calls *quock* in flight. Roosts in dense thickets during the day. Feeds at night; hunts by standing motionless, waiting for prey. Diet includes fish, aquatic invertebrates, amphibians, snakes, small mammals, birds, and some vegetation. Breeds mid-April to mid-September in colonies, nesting in trees, shrubs, or marsh vegetation, usually near water. Nest is a platform of sticks or reeds. Lays 1–7 eggs; incubates 24–26 days. May occur in lower Montane Meadows, Eastern Sierra Riparian Woodlands, Desert Marshlands, and Alkali Meadows. In the Eastern Sierra usually breeds and winters from El Dorado Co. north, less commonly south through Inyo Co.; resident on the west slope. *Distribution* Breeds in south-central and eastern Canada, locally through much of the United States, and Central and South America. Winters locally in the southern two-thirds of the United States and southward through most of its American breeding range. Also occurs on other continents.

VULTURES: FAMILY CATHARTIDAE

Cathartids are adapted to a specialized ecological niche: They are scavengers, eating both fresh and putrid carcasses. This specialization has led to several interesting traits. Their naked legs and heads (and necks in some species) allow them to eat a messy diet with minimal fouling of feathers. The exposed head skin may be black, red, orange, or yellow, varying with the species. The wings of vultures are long and broad, enabling them to soar and glide for hours on *thermals* (warm-air updrafts) in search of dead animals. They locate potential food by their keen eyesight and, in the case of Turkey Vultures, by the ability to smell certain chemicals associated with decaying meat.

Vultures do not usually kill prey. Lacking powerful talons or bills, they are unable to cut into a fresh, large kill. In fact, their feet are not adapted for grasping; the front toes are long and slightly webbed. Without much vocal ability, vultures use croaks, grunts, or hisses to communicate. A curious feature is their method of coping with heat stress. They urinate on their own legs, cooling themselves by evaporative water loss; their legs end up with a white coating of uric acid. **Note:** Recent studies using DNA suggest that the vultures belong in the Ciconiiformes, i.e., they are most closely related to storks.

Although only one species of vulture now occurs on the east slope, a few decades ago the **California Condor** *(Gymnogyps californianus)* ranged across California. We can only imagine the awesome sight of soaring condors, their 9-foot (2.7 m) wingspread casting shadows over the eastern slope.

TURKEY VULTURE, *Cathartes aura*   L: 26–32 in (66–81 cm). Ws: 68–72 in (1.7–1.8 m). Fig. 10.1d. Black or black-brown body plumage; naked head with red skin like that of a turkey; white tip on bill, pale legs. In flight the underwings show a distinctive pattern—light gray to the rear and dark in the front. Often, the long wings are held in a wide, shallow V as the vulture soars. Awkward on the ground, with stumbling take-off. May roost with many other vultures. Eats mostly carrion and roadkills, but may capture small animals and occasionally eat fruit. Preferred nest sites are small caves, rock ledges, stumps, or logs; 1–3 eggs are laid on bare surface. Both sexes share incubation for 38–41 days. They feed nestlings by regurgitating food. The Turkey Vulture breeds along the east slope, April to September; uncommon in winter. Soars alone or in groups over open habitat and forest in High Desert and Montane communities. Year-round resident of the west slope. *Distribution* Breeds from southern Canada throughout most of the United States, and Central and South America. Winters in the southern half of the United States and southward through its range.

GEESE AND DUCKS: ORDER ANSERIFORMES

Ducks, geese, and swans, generally referred to as *waterfowl,* represent the Anatidae, the largest family of the Anseriformes, in North America. They spend their whole life on or near water, which provides them with both food and safety. Their specialized aquatic adaptations include three front toes webbed for swimming and wide, flattened bills with rounded tips and serrated edges. After taking a mouthful of food, anatids press their thick tongue against the palate to squeeze out water through the serrations, which serve as a filter. Other characteristics enabling them to survive a life on water include heavy, well-oiled plumage, and arteries and veins in the legs and feet that run in opposite but parallel directions. Heat then flows from outgoing to incoming blood, conserving body heat on ice and in cold water. Most anatids are social, long-distance flyers, and migratory (north to breeding grounds, south to wintering grounds). In the Eastern Sierra ducks and geese are common inhabitants of lakes and ponds.

Geese are more terrestrial than ducks or swans, often feeding in flocks on grassy fields or cropland. While foraging on land, they take turns watching for predators. Migrating geese are conspicuous, forming huge V-formations or long, snaking lines. Geese tend to mate for life, and, as with other strongly monogamous species, the sexes are similar. Geese require two years to reach sexual maturity; ducks require only one year.

Most ducks fall into two major categories: dabbling ducks and diving ducks. The dabbling ducks are primarily freshwater ducks that feed from the water surface, submerging their head and breast, and tipping up the tail. Diving ducks swim below the water surface, as deep as 20 feet (6 m) or more. They overwinter on the ocean or on saltwater bays and have large nasal glands that excrete salt. Although both dabbling and diving ducks exhibit plumage differences between the sexes, the differences are particularly pronounced in most dabblers. Along with other distinctive markings, male dabblers have a large patch of color on their wings.

Females use plumage colors and patterns as well as courtship behaviors to identify males of the same species. Recognizing their own kind is important for ducks, because pairing occurs on the wintering grounds where many species mingle. The male follows the female to her breeding grounds to nest and leaves her when incubation is under way. Although the female has sole responsibility for raising the young, this task is not as arduous as it sounds, since the young are hatched covered with down and are able to follow their mother off the nest within a day. The young become attached to the first moving object they see, which in the wild is usually their mother, and follow it faithfully. Meanwhile, male ducks shed their bright, breeding plumage and look more like females. This *eclipse plumage* conceals them from predators during the critical time when flight feathers are molted. Ducks lose all their flight feathers at once; swimming is then their only means of escape. Females follow a similar molt sequence after the breeding season is over.

Dabbling ducks—including the Mallard, Gadwall, Pintail, Green-winged Teal, Cinnamon Teal, American Wigeon, and Northern Shoveler—are the most commonly sighted east slope anatids, because they are present both summer and winter. The females are a mottled brown and difficult to tell apart. Diving ducks—Ruddy Duck, Common Merganser, Redhead, Ring-necked Duck, Canvasback, Lesser Scaup, Common Goldeneye, and Bufflehead—are also abundant along the eastern slope, but usually during migration and in winter. The Ruddy Duck and Common Merganser are exceptions and breed in the Eastern Sierra.

GEESE AND DUCKS: FAMILY ANATIDAE

CANADA GOOSE, *Branta canadensis* L: 26–37 in (66–94 cm). Fig. 10.1f. Most widely distributed of all North America geese, the familiar Canada Goose has a distinctive long, black neck, black head and bill, and bright white cheek patches. Body brown-gray, darker above and pale on breast and belly; tail black and undertail white; legs and feet dark. Voice a honking *a-lank* or *ah-honk;* calls constantly while in flight. Of the 10 or 11 subspecies of the Canada Goose, several may occur in the Eastern Sierra, including the **Lesser Canada Goose (*B. c. parvipes*), Western Canada Goose (*B. c. moffitti*),** and **Cackling Canada Goose (*B. c. minima*).** Eats tender plant material, grasses, crops, grains. Forages in shallow marshes, around lakes, in meadows and fields. Breeds from late February to mid-June. Typical nesting sites are on elevated ground; the nest is a scrape lined with grass, leaves, and down. Lays

2–12 eggs; incubation lasts 25–30 days. Young remain with parents until next breeding season. Occurs in the Eastern Sierra in lakes or nearby fields in Montane and High Desert communities; breeds from Mono Co. north. Winters along both Sierra slopes. *Distribution* Breeds in Alaska, Canada, and south to central California, northern Utah; east to southern Kansas, the upper Midwest, to Tennessee and Kentucky. Winters in southern Alaska, British Columbia, and southern Canada south to northern Baja California, northern Mexico, the Gulf Coast, and northern Florida.

GREEN-WINGED TEAL, *Anas crecca* L: 13–16 in (33–40.5 cm). Typical North American subspecies is *A. c. carolinensis.* Male has red-brown head with green patch, dark belly, buffy breast and underparts, gray back and sides with white, vertical stripe in front of wing; voice is a high-pitched whistle. Female is mottled brown; quacks softly in descending notes. Both sexes have a metallic green wing patch. Forages for seeds of aquatic plants and aquatic invertebrates by dabbling. Breeds from May to late August. Nest is a bowl lined with vegetation and down, usually under good cover. Clutch size, 6–9; incubation lasts 20–23 days. In the Eastern Sierra most breed from Placer Co. north. The Green-winged Teal frequents Desert Marshlands, lakes, ponds, pools, and shallow streams in High Desert communities. Winters along both Sierra slopes. *Distribution* Breeds in Alaska, Canada, the northern half of the United States. Overwinters in the western, southern, and eastern coastal regions of the United States. Also occurs on other continents.

MALLARD, *Anas platyrhynchos* L: 20–28 in (51–71 cm). An abundant winter duck in California. Male has yellow bill and dark green head with narrow, white collar. Brown breast, dark gray on back, pale below; white tail with central black, upcurled tail feathers; orange legs. Male voice is *reb-reb.* Female is mottled brown with orange-and-black bill, orange feet, dark cap on head, and dark stripe through eye. Voice is familiar *quack-quack.* Both sexes have a metallic blue wing patch with white border. Consumes tender shoots, grains, seeds, and invertebrates. Feeds in shallow water, on lake surfaces, and in fields. Breeds from early April to late August. Builds nest, well-formed bowl lined with down, in tall vegetation. Rarely, nests in trees, using the abandoned nests of other species. Lays 5–14 eggs, incubates for 26–30 days. Breeds and winters along both Sierra slopes. In the Eastern Sierra, frequents Desert Marshlands, ponds, lakes, flooded fields, and agricultural areas of the High Desert. *Distribution* Breeds in Alaska, most of Canada, and south to the Southwest and Oklahoma, east to Virginia, and south through the highlands of Mexico to Puebla. Winters in southern Alaska, southern Canada, and south to Veracruz. Also occurs on other continents.

NORTHERN PINTAIL, *Anas acuta.* L: 20–29 in (51–73.5 cm). Fig. 10.1e. Graceful, slim; a common winter duck in California. Male has brown head and upper neck, gray body with white patch on flanks, gray tail and dark undertail. White on breast and front of neck terminates in a point on the brown neck. Two middle tail feathers extend well beyond the tail. Voice of male is a weak *gee* or whistle. Female is mottled brown; call is a series of hoarse quacks. Both have a metallic brown wing patch. Forages in surface water, shallow water, and on pond bottoms for invertebrates; rarely in fields for seeds and grain. Breeds from April to September. Nests are scrapes in the ground lined with grasses and down. Lays 3–12 eggs, incubates 22–24 days. In the Eastern Sierra, may be sighted on ponds and lakes in High Desert communities and in Desert Marshlands; breeds from Inyo Co. north. Winters along both Sierra slopes. *Distribution* Breeds in Alaska, Canada, the northern two-thirds of the United States. Winters on the east and west coasts, in the southern United States southward to parts of South America. Also occurs on other continents.

CINNAMON TEAL, *Anas cyanoptera* L: 14–17 in (35.5–43 cm). The handsome male is cinnamon brown over most of body; bill dark; voice a low *chuk-chuk-chuk.* Female is mottled brown; call a soft quacking. Both have a large blue wing patch edged with metallic green. Usually feeds by tipping up or searching bottom sediments for aquatic insects and other invertebrates, and the seeds, stems, and leaves of aquatic plants. Breeds from late April to late July, nesting on dry sites densely covered by vegetation. The nest is a scrape lined with grasses and down; 4–16 eggs, incubation lasts 21–25 days. Breeds throughout the Eastern Sierra in Desert Marshlands, ponds, slow-moving streams, and lake margins in High Desert communities, but does not winter here. Visits the west slope during migration. *Distribution* Breeding range includes southwestern Canada, the western United States, and western Mexico. Winters from the Southwest southward to northern Venezuela and northern Ecuador.

GADWALL, *Anas strepera* L: 19–23 in (48–58.5 cm). A nondescript duck, the male has a plain brown head and neck, dark bill, gray body, black rump, light belly, yellow legs and feet. Male makes quacking calls. Female resembles Mallard female but has white belly and is very vocal: *kaak-kak-kak-kak.* Both sexes have a small white wing patch. Eats vegetation, seeds of aquatic plants, insects, aquatic invertebrates, and fish. Searches for food in shallow water and marsh bottoms, on the water surface, and occasionally on land. Breeds from early June to late September in areas with good cover some distance from water or on islands in lakes. Builds bowl-shaped nest of grass, lined with down. Lays 9–11 eggs; incubates 26–28 days. Occurs year-round in the Eastern Sierra from Inyo Co. north and on the west slope. Prefers Desert Marshlands, and lakes, ponds, and boggy meadows in High Desert communities. *Distribution* Breeds in southwestern Canada and northern United States. Winters in the southern United States and Mexico. Also occurs on other continents.

The **American Wigeon** *(Anas americana)* and the **Northern Shoveler** *(Anas clypeata),* two additional dabbling ducks, are common winter visitors to the Eastern Sierra. Diving ducks that may commonly occur along part or all of the eastern slope during winter or migration include the **Canvasback** *(Aythya valisineria),* **Redhead** *(A. americana),* **Ring-necked Duck** *(A. collaris),* **Lesser Scaup** *(A. affinis),* and **Bufflehead** *(Bucephala albeola).*

COMMON MERGANSER, *Mergus merganser*   L: 21–27 in (53–68.5 cm). Slender diving duck. Male has dark green head; long, thin, red serrated bill; black back and tail; white sides and underparts. Calls are low croaks. Female has a red-brown head with crest on back of head and neck, red bill, light underparts, and a gray back; call is a harsh *karr, karr.* Both sexes have a white wing patch. Forages on fish and fish eggs, amphibians, aquatic invertebrates, and some aquatic plants. Searches for prey from the water surface, then dives forward quickly in pursuit. Will also probe around bottom rocks to flush prey. Breeds from late March to late September. Usually builds nest in tree cavities, on ledges, or in crevices under boulders, lining it with vegetation and down. Lays 6–17 eggs per clutch; incubates 28–32 days. Breeds and winters on both slopes. Breeds on the east slope from Mono Co. north; prefers lakes and rivers in High Desert and Montane communities. *Distribution* Breeds in central Alaska and Canada south into the forested and montane regions of the United States. Winters in southern Canada, the United States, and northern Mexico. Also occurs on other continents.

RUDDY DUCK, *Oxyura jamaicensis*   L: 14–16 in (35.5–40.5 cm). A member of the "stiff-tailed" group. While swimming, often holds the dark tail almost upright; uses it as a rudder when diving. Male is mostly rusty red with black cap, dark wings, light blue bill, and large, white cheek patches. Ruddy ducks are quiet; the male makes *chucking* calls during courtship. Female has blue bill, dark cap, dark brown back, lighter underparts, and white cheek patches with a dark horizontal streak. Feeds mostly on plants; also takes algae, insects, and crustaceans. Either filters food from surface or dives to forage in bottom sediments. Breeds from late April until late August. Builds nest on emergent vegetation over water. Clutch size ranges from 5–15 eggs; incubation takes 23 days. Breeds along the east slope; common species in fall and winter, inhabiting Desert Marshlands. Rarely comes to land. Overwinters on both Sierra slopes. *Distribution* Breeds in western Canada and most of the United States. Winters in parts of the United States southward through Central America.

HAWKS AND FALCONS: ORDER FALCONIFORMES

The eyesight of falconiform species may be the keenest among vertebrates. Unlike those of most birds, their eyes are positioned at the front of the head, giving them a wider range of binocular vision, which produces depth perception. Two sensitive areas on the retina provide visual acuity. Strong muscles adjust the curvature of the lens for precise focusing. This keen eyesight enables them, from great heights, to spot food over a large area. In addition, fish eaters are able to compensate for differences in light refraction between air and water; they aim where a fish actually is and not where it appears to be. Many falconiform species have suffered serious population declines over the last several decades. Causes include habitat destruction, human disturbance, illegal hunting, and eggshell thinning from pesticides accumulated in their animal prey. All falconiform species, usually referred to as *birds of prey,* are protected by federal law.

Two families in this order—Accipitridae and Falconidae—are represented in the Eastern Sierra. Female accipitrids and falcons are larger than the males and dominate them. Size differences are particularly exaggerated in the bird hawks and falcons. Among the explanations proposed for this phenomenon are these: The birds are so fierce and dangerously armed that a larger female can better withstand the aggressive male overtures during courtship; or differently sized sexes may avoid food competition by choosing different prey sizes; or, since the male must feed the incubating and brooding female, if he is small and agile, he can capture more prey.

HAWKS AND EAGLES: FAMILY ACCIPITRIDAE

Some accipitrids are as small as jays, others are nearly as large as eagles. In fact, eagles are actually *very* large hawks. All have a strong, hooked bill. The neck is short, legs and feet are typically yellow, and the feet are strong, with large, curved talons. A bony ridge protects each eye. These hunters have broad, rounded wings that serve for soaring and diving. Most accipitrids search for prey by circling at high altitudes on thermals, watching for mice, rabbits, snakes, and ground squirrels. When prey is sighted, they dive toward the victim and extend their talons to take it. Gripped by strong feet and talons, the prey is subdued and killed.

Courtship and territorial defense in many hawks and eagles are spectacular, involving aerial acrobatics and fights. Accipitrids' pair bond is strong; eagles mate for life. Among the larger broods of eagles, the weakest chick may be killed by a sibling or may starve, losing out in the scramble for food. Why some eagles and other species have so many young that they compete for food and some die is an important question in the fields of animal behavior and evolution. One proposed explanation is that the extra egg(s) or chick(s) is insurance in case an egg fails to hatch or chicks die from other causes.

On the eastern slope, representative accipitrids are the bird hawks—Northern Goshawk and Cooper's Hawk (genus *Accipiter*); the soaring hawks—Red-tailed Hawk and Swainson's Hawk (genus *Buteo*); and the eagles—Bald Eagle and Golden Eagle. The bird hawks soar less than the others; their shorter, rounded wings enable them to manuever in forested areas when hunting birds. The soaring hawks and eagles have longer, broad wings and broad tails.

The **Osprey** *(Pandion haliaetus),* like the Bald Eagle, is primarily a fish eater that frequents lakes. Dark brown above, white below; narrow wings have black patch at "bend." Uncommon to rare. Known for spectacular dives with wings held above back. During summer occurs on the east slope, nesting primarily from Plumas Co. north but also in scattered locations south through Inyo Co. and on the west slope.

BALD EAGLE, *Haliaeetus leucocephalus*   L: 34–43 in (86.5–109 cm). Ws: 72–90 in (1.8–2.3 m). Listed by federal and state governments as an endangered species. Adults of both sexes have white head and tail, and dark brown body. Large yellow bill, yellow eyes. Very sensitive to human disturbance. Eats mostly fish but also squirrels, rabbits, Muskrats, waterfowl, and carrion. From soaring flight or a high perch, swoops to catch prey from the water. Also captures prey on land and waterfowl in flight. Pirates fish from Ospreys (see above). Breeds February to September, nesting on cliffs or in large, live trees, particularly Ponderosa Pine on the east slope. Builds one of the largest nests known, a platform of sticks 7–8 ft (2.1–2.4 m) across and sometimes equally deep. Clutch size 1–3 eggs; incubation period about 35 days. Populations declined because of habitat loss, disturbance, hunting, and pesticides, but numbers are now increasing. May be sighted in winter along the east slope; nests from Plumas Co. north. Found near streams, rivers, and large lakes in Montane and High Desert communities. Often sighted on the west slope in winter and nests in Butte, Plumas, Tehama, and Shasta counties. *Distribution* Breeding range includes most of Alaska and Canada with scattered, local breeding in the United States. Winters in parts of Canada but primarily in the United States.

NORTHERN HARRIER, *Circus cyaneus* L: 16 in (40.5 cm). Ws: 42 in (1.1 m). Formerly known as the Marsh Hawk. A slim hawk with long tail, yellow eyes, owl-like face, and conspicuous white rump patch. Male light gray; female mostly brown above and buffy below, brown barring on wings. Both have banded tails. Hunts by hovering or flying low over fields and meadows; several wingbeats, then a glide. Diet includes mostly mice, rats, frogs, snakes, lizards, insects, small birds, and carrion. Perches on ground, low stump, or post. Breeds from April to September, nesting on the ground in grass or marsh, or in a low shrub. Female builds a platform of sticks and grasses, lays 3–8 eggs, and incubates 28–36 days. Numbers have declined because of habitat destruction and pesticides. Breeds and winters along the east slope, although not common. Preferred habitat includes Desert Marshlands, Alkali Meadows, farmland, Mountain Meadows. May occur at high elevations, particularly during late summer and fall migration. Rare along the west slope. *Distribution* Breeding range includes Alaska, Canada, and parts of the United States. Winter range includes southern Canada south to Colombia and Venezuela. Also occurs on other continents.

The **Sharp-shinned Hawk** *(Accipiter striatus),* the smallest accipiter, occurs on both Sierra slopes. A rare breeder and uncommon migrant and winter resident on the east slope. Numbers have declined in recent decades. Resembles the Cooper's Hawk.

COOPER'S HAWK, *Accipiter cooperii* L: 14–21 in (35.5–53 cm). Ws: 27–36 in (68.5–91 cm). Crow-size, otherwise similar to Northern Goshawk. Blue-gray above, black cap, white underparts with rusty red barring. Long tail has distinct black bands; eye yellow to red. Waits on perch in dense cover and suddenly takes off after prey—small birds, nestlings, reptiles, and amphibians. Also hunts while soaring and gliding. Sometimes kills prey by drowning. Breeding season is late March to August. Nest is a platform of sticks built on a large limb in a conifer or crotch in a hardwood. Clutch size ranges from 1–7 eggs; usually the female incubates, 34–36 days. Breeds and winters throughout both Sierra slopes. In the Eastern Sierra, uncommon in Montane communities; prefers Montane Riparian Woodland for hunting. More common in late summer and fall, when migrants pass through from northern breeding grounds. Numbers have declined in recent decades. *Distribution* Breeds from southern Canada to northern Mexico. Winters in most of the United States and south through Central America.

NORTHERN GOSHAWK, *Accipiter gentilis* L: 19–27 in (48–68.5 cm). Ws: 40–47 in (1–1.2 m). The largest accipiter. Its name is derived from "goosehawk." Male and female adults have blue-gray back, white underparts finely barred with gray, black crown, white stripe over red eye, short round wings, long tail banded with gray-brown. In flight, steady wingbeats alternate with glide. Inhabits deep, shaded forest; often flies low through the trees, maneuvering expertly. Voice a high *kee-a-ah*. Eats rabbits, hares, squirrels, large birds, some carrion, and insects. Carries prey to favorite perches for plucking. Nest is a platform of sticks on a large limb or tree crotch. Lays 2–5 eggs; female incubates 36–38 days. Winters and breeds on both Sierra slopes. On the east slope, breeds from March to September in montane communities. Uncommon to rare. *Distribution* Breeding range includes most of Alaska, Canada, the montane regions of the United States. Winters in the southern part of the breeding range, much of the United States, and northern Mexico. Also occurs on other continents.

SWAINSON'S HAWK, *Buteo swainsoni* L: 18–20 in (45.5–51 cm). Ws: 49 in (1.2 m). Both sexes have dark brown head and upperparts, brown eyes, white throat, and pale belly with red-brown band across breast. Long, broad wings; underneath, white on shoulders and dark on hindwings. Light tail with narrow dark bands; often fan-shaped in flight. Some individuals have dark underparts with light, banded tail. Voice is a shrill *kree-e-e*. Soars or perches while hunting prey: small rodents, birds, reptiles, amphibians, and insects. Sometimes catches flying insects and bats on the wing; flocks of Swainson's Hawks may catch insects swarming in the air or on the ground. Nest is a platform of sticks and twigs in a tree or bush, or on a utility pole or low cliff. Female lays 2–4 eggs; both sexes incubate eggs 34 or 35 days. Hunting and habitat destruction have reduced numbers. In the Eastern Sierra, breeds from Inyo Co. north, April to August, in High Desert Riparian Woodland and open High Desert communities. A common migrant in late summer and fall. On the west slope an uncommon breeder and migrant. *Distribution* Breeding range includes Alaska, western Canada, western United States, and northern Mexico. Migrates in large flocks to the southern part of South America to winter.

RED-TAILED HAWK, *Buteo jamaicensis* L: 19–25 in (48–63.5 cm). Ws: 46–58 in (1.2–1.5 m). Fig. 10.1a. Widely occurring and best known buteo in North America. Most common hawk in the Eastern Sierra. All adults have a bright red-brown upper tail. Both sexes are dark brown above and mostly white below, with a brown belt across the white belly. Long, broad wings; tail fan-shaped in flight. A few individuals have a dark breast, belly, and fore-wings (dark phase). Voice is a hoarse, descending *kree-e-e*. Eats small mammals, particularly rodents, small birds, reptiles, insects, and some carrion. Soars high, searching for food, or watches from a high perch. Breeds from March to September; needs trees or cliffs for nesting. Nest is a large platform of sticks and twigs on a branch near trunk. Clutch contains 2–5 eggs, female incubates 28–35 days. Winters and breeds on both Sierra slopes. Broad habitat requirements: open country interspersed with woods or isolated groups of trees in both Montane and High Desert communities. At higher elevations in summer and fall. *Distribution* Breeds in most of Alaska, and western and southern Canada, south to Central America. Winters in southern Canada south through its breeding range.

Two other large buteos, the **Ferruginous Hawk *(Buteo regalis)*** and **Rough-legged Hawk *(B. lagopus)*,** overwinter in the Eastern Sierra.

GOLDEN EAGLE, *Aquila chrysaetos* L: 33–38 in (84–96.5 cm). Ws: 78–90 in (2–2.3 m). Magnificent bird of prey with long, broad wings. Adults of both sexes are solid dark brown with brown eyes; feathers of head and back golden brown. Fan-shaped tail has dark bands. Legs are feathered almost to the feet. Usually silent. In flight, alternates powerful wingbeats with glide. Diet consists primarily of rabbits and rodents; also large birds, reptiles, amphibians, fish, insects, and carrion. Searches for prey from a tall perch or by soaring. Contrary to claims of ranchers, kills virtually no livestock. Breeds from January to September, building a huge platform nest of sticks and twigs, lined with softer materials. Nest sites are typically very tall trees or cliffs. Pairs for life; lays 1–4 eggs and usually incubates 43–45 days. Year-round resident on both Sierra slopes. May be sighted flying over grassland and shrub habitats in Montane and High Desert communities. Hunts in Alpine Rock and Meadow Communities, Mountain and Alkali meadows, and Sagebrush Scrub. *Distribution* Breeds in Alaska, most of Canada, western United States, small regions in the eastern United States, and northern Mexico. Winters in southern Canada and much of the breeding range south to northern Mexico. Also occurs on other continents.

FALCONS: FAMILY FALCONIDAE

These swift hunters prey on other birds, usually taking them on the wing but sometimes on the ground. The larger falcons make spectacular dives, wings held near the sides, plunging at speeds of 100–275 mph (160–440 kph). Falcons strike feet-first, killing or stunning their victim by the force of impact. They grab small birds with their talons, break the neck of stunned prey after landing, and then carry them to a tree or cliff perch for plucking.

Like the accipitrids, falcons have hooked bills and strong, taloned feet that are very large in proportion to the body. Unlike the bill of accipitrids, the falcon bill is notched near the tip and toothed on each side. The head is typically dark-capped with dark facial markings. Built for speed, the body is trim, with broad, strong shoulders, long pointed wings, and tail narrower at the tip. The sexes are usually similar, but the female is much larger. In falconer's terminology, the female is the *falcon,* and the male is the *tercel.* Most falcons do not build nests. They either lay eggs on bare rock cliffs or use the abandoned nests of other birds, sometimes in trees. The pair bond is strong, enduring until one mate dies.

AMERICAN KESTREL, *Falco sparverius*   L: 9–12 in (23–30.5 cm). Ws: 20–24 in (51–61 cm). Fig. 10.1b. Smallest and most common North American falcon. Both sexes have a bold head pattern with dark crown, black stripes, and white patches between facial stripes; red-brown back and tail, and white underparts with spots or streaks. Male has dark blue wings, female has brown wings. Frequently perches along Hwy 395 on trees, utility poles, and wires, often flicking its tail. May survey an area by hovering. Eats small birds, bats, mice, insects, reptiles, and amphibians. Will hide food for later use. Able to survive in arid regions because all needed water is obtained from prey. Breeds from early April to August, nesting in tree cavities excavated by woodpeckers or in rock crevices. Female incubates 3–7 eggs, 29–31 days. Occurs in diverse habitats—farmlands, Alkali Meadows, Sagebrush Scrub, Mountain Meadows, and Alpine Rock and Meadow Communities. Moves up to alpine elevations in summer. A common year-round resident on both Sierra slopes. *Distribution*   Breeding range extends through much of North, Central, and South America. Winters from southwestern and southeastern Canada south through its breeding range.

The **Merlin** *(Falco columbarius)* is a small gray falcon that is an uncommon winter resident and migrant on both Sierra slopes.

PEREGRINE FALCON, *Falco peregrinus*   L: 15–20 in (38–51 cm). Ws: 43–46 in (1.1–1.2 m). This beautiful falcon is listed as an endangered species by federal and state governments. Sexes similar; dark head markings like football helmet, with flaps down sides of head. Slate-blue upperparts with brown barring, white neck and underparts with the latter barred brown. Tail long, rounded at tip, slate-gray banded with black. A pair roosts together, hunts cooperatively, and has elaborate courtship displays. Mostly hunts birds, as large as waterfowl, and fish. Nests early March to late August, on high cliffs near water; 2–6 eggs. Incubation, mostly by the female, lasts 33–35 days. Populations have suffered drastic declines from habitat loss, disturbance, pesticides, and nest-robbing by falconers. Nest sites need protection. In several regions reintroduction to the wild by captive breeding programs has been successful. Rare breeder and winter resident in the Eastern Sierra from Mono Co. north, at Montane and High Desert elevations near lakes, rivers, and marshes. Rare breeder and migrant on the west slope. *Distribution* Breeding range includes northern Alaska, Canada, parts of western and southern United States, northern Mexico, and western South America. Winter range includes coastal British Columbia, central and southern United States, southward through South America. Occurs on other continents.

PRAIRIE FALCON, *Falco mexicanus* L: 17–20 in (43–51 cm). Ws: 40–42 in (1–1.1 m). Pale, large falcon with light brown upperparts and light, brown-spotted underparts. Sexes similar. Dark cap on head with 2 dark vertical streaks—below eye and on side of head—brown bars on tail, and black patches in "armpits" where wings join the body. Hunts by diving at prey or pursuing winged prey in the air; prey includes birds, mammals, insects, reptiles, amphibians. Breeds March to August. Nests on high ledges, laying 3–6 eggs; female incubates 29–31 days. Relatively common on the east slope. Prefers grasslands: Alkali Meadows and lower elevation Montane Meadows during much of the year; Subalpine Meadows and Alpine Rock and Meadow Communities during summer and early fall. Uncommon on the west slope. *Distribution* Breeding range extends from southwestern Canada through the western United States and northern Mexico. Wintering range is similar, but the limits are farther south.

GROUSE AND QUAIL: ORDER GALLIFORMES

Chickenlike birds, members of the Galliformes are good walkers, do not swim, and—with their short, round wings—can fly only short distances. With small, down-curved bills, these plump birds sift through soil and litter; their legs are sturdy, with four-toed feet and strong claws for scratching for food. They eat vegetation and seeds. In order to extract more nutrients from a diet of plant materials, the Galliformes have well-developed pouches between the small and large intestines. Such pouches tend to be large in herbivorous vertebrates; they act as "holding tanks" for plant material undergoing microbial digestion. Two families, the Phasianidae (grouse) and Odontophoridae (quail), are represented in the Eastern Sierra.

Native North American species of these families are the grouse, ptarmigan, prairie chickens, turkey, and quail. They occur in many habitats: scrub communities, forests, prairies, plains, and tundra. The grouse, prairie chickens, and ptarmigan have several distinctive features, including nostrils hidden by feathers, legs completely or partly feathered, and no hind spurs. Alpine and arctic species of ptarmigan grow feather "snowshoes" on their feet in fall and early winter. Northern species of grouse accomplish the same with a row of tissue projections on the sides of toes. The sexes in the Galliformes tend to differ in appearance, most noticeably during courtship. In many species the males are larger than females and rakishly colored, plumed, or tailed. Male grouse and prairie chickens have brightly colored patches of naked skin above the eye and bare, colorful inflatable areas on the neck that help produce courtship sounds. Male quail have bold, contrasting facial markings.

Although many phasianids are monogamous, the polygynous species (such as the Sage Grouse) are known for their group courtship displays. At the beginning of the breeding season, the males from a local population will gather in traditional clearings in early morning hours. As females appear, the males strut about in courtship posture — crests raised, tails fanned, vocalizing — as they attempt to attract a female. Studies show that only a few males out of many account for most of the matings that occur; their appearance, voice, and displays attract the females. After mating, the females lay eggs, incubate, and raise the chicks on their own. In contrast, all New World quail are monogamous and usually form flocks of family groups or *coveys* in fall.

Introduced Species   Three species have been introduced for sport hunting. The **White-tailed Ptarmigan** *(Lagopus leucurus),* an alpine species, was introduced at Mono Pass by the California Department of Fish and Game. A breeding population has been established, and the ptarmigan is expanding its range. The **Ring-necked Pheasant** *(Phasianus colchicus)* frequents farmlands and rangelands in Inyo, Mono, and Sierra counties. The **Chukar** *(Alectoris chukar)* inhabits arid foothill communities, such as Pinyon-Juniper Woodland and Sagebrush Scrub, from Mono Co. south. The chukar and pheasant are native to Asia, the ptarmigan to Canada, Alaska, and the Rocky Mountains south to northern New Mexico.

GROUSE: FAMILY PHASIANIDAE

BLUE GROUSE, *Dendragapus obscurus*   L: 18–21 in (45.5–53 cm). Male larger than female, mostly black-gray above and below with white along sides and under chin. Yellow patch of skin above eye. Hidden red air sac on each side of neck, surrounded by white patch when inflated. Female mottled brown and white, with more white below. Both sexes have a wide, dark tail with gray band at tip. Breeds from early April to late August. To court, male inflates red air sacs, makes deep hooting sounds with fanned tail held high, and struts about quickly in front of female. After mating, the male migrates to higher elevations. Female builds nest, a depression lined with vegetation, next to a tree or rock, or under branches, and lays 2–10 eggs. She incubates 25–28 days, raises the chicks, and then migrates with young to higher elevations. Winter diet is primarily conifer needles and buds. At other times the Blue Grouse eats berries, fruits, flowers, twigs, leaves, and insects. Prefers Montane and Subalpine forest, particularly areas adjacent to Montane Chaparral, Mountain Meadows, Montane Riparian Woodland, usually near water. Migrates up to dense, coniferous forest in winter, downslope in spring to breed. Year-round resident along both Sierra slopes. *Distribution* Range includes western Canada and the western United States south to Arizona and New Mexico.

*SAGE GROUSE, *Centrocercus urophasianus*   L: 21–30 in (53.5–76 cm). Fig. 10.2a and b. Largest grouse in North America. Male larger than female. Both are a mottled brown that blends with the sagebrush and have a black belly and pointed tail. Male has a black throat, black V on white breast, and a yellow patch above the eye. Breeds mid-February to late August. From mid-February through April, typically 20–70 males gather at traditional strutting grounds—open areas surrounded by cover—to court females. Displaying males strut with neck and breast feathers fluffed and tail feathers fanned. They inflate and deflate yellow air sacs on their breast, making loud "pops." The nest is a scrape lined with plant material, usually under sagebrush. Females lay 7–15 eggs and incubate 25–27 days. The Sage Grouse eats the leaves and shoots of sagebrush in winter and forbs, flowers, and some insects in summer. Prefers Sagebrush Scrub. Occurs only in the Eastern Sierra, from northern Inyo Co. north. *Distribution* Range includes southern Alberta and Saskatchewan south to northern New Mexico.

QUAIL: FAMILY ODONTOPHORIDAE

CALIFORNIA QUAIL, *Callipepla californica*   L: 9.5–11 in (24–28 cm). Common, small, plump birds with familiar *Chi-ca-go* call. Both sexes brown on back and wings, with forward-tilted black head plume. White scaled pattern on belly. Male has blue-gray on breast, black patch bordered with white under chin, and white stripe over eye and across forehead. In fall, family groups form coveys of 10–200 birds that roost together. Breeds from early April to mid-September. Male mates with a single female, but new pairs form each year. Nest is a depression lined with vegetation, under cover of shrubs, grass, or other low vegetation. Female lays 6–16 eggs and usually incubates, 18–23 days. Both parents attend the chicks. Eats a mixture of seeds, green vegeta-

Fig. 10.2.   Birds. (a) and (b) Sage Grouse, *Centrocercus urophasianus:* (a) male; (b) female. (c) Wilson's Phalarope, *Phalaropus tricolor,* female. (d) Spotted Sandpiper, *Actitis macularia.* (e) Killdeer, *Charadrius vociferus.* (f) Great Blue Heron, *Ardea herodias.* (g) California Gull, *Larus californicus*

tion, and some insects. Preferred habitat—High Desert Riparian Woodland, Montane Chaparral, and Sagebrush Scrub—includes shrubs or trees for cover, and water. Year-round resident on both Sierra slopes. *Distribution* Ranges from southern British Columbia south through the Pacific states to Baja California.

MOUNTAIN QUAIL, *Oreortyx pictus* L: 10.5–11.5 in (26.5–29 cm). Largest North American quail; secretive. Has long, straight, black head plume; blue-gray on neck, breast, and upper back; brown on lower back and wings. Chestnut patch bordered with white under throat and on belly and sides; belly patch has series of white streaks on each side. Sexes similar but female duller with shorter plume. Typical call is a chickenlike *took-took-took;* during breeding season, males loudly call *quee-ark* or *kyork.* Diet includes a variety of plant materials—foliage, buds, flowers, seeds—as well as insects. Scratches and digs for food. Breeds late March to late August. Nest is a scrape lined with grasses and pine needles under shrub, rock, or log, and near water. Monogamous; lays 6–14 eggs; female, or sometimes male, incubates 21–25 days. Family groups form coveys of 3–20 birds after breeding. Requires water and shrub understory—as in Eastern Sierra Riparian Woodlands, Montane Forest mixed with Montane Chaparral or Sagebrush Scrub, or scrub habitats alone. Migrates by walking downslope in fall to avoid snow. Year-round resident on both slopes. *Distribution* Range includes the Pacific states, from southwestern British Columbia south to northern Baja California.

RAILS AND COOTS: ORDER GRUIFORMES

The Gruiformes include a curious collection of birds from different parts of the world—sun-grebes, cranes, bustards, rails, and button-quail—all loosely considered marsh birds. They are close relatives of the shorebirds (Charadriiformes, see below). Several species belonging to the Raillidae and one species of the crane family Gruidae occur in the Eastern Sierra.

RAILS AND COOTS: FAMILY RALLIDAE

The North American members of this family include the coots, gallinules, and rails. Most rallid species are slender and chickenlike with rounded wings, short tails, long, strong legs, and long toes. Bills may be short or long and pointed. Plumage is dark or brown, blending with surrounding vegetation. Males and females are similar and monogamous. Although rails are weak flyers, they may migrate long distances.

The coots and gallinules, the most chickenlike, have earned the nickname of mud hens or marsh hens. Their distinctive features include small, brightly colored bills, small forehead plates above the bill, and red eyes. Coots occupy an ecological niche similar to that of ducks; they swim and dive, aided by lobed toes on their feet like those of grebes. They often form large flocks. Gallinules, in contrast, have long, unlobed toes that enable them to walk on floating aquatic vegetation, and they are less social. Rails are secretive birds that dwell in dense marsh vegetation. Their narrow bodies enable them to move quickly through tangles of stems; they climb by using the claw at the bend of each wing as a grapple. Long toes prevent them from sinking into mud. Striped or streaked, their plumage camouflages them in the marsh vegetation. Rails rarely venture out of cover; however, their distinctive calls signal their presence.

The **Virginia Rail** *(Rallus limicola)* occurs only in marshes with dense vegetation. A secretive species, rarely observed. Calls vary; most common is a descending series of little grunts, given in duet by members of a pair. Breeds on both Sierra slopes.

The **Sora** *(Porzana carolina)* prefers Desert Marshlands or shallow areas of ponds and lakes with marsh vegetation. Like most rails, it is rarely observed. Call is *er-we,* rising in inflection, or *keek.* Breeds in west slope valleys and on the east slope from Inyo Co. north.

AMERICAN COOT, *Fulica americana*   L: 13–16 in (33–40 cm). Ws: 23–28 in (58.5–71 cm). The common "mud hen" of lakes and ponds. Rides ducklike on water. Black head and neck, dark gray body, white patch under tail, red eyes, short white bill, green legs and feet. Eats aquatic plants and invertebrates, grass, seeds, and small fish. Forages by tipping up in shallow water or diving in deeper water; may also forage in meadows or in grain fields. Breeds from April to early September. Builds a nest of stems on a platform of marsh vegetation anchored to nearby plants. Also builds display and resting platforms for adults and for young. Clutch contains 5–17 eggs; both sexes incubate 23–24 days. Common year-round resident of both Sierra slopes. Often forms flocks in Desert Marshlands, and in lakes and ponds of High Desert communities; sometimes occurs at higher elevations.   *Distribution* Breeding range includes the Yukon, part of Mackenzie, southern Canada, and the United States south to Costa Rica. Winters from southeastern Alaska and British Columbia through the Pacific states, the Southwest, the Southeast, and through Central America.

A member of the family Gruidae, the **Sandhill Crane** *(Grus canadensis)* breeds in marshy areas in Sierra Co.; occurs in agricultural fields of west slope valleys.

SHOREBIRDS AND GULLS:
ORDER CHARADRIIFORMES

This order includes plovers, sandpipers, avocets, stilts, gulls, and terns. When young of the Charadriiformes hatch, they are covered with down. Shorebird young are capable of searching for food within hours. Species that occur in the eastern Sierra are in the families Charadriidae, Scolopacidae, Recurvirostridae, and Laridae.

PLOVERS: FAMILY CHARADRIIDAE

Plovers are small to large shorebirds with thick necks and rounded heads, short, straight bills slightly enlarged near the tip, and long, tapered wings. While foraging, they run a few steps and pause. They fly swiftly; some species migrate long distances, traveling in flocks. Most plovers are disruptively colored—dark back, white neck, throat, and belly, with one or more black breast bands—which may camouflage them as they forage, much as the stripes of a zebra break up its body outline from a distance. In flight, the upper wings and tail are also boldly patterned with dark-and-white stripes along the trailing edges; this pattern may facilitate flock movements. Monogamous, the sexes are similar in size and color pattern. Plovers nest on the ground, some species in the open but others under grassy cover. The nest is usually little more than a slight depression lined with vegetation and shells or rocks. Both parents use the broken-wing display to distract a potential predator—flying toward the intruder, landing, dragging one wing, and walking away from the nest as if injured and unable to fly. Distraction displays, which occur in other groups of birds as well, are associated with ground-nesting.

SNOWY PLOVER, *Charadrius alexandrinus*   L: 6–7 in (15–18 cm). Ws: 13.5 in (34 cm). Small, pale plover. In summer, dark mark on front of crown, behind the eye, and near shoulder; black bill and dark legs. Forms flocks except during breeding season. Eats insects and aquatic invertebrates, taken from sandy areas, salt flats, and shorelines. Nests from early April to late August in colonies at Owens, Mono, and Crowley lakes, and in smaller numbers at other lakes and reservoirs. Nests are merely scrapes on open ground and contain 2 or 4 eggs. Both sexes share incubation, 25–32 days. Eggs are vulnerable to overheating; parents wet breast feathers and cool eggs by wetting them. Adults cool off by standing in water. Sensitive to disturbance by people. Breeds along both Sierra slopes; in the Eastern Sierra breeds and forages around brackish or saline lakes in High Desert communities.   *Distribution*   Breeding range includes the western United States, Midwest, Gulf

Coast, and the Pacific Coast to Peru and Chile. Winters along the Gulf Coast and Pacific Coast south to Chile. Also occurs on other continents.

The **Semipalmated Plover** *(Charadrius semipalmatus)* is a locally common migrant on both Sierra slopes, stopping at mudflats and lakeshores.

KILLDEER, *Charadrius vociferus*   L: 9–11 in (23–28 cm). Ws: 19–21 in (48.5–53.5 cm). Fig. 10.2e. During the breeding season, the most widely distributed shorebird on both Sierra slopes; less common in winter. Often occurs far from water, usually solitary. Gray-brown back and top of head, black bill, black ring from bill around back of head, two black breast bands. Rump and lower tail bright rufous; legs flesh-colored or yellow. Loud, piercing call in flight, with repeated *kill-deer* or just *deer*. Takes mostly insects, also aquatic invertebrates, worms, and some seeds. Breeds from mid-March to mid-August, nesting in open areas near water. Lays 3–5 eggs, and both sexes share incubation for 24–28 days. Adults protect eggs and young by flying into the faces of approaching livestock. In the Eastern Sierra frequents Mountain Meadows, Alkali Meadows, and the shores of ponds and lakes. *Distribution*   Breeds in much of Canada, the United States, Baja California, Central Mexico, and western South America to northern Chile. Winters from southern British Columbia south along the Pacific Coast; the east coast, parts of the central and southern United States, to northern South America.

AVOCETS AND STILTS: FAMILY RECURVIROSTRIDAE

Avocets and stilts are slender, wading shorebirds with long legs and a long, slender bill—up-curved near the tip in avocets, as the family name suggests. They use the thin bill to probe mud, explore beneath shallow water, or pick up insects from the mud surface. Their wings are long and pointed for strong, swift flight, and their feet are partially to completely webbed for swimming. The monogamous males and females are similar and contribute equally to raising young. To keep the young warm and dry, the parents *brood* (cover) them frequently. This parental behavior is so strong that young birds have been observed trying to brood their siblings!

BLACK-NECKED STILT, *Himantopus mexicanus*   L: 13.5–15.5 in (34.5–39.5 cm). Smaller and slenderer than the American Avocet. Black above and white below, with pink legs, red eyes, and a very thin, pointed bill that is straight or very slightly up-curved. Calls *ip-ip-ip* in flight. Diet includes aquatic invertebrates, insects, snails, small fish, and some seeds. Nests from late April to August; lays 3–5 eggs on scrape in ground. Both parents incubate 22–26 days. Both slopes; in the Eastern Sierra from Inyo Co. south and from Nevada Co. north, in High Desert communities on the shores of alkaline lakes and ponds, and in wet meadows. *Distribution* Breeds along the Atlantic coast from New Jersey south to southern Florida; Oregon, Idaho, and Colorado south through the Southwest; Kansas, the Gulf Coast, and south through South America. Winters from central California south along the Pacific coast; from the Gulf Coast and Florida south along the coast to the limits of the breeding range.

AMERICAN AVOCET, *Recurvirostra americana* L: 17–18.5 in (43–47 cm). Ws: 27–38 in (68.5–96.5 cm). White with black-and-white wings and back; head and neck turn red-brown during breeding season. Long bill has a strong up-curve, particularly on female. In flight, shows strong black-and-white wing pattern. Long, trailing blue-gray legs are distinct field marks. Hunts for insects, aquatic invertebrates, seeds, and an occasional fish. Will snatch flying insects from the air as well as feed with head below water while swimming. Sweeps bill from side to side in soft mud to stir up prey. A flock of avocets sometimes feeds cooperatively—walking in a line and thrusting their bills under water with each step. Breeds from late March to late July. Courtship behavior is elaborate, including ritualized displays by both sexes. Nest is a grass-lined scrape in meadow, mudflat, saline flat, or marsh. Lays 3–5 eggs; both sexes incubate, 23–29 days. Both Sierra slopes. Common breeder from northern Inyo Co. north, occurring in various wetland habitats, such as Alkali Meadows, Desert Marshlands, mudflats, lakeshores, ponds, and streambanks. *Distribution* Breeding range extends from southwestern Canada through the western United States to northern Mexico. Winters along the Pacific coast from California south and from the Texas coast south through Central America.

SANDPIPERS: FAMILY SCOLOPACIDAE

The Scolopacidae includes sandpipers, phalaropes, woodcock, and snipe. All lack the bold, dark markings of the plovers. They tend to be brown or gray above and paler or white below in winter. They assume darker colors and chestnut markings in the breeding season. In several species, including the Spotted Sandpiper and phalaropes, females are larger than males, brighter in coloration, and more aggressive. It is the female who defends the territory. The female Spotted Sandpiper is *polyandrous* (mating with more than one male). She leaves her first mate a clutch of eggs for incubation and goes on to nest with one or more additional males. Only one parent is required to raise the chicks, because the young can feed themselves almost immediately after hatching.

Sandpipers and phalaropes are highly gregarious, slender, long-legged waders; woodcock and snipe, solitary species, are more chunky. All scolopacids have a fourth, backward-pointing toe and long, slender bill. Like the plovers, they have long, pointed wings for rapid flight. Some species migrate great distances. Although many species may flock together on tidal mud flats and sandy beaches during migration along the Atlantic and Pacific coasts, they seldom compete with each other for food. They take different prey species, depending on how deeply they probe the mud and on the shape of their bill. For example, a long, down-curving bill enables the Long-billed Curlew to take prey ranging in size from toads to small insects and to feed in varied habitats, from grasslands to salt marshes. In contrast, the Spotted Sandpiper, with a shorter bill, often forages in shallow water and is adept at catching insects, young fish, and small invertebrates. Several sandpipers other than those described below visit Eastern Sierra wetlands, but only during migration.

The phalaropes are small, graceful birds with long, straight, thin bills. Unlike other sandpipers, they forage for food while swimming. In fact, their toes are lobed like those of grebes and coots, and some species may feed and sleep on the water. Phalaropes have an odd way of foraging for food: They sometimes swim rapidly in circles, spinning like a top, stabbing at the water with their bill, seemingly stirring up food from the bottom sediments. Female phalaropes assume bold, red-brown markings during the breeding season and defend their mates against other females. Males incubate the eggs and care for the chicks; they develop *brood patches* (a bare area on the abdomen for applying body heat to eggs). Soon after laying their eggs, the females leave the nesting grounds.

The **Willet** *(Catoptrophorus semipalmatus),* a large gray sandpiper with white wing patches, breeds in wet meadows on the east slope from El Dorado Co. north. Winter visitor on west slope.

SPOTTED SANDPIPER, *Actitis macularia*   L: 7–8 in (18–20 cm). Ws: 13–14 in (33–35.5 cm). Fig. 10.2d. Widespread shorebird but not locally common; occurs singly or in pairs. In winter, drab brown above and white below; in summer, brown above, black-and-brown spots on underparts. Call a shrill *peet-weet.* Sexes similar, female slightly larger. Often teeters, holding head down and bobbing tail up and down. Feeds on insects and an occasional fish. Breeds from April to August, laying 1–5 eggs per clutch. Nest is a shallow scrape lined with vegetation. Male usually incubates, 20–24 days. Breeds, but rarely winters, along both Sierra slopes. Occurs on lakeshores and sandbars of streams in both High Desert and Montane communities; also breeds and forages in Alpine and Subalpine meadows.   *Distribution* Breeding range includes much of Canada and Alaska and parts of the western, southwestern, and southeastern United States. Winter range includes parts of the U.S. breeding range, south through Central and South America to northern Chile, northern Argentina, and Uruguay.

The **Long-billed Curlew** *(Numenius americanus)* is a large reddish-tan sandpiper with a very long, down-curved bill. Breeds in wet meadows and grasslands in Sierra and Plumas counties, and less commonly in Inyo Co.; winter visitor on west slope.

COMMON SNIPE, *Gallinago gallinago*   L: 10–11 in (25.5–28 cm). Ws: 18–20 in (46–51 cm). Stocky sandpiper with long, straight bill. Brown streaks on head, upperparts, and breast; white below. In flight, exposes brown rump and orange tail; flight pattern a distinct zigzag. Eyes set back on head, allowing forward and rear vision. Diet includes insects, larvae, mollusks, crustaceans, and worms. Probes with bill for food in mud, wet soils, and shallow marshes; tip of flexible bill is highly sensitive and can manipulate objects. Breeds mid-April through August. Female selects a dry nest site in marsh or meadow; nest is a scrape lined with vegetation, containing 3–5 eggs that female incubates for 18–20 days. Both sexes care for chicks and will use distraction displays if predators threaten. Common Sierra resident. On the east slope prefers wet habitats such as Desert Marshlands, Alkali and Mountain meadows. Moves to higher elevations in spring and summer.   *Distribution*   Breeding range includes most of northern and eastern Canada; the southwestern, midwestern, and northeastern United States; and parts of South America to Tierra del Fuego. Winters in southern British Columbia, the western and central United States, south through Central and South America to Tierra del Fuego. Also occurs on other continents.

WILSON'S PHALAROPE, *Phalaropus tricolor*   L: 8.5–9.5 in (21.5–24 cm). Ws: 14.5–16 in (37–41 cm). Fig. 10.2c. Plumage is pale gray on crown, back of neck, and other upperparts; white below, gray stripe through eye, wings dark. In flight, exposes white rump. Female assumes bright breeding plumage: heavy black stripe through eye and continuing down neck, cinnamon streak on neck, and cinnamon streaks on back. Toes have lateral flaps for swimming, although toes of other phalaropes are lobed. Feeds along shore, wades through shallow water, or swims in deep water, looking for aquatic invertebrates, insects, and seeds. Arrives for breeding by early May and leaves by mid-August. Nests in loose colonies in tall grass; female usually builds the grass-lined nest scrape. Lays 3–6 eggs that the male incubates for 18–27 days. Up to 80,000 dainty Wilson's Phalaropes that nest locally and in the north gather on Mono Lake in late July after breeding to molt and fatten up for their southward migration, a trip of more than 5000 mi (8000 km). Most individuals leave the lake by the first week in August. Breeds from northern Inyo Co. north on the east slope; rare summer visitor on the west slope. Frequents High Desert wetland habitats, such as Desert Marshlands, Alkali Meadows, lakes, and ponds. *Distribution* Breeding range includes parts of southern Canada and the western and midwestern United States. Winter range includes much of western South America.

The **Red-necked Phalarope** *(Phalaropus lobatus),* also known as the Northern Phalarope, is a common migrant in the Eastern Sierra; occurs on west slope. In August more than 65,000 birds stop over on Mono Lake to feed, en route from their northern breeding grounds to the southern oceans where they winter. The first birds arrive in mid-July, the last depart in September.

GULLS AND TERNS: FAMILY LARIDAE

The Laridae is a large family that includes gulls, terns, skuas, jaegers, and skimmers—species generally considered to be seabirds. Actually, many are waterbirds of coastal and inland habitats. Larids are typically monogamous, sometimes pairing for life. Males and females are similar. Many species use calls and displays to advertise and defend territories, and to attract mates. Both terns and gulls usually nest in large, traditional colonies. Great-horned Owls, rats, weasels, Coyotes, Raccoons, and foxes prey on the eggs and chicks.

Gulls and terns of the Eastern Sierra are superficially similar with their long wings, gray upperparts, and white underparts, but they differ considerably in their shape and their foraging behavior. Gulls are stocky with thick, slightly hooked bills. They are good swimmers with large, webbed feet. Glands that excrete salt, high on the sides of the head, enable them to drink salty and brackish water. Gulls generally forage by circling and hovering over water and then picking up food from the surface, or by swimming and taking prey from the water. They have notoriously opportunistic food habits, feeding on garbage, dead animals, and eggs and chicks of other species. Highly gregarious, they roost in flocks.

Terns, in contrast, are slender and graceful, and have a slim bill pointed at the tip. Their tails are forked, and their wings have a sharp bend and narrow tips. Their feet are webbed but small; swimming ability is poor. In summer, terns have a jaunty black cap. Unlike gulls, terns forage by diving into the water, sometimes from spectacular heights. They feed primarily on fish.

The *Ring-billed Gull (Larus delawarensis)* is similar in appearance to the California Gull but paler gray above and with a black ring on the bill. Nests on islands in lakes from about Lassen Co. north. In winter, occurs as far south as Inyo Co. A summer visitor to the west slope; winters in the Central Valley.

CALIFORNIA GULL, *Larus californicus*   L: 21–22 in (53.5–56 cm). Ws: 48–54 in (1.2–1.4 m). Fig. 10.2g. Adults white with gray across the top of the wings and back; black-and-white wingtips, sturdy yellow bill with a red-and-black spot on the lower mandible, and yellow-green legs. First-year birds are brown, and second-year birds are intermediate. Various loud, screaming, and chuckling calls. Eats whatever is available: fish, arthropods, small vertebrates, occasionally eggs and chicks of other species as well as its own, grain, fruits, foliage, and garbage. Forages in water and farmlands, at lakeshores and garbage dumps. Gulls arrive in March and form breeding colonies of tens to thousands of pairs. The nest is a scrape on the ground, lined with feathers and vegetation. Both sexes incubate 1–5 eggs, 23–27 days. Mono Lake hosts the largest California Gull colony in California, second only to the Great Salt Lake/Utah Lake colony. Up to 50,000 gulls have recently nested on the large and small islands of the lake. Most gulls leave by mid-August, but about 8000 remain through early fall. In the Eastern Sierra a common breeder from Mono Co. north, using islands in High Desert lakes, marshes, and ponds for nesting. A common summer visitor and migrant along the west slope.   *Distribution*   Breeds from Mackenzie south through parts of western Canada and the western United States. Winters from southern Washington along the Pacific coast south to Colima, Mexico.

CASPIAN TERN, *Sterna caspia*   L: 19–23 in (48.5–58.5 cm). Ws: 50–55 in (1.3–1.4 m). Large, heavy-bodied tern, with wide, pointed wings and shallowly forked tail. Body mostly white, back and wings gray with darker wingtips. Black cap in summer; mottled cap in winter. Heavy, dark red bill; black legs. Eats mostly fish, and rarely eggs and chicks of other species. Dives for fish; also feeds from water surface. Nests on islands or peninsulas in lakes and rivers in colonies. An unlined scrape on the ground serves as a nest, containing 1–5 eggs. Both sexes incubate, 20–22 days. In the Eastern Sierra, breeds from late April to August, from Mono Co. north along the eastern edge of California, near lakes and large rivers in High Desert communities. Uncommon visitor on the west slope.   *Distribution*   Widely distributed. Breeding range includes parts of the western United States south to northern Mexico, interior Canada from Mackenzie south through the Midwest, localities on the east coast of Canada and through the eastern United States. Winters along the coasts from central California south and from North Carolina south to northern Colombia and Venezuela. Also occurs on other continents.

FORSTER'S TERN, *Sterna forsteri*   L: 14–16.5 in (35.5–42 cm). Ws: 30 in (76 cm). Slender bird, white with gray wingtips. In summer, black cap, red-orange bill with black tip, orange legs and feet. In winter, no cap but dark bar through eye, and duller bill, feet, and legs. Diet includes small fish, insects, and amphibians. Dives into water for fish; also snatches insects in the air and from the water surface. Builds nests on islands in lakes or, more commonly, on emergent aquatic vegetation and on sandbars, in loosely structured colonies. Nests contain 1–5 eggs; both sexes take turns incubating over the 23-day period. On the east slope during the breeding season (late April to mid-September), typically from central Mono Co. north. Forages in lakes and ponds of High Desert and lower Montane communities and in Desert Marshlands. Local breeding and wintering on the west slope.   *Distribution*   Breeding range includes interior southern Canada, most of the interior United States, and south to Tamaulipas, Mexico. Winter range is confined to both coasts, beginning at central California in the west and Virginia in the east, continuing south to southern Mexico.

The **Black Tern** *(Chlidonias niger)* breeds in Desert Marshlands in the Eastern Sierra primarily from El Dorado Co. north; may be sighted farther south in summer. Also in the Central Valley. Numbers have been declining because of human disturbance, reduction of marshlands, and pesticides.

PIGEONS AND DOVES: ORDER COLUMBIFORMES

All birds of this order have the same plump, short-legged silhouette with short neck and small head.

PIGEONS AND DOVES: FAMILY COLUMBIDAE

These gentle, grain- and fruit-eating species have short bills with swollen tips. Soft, thick plumage, sometimes iridescent or metallic, is loosely attached to the skin and pulls out easily. Males and females differ little in appearance. Pigeons and doves have a specialized crop that enables them to nourish their helpless nestlings, much as mammals nurse their young. Although many birds have a *crop* (an expandable part of the esophagus), the crop in the Columbidae is a two-chambered sac that sloughs off a fat- and protein-rich "pigeon milk" from the walls. The milky liquid is regurgitated from the crop as a nestling thrusts its bill into a parent's mouth.

Many pigeons and doves are migratory and have excellent navigational skills. By studying the behavior of homing pigeons, researchers have discovered that pigeons may use several kinds of cues to orient themselves. They can navigate by sun compass, which means that they can compensate for the movement of the sun by time of day and by latitude. They also use familiar landmarks and perhaps familiar scents as well when nearing their home loft. In addition, pigeons use the magnetic field of the Earth, an ability that is difficult for us to understand. This ability was confirmed by several ingenious, carefully controlled experiments in which miniature electric coils or magnets were attached to the birds to determine if homing ability was disrupted.

Introduced Species The **Rock Dove** *(Columba livia)* is a large pigeon that lives in urban areas. Prominent white rump patch. Frequents the lower-elevation populated areas of both Sierra slopes.

BAND-TAILED PIGEON, *Columba fasciata* L: 14–15.5 in (35.5–39.5 cm). Large pigeon. Tail has prominent dark band and broad gray tip. Back of neck has narrow white band above patch of iridescent green. Owl-like call, a low-pitched *whoo-whoo-hoo* or *whoo-uh*. Except for breeding season, often travels in large flocks. Forages both on the ground and in trees and shrubs for acorns, fruits, and pine seeds. Breeds from early April to mid-September, often raising 2 broods. Builds a platform of twigs 30–35 ft (9–11 m) high in a tree near water. Lays 1 or 2 eggs; both sexes incubate for 18–20 days. On the east slope occurs in Montane Forest and Montane Riparian Woodland with a mix of oaks and conifers, from northern Inyo Co. north. More common on the west slope. *Distribution* Breeds from southern British Columbia south along the west coast to Baja California, Utah and Colorado, the Southwest through montane areas of Mexico into Central America. Winters in southern California and the Southwest south through Mexico and Central America.

MOURNING DOVE, *Zenaida macroura* L: 11–13 in (28–33 cm). Ws: 17–19 in (43–48.5 cm). Widely distributed, graceful dove. Long, narrow, pointed tail bordered in white; slender, pointed wings. Brown below and gray-brown above with scattered black spots. Call is a repetitious, mournful *oo-ah-ooo-oo-oo*. Eats seeds, grain, occasionally insects and snails. May mate for life. Breeds from March to late September; raises 2–6 broods per year; 2 eggs per clutch. Nest is a platform of twigs on a tree branch or on the ground. Both sexes incubate, the male during the day and the female at night, for 14 days. In winter, travels in flocks. Common on both Sierra slopes. On the east slope occurs at lower elevations in open areas with scattered trees, such as lower Montane Forest, Pinyon-Juniper Woodland, and High Desert Riparian Woodland. Thrives in disturbed areas. *Distribution* Breeds in southeastern Alaska, southern Canada, the United States, and Mexico to western Panama. Winters in much of the United States and Mexico, south to western Panama.

CUCKOOS: ORDER CUCULIFORMES

Cuckoos and their relatives are landbirds with unusual feet; two toes are directed forward and two toes backward. Most species are slim and long-tailed, resembling the songbirds (Passeriformes). Cuckoos from the Old World are known as "brood parasites" because they lay eggs in other birds' nests.

CUCKOOS: FAMILY CUCULIDAE

The two roadrunner species of North America are ground-dwelling cuckoo forms that differ considerably in appearance and habits from most species in this family.

GREATER ROADRUNNER, *Geococcyx californianus* L: 20–24 in (51–61 cm). Ground-dwelling cuckoo. Gangly with long, sturdy, pale blue legs, long brown tail with white edging, and large head. Black bill is as long as the head; short crest of brown feathers; 2 patches of colored skin behind each eye—1 blue and 1 red. White plumage on undersides, with brown-streaked, white plumage above. Short wings show white crescents in flight. Voice is a dovelike series of *coos*. Rarely flies but runs and jumps in pursuit of prey. Eats insects, small mammals, small lizards and snakes, scorpions, fruits, and seeds. Mates for life. Breeds from March to late August. The nest, a platform of twigs in a low tree or shrub thicket, usually contains 3–5 eggs. Incubation takes 17–20 days. In the Eastern Sierra occurs in lower elevation, shrubby habitats near water, including Sagebrush Scrub, Pinyon-Juniper Woodland, and Creosote Bush Scrub as far north as Inyo Co. Occurs at lower elevations throughout the west slope. *Distribution* The Southwest and south into Baja California and southern Mexico.

OWLS: ORDER STRIGIFORMES

Powerful, taloned, hook-billed hunters, owls—like hawks and falcons—are birds of prey. In addition to large, strong feet and hooked bills, owls have acute vision and hearing that are adapted to nocturnal hunting. As in the Falconiformes, the large eyes are set forward for binocular vision and good depth perception. In order to see to the sides, owls must turn their heads; their flexible necks permit a turning radius of over 180°. The high concentration of rod cells in their eyes enables them to see in limited light.

In extreme darkness—for example, on moonless nights—owls must rely on hearing to track prey. Their range of hearing is comparable to that of humans, but their ability to home in on prey is far superior because of several unique adaptations. The two ear openings are large, different in size and shape, and asymmetrically positioned on the sides of the head. Sound waves reaching the ears at different times permit owls to triangulate the source. In addition, nocturnal owls have a *facial disc* (a ring of flattened feathers around the face) that may act as a parabolic reflector, collecting and focusing sounds. Other feather structures are unique to owls. Some species have "ear tufts" on the head that are formed from feathers, and all species have modified wing feathers that result in silent flight.

Owls regurgitate undigestible portions of their prey—such as fur, feathers, bones, and teeth—in compact pellets. Pellets accumulate under roosting and resting places. Researchers use these pellets to learn what owls are eating. Parts of skulls and teeth are often sufficiently intact to identify which species of small mammals and other vertebrates the owls are preying upon. In the daytime songbirds may discover a roosting owl, give alarm calls, and gather in a flock to harass the unlucky predator. This *mobbing* behavior often involves "dive-bombing" the owl, making frenzied flights around the perch site, or simply perching and calling near the owl. The songbirds stop this harassment when the owl finally leaves to find a more restful perch. Like most other birds of prey, owls are territorial. Their strange and diverse calls advertise their territories and also attract mates. Females in most species are larger than the males. During nesting, the female usually incubates the eggs, while the male brings food. All owls are protected by law from hunters.

BARN OWLS: FAMILY TYTONIDAE

This worldwide family contains the barn owls and bay owls, which are different from the "typical" owls. Members of the family have triangular or heart-shaped facial discs, small dark eyes, and small bills. Tails are short and square, wings are long, and the legs, but not the feet, are feathered; the feet have two toes directed forward and two backward.

BARN OWL, *Tyto alba*   L: 14–20 in (36–51 cm). Ws: 43–47 in (1.1–1.2 m). Pale, slender owl; white below, buffy to rust above with scattered dark spots. Dark edge to heart-shaped facial disc. Calls a mixture of wheezes, hisses, and chuckles. When perched, may lower head and move it back and forth. Eats voles, pocket gophers, kangaroo rats, other rodents, frogs, birds, and insects. May mate for life. Breeds from January to November, raising 2 broods. Does not build a nest. Raises young in barns or other structures, burrows in cliffs or in the ground, caves, or tree hollows. Lays 3–10 eggs, incubates 29–34 days. Present on both Sierra slopes, but populations have declined in recent years. Hunts at lower elevations in grassy, open areas, such as lower Montane Meadows, Alkali Meadows, Sagebrush Scrub with open areas, and farmland. *Distribution*   Ranges throughout most of the United States and south to Tierra del Fuego, and on other continents.

Except for the barn and bay owls, all owls—about 140 species worldwide—belong to the Strigidae. There are 17 species in North America, ranging in size from the tiny Elf Owl, 5–6 inches (13–15 cm) long, to the Great Gray Owl, 24–33 inches (61–84 cm) long. The strigids have four toes, including an outer toe that can grip forward or backward. The leg feathering usually extends out to the feet.

The **Flammulated Owl** *(Otus flammeolus),* small with tiny ear tufts, breeds locally on both slopes. On the east slope, particularly in Pinyon-Juniper Woodland and Montane Forest. Eats mainly insects. Winters in Mexico and Central America.

WESTERN SCREECH-OWL, *Otus kennicottii* L: 7–10 in (18–25.5 cm). Ws: 18–24 in (46–61 cm). A small owl with ear tufts. Brown and white with black barring above; lighter below. Monotonous call of repeated whistles on same pitch that begin slowly and speed up. Roosts in tree hollow or dense foliage during the day. If threatened, may stretch body and freeze, appearing like a broken tree branch, camouflaged by barklike plumage pattern and color. Catches small rodents, songbirds, reptiles, fish, and amphibians by pouncing. Takes insects from foliage or by aerial fly-catching. Breeds from February to mid-June in tree holes, without nesting material. Clutches range from 2–8 eggs, incubation lasts about 26 days. Both Sierra slopes. On the east slope, prefers lower Montane Forest with oaks, Pinyon-Juniper Woodland, and Eastern Sierra Riparian Woodlands. Distribution Ranges from southeastern Alaska along the west coast through the western United States, Baja California, and Mexico south to Mexico City.

GREAT HORNED OWL, *Bubo virginianus* L: 18–25 in (46–64 cm). Ws: 36–60 in (0.9–1.5 m). Fig. 10.3e. Common, large, powerful owl with ear tufts. Brown, marked with darker brown on back and upper breast. Barred brown-and-white underparts, with white throat and feet. Call is a deep *whoo hu-hoo-hoo-hoo-hoo* that has earned it the name Hoot Owl. Often mobbed in the daytime, when discovered by small birds. When threatened, will raise back feathers. Hunts at dusk and at night for rabbits and rodents, also birds, reptiles, and amphibians. Breeds mid-January to mid-June, using abandoned hawk, eagle, or crow nest, tree hollow, or cave. Lays 1–6 eggs; both sexes may incubate, 34–36 days. Inhabits most plant communities, particularly where trees are interspersed with open, grassy areas, as in Subalpine and Mon-

Fig. 10.3. Birds. (a) Common Nighthawk, *Chordeiles minor*. (b) Calliope Humming-bird, *Stellula calliope*. (c) White-throated Swift, *Aeronautes saxatalis*. (d) Belted Kingfisher, *Ceryle alcyon*. (e) Great Horned Owl, *Bubo virginianus*. (f) Red-breasted Sapsucker, *Sphyrapicus ruber*. (g) White-headed Woodpecker, *Picoides albolarvatus*

tane forests bordering meadows, Eastern Sierra Riparian Woodlands, and Pinyon-Juniper Woodland. Occurs on both Sierra slopes. *Distribution* Ranges through North, Central, and South America, except above treeline. The northern populations winter in southern Canada and the northern United States.

NORTHERN PYGMY-OWL, *Glaucidium gnoma*   L: 7–7.5 in (18–19 cm). Ws: 15 in (38 cm). Robin-size "owlet," mostly rust or gray-brown with white-streaked underparts and black streaks on flanks. Tail barred with white; black patch on back sides of neck. Unlike most owls, it has a tail that protrudes beyond the folded wings. Call is a sharp whistle or series of coolike sounds. Hunts by day or in the evening for mice, songbirds, small lizards, snakes, and insects. Breeds mid-April to late August, nesting in abandoned woodpecker hole or natural tree hole. Lays 2–7 eggs, incubates 28 days. Inhabits Montane Forest next to meadows. Both Sierra slopes. *Distribution* Ranges from southeastern Alaska through western North America to Guatemala.

BURROWING OWL, *Athene cunicularia*   L: 9–11 in (23–28 cm). Ws: 20–24 in (51–61 cm). Only ground-dwelling owl in the New World. Small, earless, sandy brown above and white below. Long, nearly bare legs. During the day, may stand in open areas by burrow or perch on fence post; bobs up and down. Call is a sad *coo-coo-roo.* May occur as a solitary pair or in colonies of 10 or more pairs. Hunts small mammals, frogs, fish, birds, insects, and carrion. Hovers over prey, like the American Kestrel, and also fly-catches from a perch. Uses abandoned burrows of ground squirrels. Breeds from mid-April to mid-August, laying 4–11 eggs in a burrow lined with plant material and feathers. Both sexes incubate 26–31 days. Inhabits both Sierra slopes; on the east slope occurs in open, grassy areas and in scrub habitat, such as Sage-brush Scrub and Pinyon-Juniper Woodland. *Distribution* Ranges from southern Canada and western United States as far east as Minnesota and south to Louisiana; south through Mexico, and Central and South America. A disjunct population in Florida. Canadian and northern U.S. populations migrate south for the winter.

Other owls that occur on both Sierra slopes include the **Spotted Owl** *(Strix occidentalis),* which inhabits old-growth coniferous forest from Alpine Co. north on the east slope and Tulare Co. north on the west slope, and the rare **Great Gray Owl** *(Strix nebulosa),* found in Subalpine Forest on the east slope from Mono Co. north. The **Long-eared Owl** *(Asio otus)* prefers Montane Forest adjacent to meadow, and the **Short-eared Owl** *(Asio flammeus)* prefers lower-elevation grassland, Desert Marshlands, and Alkali Meadows; both species range throughout the east slope, but their numbers have declined in recent years. Another "owlet," the **Northern Saw-whet Owl** *(Aegolius acadicus)* also occurs throughout the east slope in Montane and Subalpine forest, Sagebrush Scrub, and Riparian Woodlands. It stores excess prey on branches.

GOATSUCKERS: ORDER CAPRIMULGIFORMES

The name of this worldwide order is based on the folktale that the birds sucked milk from goats at night, a belief fostered by these birds' nocturnal habits, strange songs and cries, and unusual appearance. In fact, some of the species are referred to as "nightjars" for their jarring, nocturnal cries. All caprimulgiform species have soft, loose, cryptically colored plumage, mottled with shades of brown, black, and gray, and startling patches of white. The head is flat with no distinct neck, and the dark eyes are large. The bristled *gape* (the width of the mouth) is enormous, but the bill is small. Small legs and weak feet are barely visible.

NIGHTHAWKS: FAMILY CAPRIMULGIDAE

Swift, silent fliers with long wings, the nighthawks and their relatives catch insects in flight, aided by wide gapes and bristles. An unusual behavior is their tendency to perch lengthwise on branches rather than crosswise as most birds do—a behavior that, with their mottled plumage, makes them very difficult to see. Other characteristic behaviors include laying eggs directly on the ground and swallowing stones, which help crush insect parts in the gizzard.

Physiological adaptations to cope with heat and cold—maintaining body temperature and energy balance—are found in some members of the family. Since their open nests and daytime roosts are not shielded from the afternoon sun, the birds may overheat. They will then flutter the upper region of the throat, an action similar to a dog's panting, which results in evaporative cooling. In cold weather, several caprimulgids, including the Common Poorwill and the Common Nighthawk, can become torpid, conserving body energy reserves. The state of torpor is similar to hibernation: Heart, breathing, and metabolic rates drop, as well as body temperature. Torpor may occur each night or continously for several weeks. Seven species of caprimulgids occur in North America.

The *Lesser Nighthawk *(Chordeiles acutipennis)* breeds in Mojave Desert communities—Creosote Bush Scrub and Joshua Tree Woodland—from Inyo Co. south. Similar in appearance and foraging habits to Common Nighthawk, but smaller. Call is a low *chuck chuck* or a soft trill. Usually absent from the west slope.

COMMON NIGHTHAWK, *Chordeiles minor* L: 8.5–10 in (21.5–25.5 cm). Ws: 21–24 in (53–61 cm). Fig. 10.3a. Typically seen in flight at dusk or at night. In urban areas may hunt insects around bright streetlights or spotlights. Wings are long, narrow, and pointed with a broad, white bar; white patch at throat. Male has a broad, white tail bar. In flight, leisurely repeats loud, nasal calls of *peent* or *beer*. Forages in open areas in all communities of the Eastern Sierra to above treeline. Flies over treetops and roofs or high in the sky pursuing insects, which it engulfs in its wide, open mouth. During males' steep courtship dives, wing feathers vibrate, producing a loud noise. Female lays 2 eggs in open areas on rocks, soil, or roofs of buildings; usually incubates 18–20 days. Migrates in flocks of up to 1000 birds in fall. Breeds along both Sierra slopes from early June to late August. *Distribution* Breeding range includes all but the extreme north of Canada, the United States south through parts of western Central America. Winters in South America as far south as northern Argentina.

COMMON POORWILL, *Phalaenoptilus nuttallii* L: 7–8 in (18–20.5 cm). A bird more often heard than observed. Black on throat and lower face, and above eye, with white neck band. White on tips of outer tail feathers. Rounded wingtips. Named for call *poor-will* or *poor-will-ee*. By day, roosts on the ground, concealed by vegetation. Catches moths and grasshoppers by flying up from the ground or from a perch; sometimes catches insects on the wing. Breeds from late March to late August; lays 2 eggs; both sexes incubate, about 20 days. In forests, nest may be next to a log or hidden by shrubs. Breeds along both Sierra slopes. Occurs in clearings in Montane Forest and in most High Desert communities, including Pinyon-Juniper Woodland and Sagebrush Scrub. *Distribution* Breeding range includes southwestern Canada, the western United States, and south to central Mexico. Wintering range includes the Southwest to central Mexico.

SWIFTS AND HUMMINGBIRDS: ORDER APODIFORMES

This order contains two very distinct groups of birds, the widely distributed swifts (Apodidae) and the New World hummingbirds (Trochilidae). Contrary to the name of the order, which means "without feet," the species do have small, inconspicuous feet.

SWIFTS: FAMILY APODIDAE

Swifts earn their names from their rapid, highly maneuverable flight. They spend their days foraging at both low and high altitudes for insects. Observers claim that swifts may fly as fast as 100 mph (160 kph). All species are streamlined: cigar-shaped, with long, pointed wings. Because they spend entire days in flight and migrate in both fall and spring, swifts may fly hundreds of thousands of miles in their lifetimes; long-lived individuals (nine years or older) may fly more than 1 million miles. Their flight is characteristic: several rapid, choppy wingbeats followed by a glide. Although they resemble swallows in shape and feeding behavior, swifts may be distinguished by their way of flying.

The tiny feet of some swifts have four toes that may all point forward or move to the sides in pairs for grasping nest, perch, or roost sites. Each toe is clawed, enabling swifts to cling to the vertical surfaces of caves, cliffs, or chimneys while roosting or nesting. Swifts of the genus *Chaetura*, such as the Vaux's Swift, are highly social and may gather in flocks of several hundred to thousands before roosting. Beginning at dusk, these swifts circle around a common roost site, such as a hollow tree or chimney, for up to an hour, all flying in the same direction. A few birds, as they approach the roost, may suddenly depart from the circle and enter. This initiates a stunning chain reaction: As each bird in the whirling flock approaches the roost, it heads into the opening.

Swifts have large salivary glands that produce a gluey secretion used to bind nest material together and attach the nest to a vertical surface. To feed nestlings, adults forage for long periods, gathering a ball of several hundred insects in the throat. During cold weather, when insects are not flying, individuals of some swift species will become torpid at night to conserve energy (see Caprimulgiformes).

Two swifts are uncommon breeders along both Sierra slopes. The **Black Swift** *(Cypseloides niger)* nests in wet rock crevices or small caves next to or behind waterfalls. **Vaux's Swift** *(Chaetura vauxi)* inhabits Montane Forest, nesting in hollow trees and, occasionally, chimneys.

WHITE-THROATED SWIFT, *Aeronautes saxatalis* L: 6–7 in (15–18 cm). Ws: 14 in (36 cm). Fig. 10.3c. The only swift with a bold black-and-white pattern. Sooty black above with white throat patch tapering along breast to belly; white flank patches, visible from above and below. Tail slightly forked. Forages on the wing for insects above forest canopy and lakes; even drinks in flight, by swooping down to water. Breeds from May to early August. Courts and copulates in flight, pairs pinwheeling downward while mating. Requires rocky cliffs or canyon walls for nesting. Builds nest cup of feathers and dried vegetation, held together with saliva, in a rock crevice. Incubates 3–6 eggs about 18–20 days. Breeds along both slopes. In the Eastern Sierra frequents Montane Forest, Pinyon-Juniper Woodland, and Sagebrush Scrub. *Distribution* Breeds from southern British Columbia, western United

States, Baja California, south through interior Mexico to Central America. Winters from California, Arizona, and southern New Mexico south through Central America.

HUMMINGBIRDS: FAMILY TROCHILIDAE

Because hummingbirds occur only in the Western Hemisphere, they probably evolved after the continents of Africa and South America drifted apart about 100 million years ago in the mid-Cretaceous period. The group most likely originated in South America, eventually spreading to Central and North America. There are about 340 known species with only 8 common to the United States. The smallest hummingbird is about 2.5 inches (6.5 cm) long, and the largest is 8.5 inches (21 cm) long. Hummingbirds occupy a wide range of elevations and plant communities, from deserts to alpine meadows.

Hummingbirds are specialized in anatomy, physiology, and behavior for a diet of nectar, which they obtain by probing flowers. The bills of the nectar-feeding species differ in length or shape (straight or curved) in relation to the length and shape of the flowers they visit. Contrary to popular belief, hummingbirds do not siphon nectar with their long, slender bill. Instead, they move their long, brush-tipped tongue rapidly in and out of the bill tip to gather nectar. Unlike other nectar-feeding birds, hummingbirds can hover at flowers while they feed. Their rotating shoulder joints permit wingstrokes backward and forward, enabling them to hover. In normal flight hummingbirds are fast, acrobatic fliers, aided by narrow, pointed wings. They rapidly shift direction to avoid obstacles, darting confidently between branches and leaves. They have been clocked at speeds of 50 mph (80 kph) or faster.

Small, warm-blooded animals, such as hummingbirds, have high metabolic rates. Hummingbirds meet this high energy demand principally with nectar, pollen, and small insects gleaned from flowers for protein. Before roosting, hummingbirds fill their crops with nectar to provide some nighttime nourishment. When nights are unseasonably cold, hummingbirds become torpid, which lowers their metabolic rate and conserves energy. Yet, despite their high energy needs, hummingbirds spend most of the day at rest. Recent studies have discovered that hummingbirds must wait for their crop to empty between feeding bouts. These studies also show that the hummingbird intestine removes sugar from stomach contents faster than that of any other vertebrate studied to date.

Hummingbirds are colorful, with most species at least partly metallic green (which may offer some protective coloring). Males also have a throat patch of striking metallic color—such as red, blue, purple, or red-orange—that shows clearly in direct light. The throat patch is most conspicuous during territorial or courtship displays, which consist of steep dives and rapid, short swings in patterns that differ with each species. Males are fiercely aggressive but allow sexually receptive females into their territories, often mating with more than one. The females nest and raise the young entirely on their own, defending small nesting territories against other hummingbirds. While incubating eggs, they do not become torpid at night during cold spells, gaining some protection from the nest. Maintaining their normal body temperature ensures that the eggs stay warm.

A mutually beneficial relationship has evolved between hummingbirds and a number of plants. As the hummingbirds feed on the nectar, they move pollen among flowers. Flowers that depend on them for pollination tend to be tubular, bell-shaped, or have nectar-filled spurs. In the Eastern Sierra, hummingbird flowers include Indian paintbrush, penstemon, Red Columbine, larkspur, gooseberry, currant, manzanita, and Scarlet Gilia. Bees, hawkmoths, and swallowtail butterflies also obtain nectar from these flowers, but hummingbirds often chase away these competitors. Many people believe that most hummingbird flowers are red and that hummingbirds are innately attracted to the color red for that reason. However, hummingbirds feed from flowers of many colors, indicating that they do not respond only to red.

BLACK-CHINNED HUMMINGBIRD, *Archilochus alexandri*   L: 3.5–4 in (9–10 cm). The only dark-throated hummingbird. Metallic green above and white below. Male has a metallic violet-blue band below a black throat. Soft, high-pitched song. Forages among flowers for nectar, pollen, and insects; also fly-catches. Breeds from May to early August, usually lays 2 eggs, incubates 13–16 days. Often builds nest in a branch fork or on a drooping limb over water, near patches of flowers. Tiny nest, made of spiderwebs and plant down, about 1.5 in (4 cm) in diameter and 1 in (2.5 cm) deep. Breeds throughout both Sierra slopes at lower elevations and moves to higher elevations in July. Occurs in flower patches in Eastern Sierra Riparian Woodlands, Montane Meadows, Montane Forest, and Pinyon-Juniper Woodland. *Distribution*   Breeding range includes southwestern British Columbia, the western United States east to northwestern Montana, south into northern Baja California and northern Mexico. Birds that nest in southern California may move into the Sierra after breeding. Wintering range includes southeastern California and southern Texas south to southern Mexico.

CALLIOPE HUMMINGBIRD, *Stellula calliope*   L: 2.8–3.5 in (7–9 cm). Fig. 10.3b. Of all hummingbirds in the United States, the Calliope is the smallest and is found at the highest elevations. Metallic green above and white below. Male has metallic purple feathers at throat; in display, they appear as streamers against a white throat. Female has dark speckles on throat. Forages in gooseberry, currant, and manzanita flowers, among others. Diet consists of nectar, sap from sapsucker holes (see Woodpeckers: Family Picidae), insects, and spiders. Lays 2 eggs, incubates 15–16 days. Breeds from early May to early August. Places tiny nest, about 1.5 in (4 cm) in diameter, on small, dead conifer limbs, in mistletoe, or among pinecones. Builds nest of moss, bark, lichens, and other plant material bound together by cocoon and spider silk. Female may stack a new nest on top of previous nests. Breeds on both Sierra slopes. Occurs in flower patches in Montane Riparian Woodland, Montane Chaparral, Montane and Subalpine forest, and Mountain Meadows, up to 11,500 ft (3500 m) elevation. *Distribution* Breeding range extends from central British Columbia and southwestern Alberta south through the western United States, east to northwestern Wyoming, northern Baja California, and western Texas. Wintering range extends from northern Mexico and Baja California to south of Mexico City.

RUFOUS HUMMINGBIRD, *Selasphorus rufus* L: 3.5–4 in (9–10 cm). Male red-brown above and on the sides; patches of metallic green on wings, head, and tip of tail. Bright red metallic throat against white chest. Female metallic green above and white below; base of tail and sides light red-brown; throat speckled. A summer migrant along both Sierra slopes. Males leave their northern breeding grounds and migrate south beginning in June, followed in July by females and young; the last arrive in the Eastern Sierra in early September. Along the migration route, individuals stop and establish feeding territories, which they aggressively defend against all other hummingbirds for up to 2 weeks. During these stopovers, the birds build up fat in order to continue migrating. Diet consists of nectar, insects, and spiders. Occurs in flower patches in Montane Forest, Montane Riparian Woodland, Montane Meadows, and creekside or spring-fed meadows near Sagebrush Scrub and other arid communities. *Distribution* Breeds the farthest north of any hummingbird, from April to July: in southern Alaska and southern Yukon, along the coast through southwestern Canada, through the western United States to northwestern California, eastern Oregon, and central Idaho. Winters from coastal southern California east to southern Texas and south through southern Mexico.

The *Broad-tailed Hummingbird *(S. platycercus),* a Rocky Mountain and Great Basin species, was recently observed nesting from southern Mono Co. south. Mostly green; males have red throats. Anna's Hummingbird *(Calypte anna)* and Costa's Hummingbird *(Calypte costae)* may occasionally be observed along the southern portion of the east slope; both also occur on the west slope. Anna's males are green with red throat and crown; Costa's males are green with white underparts, purple crown and throat.

KINGFISHERS: ORDER CORACIIFORMES

The members of this order all have small feet with two or three toes joined along their lengths. Nesting sites are tree cavities, burrows, or crevices. Mostly carnivorous, coraciiform species eat small fish, amphibians, reptiles, small mammals, and insects.

KINGFISHERS: FAMILY ALCEDINIDAE

Kingfishers tend to be brightly colored with long, straight, sharp bills, large heads, and small feet with three toes directed forward (two of these joined) and one backward. The sexes are similar but may differ in color of underparts. Individuals are solitary outside of the breeding season; while nesting, males and females take turns incubating eggs and feeding young. New World kingfishers build their nests in burrows that they excavate along the banks of creeks and rivers. All New World kingfishers are true fishers that dive head-first from a hovering position, a branch, or a power line into water in pursuit of small fish and other small vertebrates and crustaceans. Their dives are shallow, and the birds quickly return to the air. The kingfishers are the only group of coraciiform birds widespread in the Temperate Zone. Of the eighty-six kingfisher species, only six occur in the New World.

BELTED KINGFISHER, *Ceryle alcyon*　L: 11–14.5 in (28–37 cm). Fig. 10.3d. The only kingfisher in most of the United States and Canada. Heavy, gray bill; large, slate-gray head with unkempt double crest; blue-gray back and chest band; white collar and belly. Female has rust belly band. Perches on branch over water, watching for prey, then dives, either straight or in spiral. Diet is mostly small fish, supplemented by frogs, salamanders, and insects. Sometimes beats prey against a perch, tosses it into the air, and swallows it headfirst. Regurgitates pellets of undigested material. Breeds from early April to mid-August; 5–8 eggs per nest, 22–24 days' incubation. Occurs year-round along both Sierra slopes near water—lakes, creeks, and rivers—in all plant communities except where snow and ice prevent foraging. *Distribution*　Breeding range includes Alaska and Canada, except the northernmost areas, and most of the United States. Winters in southern coastal Alaska, along the west coast of Canada, most of the United States, Mexico, Central America, and northern South America.

WOODPECKERS: ORDER PICIFORMES

Most piciform species have two toes pointed forward and two backward, which probably evolved for perching in trees and later for climbing up tree trunks. In fact, all species of the order are arboreal. Most of the piciform families are tropical. Only the family Picidae represents the Piciformes in the Temperate Zone.

WOODPECKERS: FAMILY PICIDAE

Many woodpecker traits are associated with tree-dwelling and foraging for food on tree trunks and branches. Primarily insect-eaters, woodpeckers pry off bark and dig in crevices to find insects. Their acute hearing enables them to detect insect activity in the tree. They also catch insects from a perch and eat pine seeds, nuts, or fruit. Several species peck small holes in tree trunks in order to lick up the exuded sap. The wings of woodpeckers reflect their arboreal lifestyle. They are short and broad, enabling them to maneuver among trees, but most woodpeckers are not suited to rapid or long-distance flight. Typical woodpecker flight is undulating, a pattern of several wingbeats followed by a glide.

While foraging for insects, woodpeckers often fly to the base of a tree and gradually climb up, sometimes circling the tree several times. Their clawed, long-toed feet enable them to cling to the trunk, while their strong tail with stiff, central tail feathers serves as a brace. The sturdy, pointed bill is a probing and digging tool for exposing insects. A thick-walled skull, protected brain, and strong head muscles reduce the shock of pounding blows. The long, narrow tongue may be extended several times the length of the bill. The tip of the tongue, sticky with mucus, is barbed and pointed in insectivorous woodpeckers but hairy or brushlike in sapsuckers.

The powerful bill is also used in territorial and courtship activities. Males and females peck in a distinct rhythm against a hollow tree or any other object that resonates. The rapid drumming advertises territory as well as attracts unmated females. Woodpeckers use their bill to excavate a nesting hole in a tree that is completely or partly dead, an effort that requires a week to a month of work by both sexes. For a medium-size species, the entrance is typically 1.5–2 inches (4–5 cm) in diameter; the interior is wider, to accommodate the nesting chamber, and is about 15 inches (38 cm) deep. After the eggs are laid, both sexes incubate and feed the young. Many species of birds that cannot excavate their own tree hole depend on old woodpecker holes for their nesting sites.

LEWIS' WOODPECKER, *Melanerpes lewis*   L: 10.5–11.5 in (26.5–29 cm). Named for Meriwether Lewis of the Lewis and Clark expedition. Greenish black back and wings, reddish face, broad gray-white collar, and pink belly. Female is smaller and duller. Call sounds like a harsh *churr-churr*. Requires high perches for frequent fly-catching; also forages on the ground. Will store acorns and other nuts in holes or cracks in trees. Nests from early May to July; lays 5–9 eggs, and incubates 13–14 days. Male excavates most of the nest. Resident on the east slope but numbers fluctuate; August to May, non-breeding, on the west slope. Frequents open forest or burned or logged areas, including Montane Forest, Montane Riparian Woodland, and Pinyon-Juniper Woodland. *Distribution*   Breeding range includes southwestern Canada south through the Southwest. Winters from Oregon, southern Idaho, and Nebraska south through northern Mexico.

The **Acorn Woodpecker** *(Melanerpes formicivorus),* a close relative of the Lewis' Woodpecker, occurs on the east slope only in Plumas and Lassen counties in mixed conifer-oak Montane Forest. It is noted for living and defending territories in groups, nesting cooperatively, and drilling holes in trees for storing thousands of acorns for winter food.

RED-BREASTED SAPSUCKER, *Sphyrapicus ruber*   L: 7 in (18 cm). Fig. 10.3f. Distinguished by red head and breast, yellow belly, black-and-white-barred back, wings, and tail; dark area on wing with white bar, which becomes a broad, white wing patch in flight. Immatures are mostly mottled white and brown. Eats insects, fruit, and tree sap obtained by drilling shallow holes and periodically lapping up exuded sap. Also carries sap to young in nest. Prefers trees with dead or diseased heartwood for nest hole. Nests from early May to late July; 3–7 eggs; 12–14 days' incubation. Resident on both Sierra slopes, preferring Montane and Subalpine forest and Montane Riparian Woodland. *Distribution*   Breeding range includes southeastern Alaska, western British Columbia, and west of the Cascades south through the Sierra Nevada to the mountains of southern California. Winters at lower elevations throughout the breeding range and south to northern Baja California.

The *Red-naped Sapsucker *(Sphyrapicus nuchalis),* formerly considered a subspecies of the Yellow-bellied Sapsucker, is similar to the Red-breasted Sapsucker but with red restricted to head area; black bands around bill. Resident on the east slope, occurring in open Montane Forest or Riparian Woodlands. May interbreed with Red-breasted Sapsucker.

Williamson's Sapsucker *(Sphyrapicus thyroideus)* prefers Montane and Subalpine forest on both Sierra slopes. Male yellow-bellied with black back and chest, small red patch on throat. Female has a brown head, black chest patch, yellow belly, and brown-and-white barring on back.

The *Ladder-backed Woodpecker *(Picoides scalaris)* has irregular, thin, black-and-white stripes across its back; male has a red crown. Occurs in Pinyon-Juniper Woodland, High Desert Riparian Woodland, and High Desert Scrub communities on the east slope, from Inyo Co. south.

Nuttall's Woodpecker (Picoides nuttallii) has heavy black stripes against white across its back; male has red nape. An uncommon resident in the southern portion of the east slope, but common in the foothills of the west slope. Prefers Riparian Woodlands for nesting and foraging.

The **Downy Woodpecker** *(Picoides pubescens)* is the smallest woodpecker in the United States, about 6 in (15 cm) long. Looks like a small Hairy Woodpecker. Breeds in lower-elevation forests and riparian woodlands along both Sierra slopes.

HAIRY WOODPECKER, *Picoides villosus* L: 8.5–10.5 in (21.5–26.7 cm). Black-and-white head, rows of white spots on black wings; black tail with white outer feathers; white undersides and white back. Male has red patch on back of head. Voice is a loud, sharp *peek*. Forages mainly on trunks and branches in search of insects, particularly woodborer larvae. Also eats fruit and pine seeds; feeds at sapsucker holes. Breeds mid-March to late August, laying 3–6 eggs; incubates about 14 days. Common year-round on both slopes. Frequents Montane and Subalpine forest, Pinyon-Juniper Woodland, and Eastern Sierra Riparian Woodlands, particularly recent burns. *Distribution* Breeds and winters up to treeline in Alaska and Canada, most of the United States, and south through the highlands of Mexico to Panama.

WHITE-HEADED WOODPECKER, *Picoides albolarvatus* L: 9 in (23 cm). Fig. 10.3g. The only woodpecker with a white head. Body is black, with white outer wing patch conspicuous in flight. Male has a small red patch on back of head. Voice is a squeaky *ick*. Fly-catches and also forages for insects on branches and trunks, preferring trees with thick, creviced bark, which it flakes off. Pecks off the scales of pinecones to remove seeds. Breeds mid-April to early August. Prefers pine stump for nest cavity; lays 3–7 eggs; incubates 14 days. Often migrates downslope in winter. Resident on both Sierra slopes, but uncommon on the east slope. Prefers mature Subalpine and Montane forest. *Distribution* Resident from southern interior British Columbia south through north-central Washington and northern Idaho, and south to southern California and western Nevada.

The **Black-backed Woodpecker** *(Picoides arcticus)* is an uncommon resident of both slopes; on the east slope from northern Inyo Co. north. Male is distinguished by its yellow crown.

NORTHERN FLICKER, *Colaptes auratus,* Red-shafted ssp., *C. a. cafer* L: 12.5–14 in (32–36 cm). Undersides of wings and tail are salmon-red, back and wings brown with black bars, underparts white with distinct black spots, collar black, face and throat gray, and crown brown; conspicuous white rump patch. Male has a red "moustache." Voice is loud, repetitive *wick* or shrill *kee-yer.* Forages primarily on ground by probing the soil and occasionally on tree trunks; rarely, fly-catches from a perch. Eats mostly ants; also insects, acorns, seeds, and fruit. Breeds from early May to late July; lays 3–12 eggs; incubates for 11 days. Common resident throughout both Sierra slopes. Prefers open forest of all types, including Montane and Subalpine forest, Eastern Sierra Riparian Woodlands, and Pinyon-Juniper Woodland. *Distribution* Breeds in southeastern Alaska, coastal British Columbia, and southwestern Canada south to northern Baja California, western Texas, and the interior highlands of Mexico to southern Mexico. Winters throughout its breeding range from southwestern Canada south.

The **Pileated Woodpecker** *(Dryocopus pileatus)* is an uncommon inhabitant of Montane and Subalpine forest on both Sierra slopes. Larger than the Flicker, it is distinguished by a pointed red crest, black back, boldly striped black-and-white face and neck pattern, and black-and-white wings in flight. On the east slope, occurs from Mono Co. north.

PERCHING BIRDS: ORDER PASSERIFORMES

About 5000 species—three-fifths of the world's birds—belong to the Passeriformes, the perching birds. All have a unique arrangement of the palate bones and a unique sperm structure, nine or ten primary wing feathers, and twelve tail feathers. Their "perching feet" have three toes directed forward and one backward, enabling the birds to grip thin branches and twigs. Passeriform birds tend to have complex displays and nesting behaviors. Their eggs are often colorful and intricately marked. The young hatch unfeathered and helpless.

The order is divided into two groups, the flycatchers and the true perching birds. Flycatchers are distinguished by a simple voicebox, the shape of the middle-ear bones, and a number of skeletal features. The true perching birds, known as "songbirds," represent about half the birds of the world. They have a voicebox with many muscles for control, resulting in complex songs in some groups. The differences among passeriform (Passerine) birds are small. Their body design is similar, so differences in skeletal features, bill morphology, musculature, and behavior have been used for tentatively grouping species into families. With the advent of molecular techniques, such as DNA hybridization, DNA sequencing, and mitochondrial DNA comparisons, new tools are now available for determining evolutionary relationships.

Introduced Species   The **European Starling** *(Sturnus vulgaris)* was introduced to the United States in 1890 and is now resident throughout the Sierra Nevada. In the Eastern Sierra it inhabits the lower mountain slopes and the High Desert valleys. The **House Sparrow** *(Passer domesticus),* also known as the English Sparrow, was first introduced to Brooklyn, New York, in 1850 and later to other locations; it has since spread throughout North America. It is resident on both Sierra slopes around buildings in all communities up to and including Montane Forest, rarely at higher elevations.

FLYCATCHERS: FAMILY TYRANNIDAE

The only flycatcher family represented in the United States and Canada, the Tyrannidae consists of about 370 species. Based on *tyrannus,* a Latin word for "ruler" or "monarch," the family name reflects flycatchers' aggression and fierce demeanor. Fly-catching, short flights from a perch to catch insects, is the main feeding behavior for these insect-eating birds. As they catch their prey in the air, often the bill snaps shut with an audible click, the action of special ligaments that contract as the mouth gapes. Flycatchers also glean insects from foliage or catch them on the ground. They characteristically perch upright and often erect their crown feathers. They have a flat bill with bristles around the base, ten primary flight feathers per wing, short legs, small feet, and simple, whistling songs consisting of only a few syllables. Almost all species in the United States and Canada are gray, olive green, or brown, sometimes with a yellow belly. All migrate south for the winter.

OLIVE-SIDED FLYCATCHER, *Contopus cooperi*   L: 7–8 in (18–20 cm). A robust flycatcher that prefers high, exposed perches. Large head and body with short tail. Dark, olive brown head, back, and sides; white from throat to belly. Two narrow white wing bars on dark wing; dark tail. Song resembles a loudly whistled *whip-three-beers,* and the call is *pip-pip-pip.* Perches on tall tree or snag for fly-catching, flies away, and returns to the same perch. Diet consists of insects. Nests from mid-May to late July; lays 3–4 eggs, incubates 14 days. Builds nest toward end of horizontal branch, often in clump of twigs or conifer foliage; materials include twigs, grass, lichen, moss, and foliage. Breeds along both Sierra slopes. Inhabits Montane and Subalpine forest and Montane Riparian Woodland but needs a clearing, tall trees, or burned area for foraging space. *Distribution* Breeds in Canada up to treeline and the United States, except for the Southeast. Winters from Colombia and Venezuela south to Peru.

WESTERN WOOD-PEWEE, *Contopus sordidulus* L: 6–6.5 in (15–16.5 cm). Fig. 10.4c. A dark flycatcher with olive brown head, back, and sides; whitish buffy from chin to belly; dark wings with two wing bars; upper bill dark, lower orange. Song is descending *pee-er* or *pheer.* Fly-catches in the tops of trees; distinct bill click as insect is caught. Also gleans insects and spiders from trees and occasionally eats berries. Nests from early May to early August, laying 2–4 eggs; incubates 12–13 days. Builds nest on horizontal branch using plant fibers, grass, leaves, and bark bound with spider silk and lined with grass and feathers. Commonly breeds on both Sierra slopes; on the east slope in Subalpine and Montane forest and Montane Riparian Woodland. *Distribution* Breeding range includes eastern Alaska to southern Mackenzie, south through western Canada and the western United States, as far east as western Texas, and south through Mexico to Nicaragua. Winters from Colombia and Venezuela south to Peru and Bolivia.

EMPIDONAX FLYCATCHERS, *Empidonax* species L: about 5.5 in (14 cm). These flycatchers are treated as a group because their appearance and habits are similar. They can be identified by habitat and voice. All have a brown-gray head and back, light gray chest, pale yellow-gray belly (except the Gray Flycatcher), dark brown tail, and wings with two distinct wing bars. Most have a light eye ring and a bicolored bill, black above and orange below. All fly-catch from perches, typically flicking their tail. Diet consists mostly of insects. On the east slope, nest from mid-May to late July. Build nest, 3–4 in (7.6–10 cm) in diameter, in crotch of tree or shrub. Nest is constructed of grass and stems, and lined with moss, feathers, and plant down. Female incubates the 2–4 eggs for 14–18 days.

Sierra populations of the **Willow Flycatcher** *(E. traillii)* have declined drastically because of habitat destruction and nest parasitism by the Brown-headed Cowbird. The species is now rare. It occurs in willow thickets in Montane Riparian Woodland and in streamside willows in Subalpine and Montane meadows along both Sierra slopes.

HAMMOND'S FLYCATCHER, *E. hammondii* Sings a repetitive, low-pitched *seput, tsurrt, chu-lup;* call is a sharp *pip* or *peep.* Nests throughout both slopes in dense Montane and Subalpine forest. *Distribution* Breeding range includes east-central Alaska and the southern Yukon, British Columbia and southwestern Alberta south through the western states, east to southeastern

Fig. 10.4. Birds. (a) Cliff Swallow, *Petrochelidon pyrrhonota*. (b) Violet-green Swallow, *Tachycineta thalassina*. (c) Western Wood-Pewee, *Contopus sordidulus*. (d) Clark's Nutcracker, *Nucifraga columbiana*. (e) Say's Phoebe, *Sayornis saya*. (f) Horned Lark, *Eremophila alpestris*. (g) Steller's Jay, *Cyanocitta stelleri*. (h) Black-billed Magpie, *Pica pica*

Wyoming and western Colorado. Winters from northern Mexico south to Honduras.

DUSKY FLYCATCHER, *E. oberholseri*   Has a low-pitched song with a mixture of *prll-it, prrdrrt, pseet;* the call is a soft *whit.* Breeds throughout both slopes in scrub communities, such as Sagebrush Scrub and Montane Chaparral, mixed with conifers; Montane and Subalpine forests; Montane Riparian Woodland; and Pinyon-Juniper Woodland.   *Distribution*   Breeds from southwestern Yukon through British Columbia and east to Saskatchewan, south through the western United States and east as far as South Dakota. Winters from southern Arizona south to Guatemala.

*GRAY FLYCATCHER, *E. wrightii*   Gently lowers and raises its tail. The song is a rapid *churweeoo* followed by a *cheep.* Occurs on the east slope only, from southern Mono Co. south in Sagebrush Scrub and shrubby, open Montane Forest and Pinyon-Juniper Woodland.   *Distribution*   A flycatcher of the Great Basin, breeds from Washington east to southern Wyoming and south to eastern California, Arizona, and central New Mexico. Winters from central Arizona to Oaxaca.

The **Black Phoebe** *(Sayornis nigricans)* is all black except for a white belly. On the east slope occurs only in High Desert Riparian Woodland and other High Desert wetlands, from Inyo Co. south. Occurs on the west slope in similar habitats.

SAY'S PHOEBE, *Sayornis saya*   L: 7–8 in (18–20 cm). Fig. 10.4e. A particularly active flycatcher of High Desert scrub communities. Brown head with gray-buff back and chest, rusty red belly, black tail, dark wings with 2 light wing bars on each. Raises and lowers tail slowly. Call is a descending whistle, and song is soft *chu-weer* or *pippety-chee.* Forages mostly on flying insects. Darts from perch to perch fly-catching; also dives for insects from hovering position. Breeds from mid-March to late July, laying 3–7 eggs and incubating about 12–14 days. Nests in tree cavities, rocky shelves, crevices, and under bridges. Constructs nest of moss, stems, and grass bound with spider and cocoon silk and hair. Breeds in low foothills along the western Sierra slope. On the east slope occurs primarily from southern Mono Co. south in Sagebrush, Shadscale, Alkali Sink, and Creosote Bush scrub communities.   *Distribution*   Breeding range extends from western and northern Alaska east to central MacKenzie and southwestern Manitoba, south through the western United States, except the west coast, to north of Mexico City. Winters in the Southwest to central Texas and south through Baja California, Oaxaca, and Veracruz.

ASH-THROATED FLYCATCHER, *Myiarchus cinerascens*   L: 7.5–8.5 in (19–21.5 cm). Brownish back and head, slightly crested; white throat, pale yellow belly. Two white bars on each dark wing, red-brown outer wing patch and tail. Call is *pee-reer*. Eats mostly insects, some berries. Captures insects by flycatching from exposed perches; may not return to the same perch. Breeds from late April to late July, lays 3–7 eggs, incubates about 15 days. Nests in tree holes, including woodpecker holes. Builds nest of stems, grasses, animal fur, and sometimes shed skins of lizard or snake. Breeds throughout the Sierra in lower-elevation communities. On the east slope inhabits Pinyon-Juniper Woodland and High Desert scrub communities where they meet High Desert Riparian Woodland. *Distribution*   Breeds in the western United States except northernmost regions and parts of California and Arizona. Winters from central Arizona and southern California south to Costa Rica.

WESTERN KINGBIRD, *Tyrannus verticalis*   L: 8–9.5 in (20–24 cm). Conspicuous flycatcher of arid, open habitats. Head (with hidden red patch on crown) and back gray, throat white, belly bright yellow, wings brownish, tail black with narrow white outer edges. Voice a brief *kip* or a twitter. Flycatches from a perch—tree, shrub, fence post, or power line; also forages on ground. Eats mostly insects, occasionally berries. Nests from mid-April to late July, lays 2–7 eggs, incubates about 14 days. In a tree crotch or on a horizontal branch, female builds nest of weeds and twigs, lined with feathers or hair. Breeds along both Sierra slopes in open habitats. On the east slope, occurs in High Desert scrub communities and Pinyon-Juniper Woodland, or near farms. *Distribution*   Breeding range extends from southern British Columbia east to southern Ontario, and south through the western and midwestern United States to northern Mexico. Winters from southern Mexico to Costa Rica; locally along the Gulf and Atlantic coasts.

The *Eastern Kingbird *(Tyrannus tyrannus)* is similar to the Western Kingbird, but black above and white below with a narrow band of white at tip of tail. Uncommon breeder on the east slope from Alpine Co. south.

SHRIKES: FAMILY LANIIDAE

The Laniidae are predatory songbirds known as shrikes, a name that refers to their shrill, shrieking calls. Only two occur in North America, and both inhabit the Sierra Nevada. Shrikes tend to be black, white, and gray. They are robust with large heads, thick hooked bills, and strong feet. On both sides of the upper bill is a toothlike structure, matched by notches on the lower bill, which makes the shrike bill a powerful hunting weapon.

Preferring open habitat, shrikes perch on a treetop, branch, fence, or overhead line. Known also as "butcher birds," they prey upon insects and small vertebrates, such as snakes, lizards, birds, and rodents. With their sharp eyesight they can track moving insects and mice at a distance. They may swoop down on prey on the ground, dart out and fly-catch, or hover over prey before attacking. By hopping from perch to perch within a tree or shrub, they attempt to flush small birds. Shrikes grab birds with their feet or knock them to the ground with a forceful blow and kill prey by biting their neck, severing the vertebrae. They store extra food by impaling prey on sharp twigs, thorns, and barbed wire. They carry small prey in their bills, large prey with their feet, to storage sites. Apparently shrikes recall the locations of their stored food and usually eat it within a week. They regurgitate indigestible material, such as fur, feathers, and bones, in pellets.

The **Northern Shrike** *(Lanius excubitor)* is similar to the Loggerhead Shrike, but is larger and lighter in color with a faintly barred breast. Uncommon in fall and winter on both slopes.

LOGGERHEAD SHRIKE, *Lanius ludovicianus*   L: 8–10 in (20.5–25.5 cm). Predatory songbird of open country. Gray above with black face mask and stout, hooked bill. Pale below, black wings with white patches, and black tail with white outer feathers. Song is a series of calls and trills. Forages by swooping on prey and by fly-catching. Breeds from mid-March to mid-August; female lays 3–8 eggs and incubates 15–17 days. Builds cuplike nest of twigs, lined with grass and feathers, in bush or tree. A resident of both Sierra slopes. On the east slope occurs in Pinyon-Juniper Woodland, Sagebrush Scrub, and other High Desert scrub communities. *Distribution* Breeds in interior southern Canada, around the Great Lakes, in most of the western and southern United States, and south to south-central Mexico. Winters in the southern half of the United States to south-central Mexico.

VIREOS: FAMILY VIREONIDAE

These small, slow-moving insectivorous birds live in the canopies of trees and shrubs. Most of the North American vireos have nostrils partly covered by bristles and slightly flattened bills. Their songs consist of short, bright phrases. Vireos usually forage in foliage, gleaning insects and spiders, and taking some seeds and fruit. The male and female together build a basketlike nest of plant fibers and twigs that is suspended from a branch fork.

The *Gray Vireo *(Vireo vicinior)* is gray above and white below with indistinct white eye ring and 1 faint wing bar. On the east slope occurs in Pinyon-

Juniper Woodland, Sagebrush Scrub, and other High Desert scrub communities from Inyo Co. south.

*PLUMBEUS VIREO, *Vireo plumbeus* L: 5–6 in (12.5–15 cm). A large, tame vireo of coniferous forest. Olive green above, gray head, white "spectacles," and 2 white wing bars on each wing. White below with yellow along sides. Song consists of slow, clear phrases of 2–6 notes rising and falling in pitch, with pauses in between. Lays 3–5 eggs; both sexes incubate 12–14 days. Breeds on Eastern Sierra slope from mid-April to mid-August. Inhabits lower Montane Forest, particularly with oaks, and Pinyon-Juniper Woodland. *Distribution* Breeds throughout the interior West, from central and southeastern Montana east to southwestern South Dakota and south to El Salvador and Honduras. Winters from the Southwest south to Honduras.

WARBLING VIREO, *Vireo gilvus* L: 5–6 in (12.5–15 cm). A plain vireo primarily of riparian woodlands. Gray above, pale below; indistinct white stripe over eye. Long, clear warbling song; male may sing for hours. Gleans insects and spiders from twigs and foliage of trees and shrubs. Usually nests in a deciduous tree or shrub. Lays 3–5 eggs; both sexes incubate 12–14 days. Breeds on both Sierra slopes; on the east slope from mid-April to late August in Eastern Sierra Riparian Woodlands and Montane Forest mixed with oaks. *Distribution* Breeds from southern Alaska, southern Mackenzie, and British Columbia east across Canada, south through most of the United States except the coastal Southeast, to south-central Mexico. Winters from Mexico south to El Salvador.

JAYS, MAGPIES, AND CROWS: FAMILY CORVIDAE

The bold and noisy corvids include the largest songbirds in the world, the ravens. Corvids have ten primary wing feathers, nostrils concealed by bristles, broad wings, and strong legs and feet. In addition, most members of the family have either an enlarged oral cavity behind the tongue, an expanded esophagus, or a throat pouch—all used for carrying food. The sexes are identical, and many corvids pair for life. They begin courtship and nesting earlier in the year than most songbirds; young are usually out of the nest by early summer. In almost all species the female incubates the eggs, fed by the male. Corvids, like the parrots, are long-lived and considered the most intelligent of birds. In the laboratory, they solve simple puzzles, and learn tasks and associations quickly. Compared to other songbirds, the vocal repertoire of most corvids is extensive; it includes a variety of calls for different occasions and a quiet, long, rambling male song. Some species are capable of mimicking sounds and human speech. Corvid young are dependent for a long period and may remain in family groups for several months to a year after they are independent. Individuals usually overwinter in the general vicinity of breeding areas.

Although corvids are omnivorous, food specialization does occur. The bill size reflects the eating habits of each species. All take insects, seeds, nuts, berries, eggs, nestlings, mollusks, and small vertebrates. Ravens and some crows prefer carrion. Several jay species and the nutcrackers prefer acorns, beechnuts, hazelnuts, or pine seeds. Most corvids store excess food, particularly for winter months; many bury food in the ground. Individuals may dig up and consume their food stores a few days to weeks or months later. Laboratory and and field studies of jays, nutcrackers, and crows indicate that individuals recall the precise location of their food stores. Considering that each one may store hundreds or even thousands of items, the precision and scope of their spatial memory is astounding.

The specialized seed-eaters have mutually beneficial relationships with certain forest trees. The birds obtain a year-round food supply from the trees and, in turn, disperse their seeds by burying them throughout the forest. Unretrieved seeds may germinate in spring or summer. Some relationships are highly specialized. In the Sierra, the Single-leaf Pinyon is dependent on seed dispersal by the Pinyon Jay and Clark's Nutcracker; the Whitebark and Limber pines likewise depend on the Clark's Nutcracker. These pines have large seeds that lack the woody wing typical of other conifer seeds, so they cannot be dispersed by wind. When the ripe cones of wind-dispersed conifers open, the wind blows out the seeds. However, pinyon cones hold the ripe pine seeds in place, even when open, and whitebark cones do not open even when the seeds are ripe.

STELLER'S JAY, *Cyanocitta stelleri*   L: 12–13.5 in (30.5–34 cm). Fig. 10.4g. The only crested jay in the Sierra Nevada. Black head and long, pointed crest; white streaks above black, pointed bill; black back and sooty breast. Rump, sides, and belly light blue; wings and tail dark blue barred with black. Frequents picnic grounds and campsites, usually in pairs. Has variety of noisy calls, particularly *shook-shook-shook* and a nasal *wah-wah-wah*. One call imitates the cry of a Red-tailed Hawk. Stores conifer seeds in fall. Nests from late March to early July; female incubates 3–5 eggs for 16–18 days. Builds nest of twigs with a lining of pine needles and feathers, on a horizontal limb near trunk or in tree crotch. Common year-round inhabitant of coniferous forest on both slopes. Occurs in Subalpine and Montane forest and Pinyon-Juniper Woodland; in winter many move downslope. *Distribution* From southern Alaska south through British Columbia and southwestern Alberta, south throughout the Rocky Mountains, coastal ranges and Sierra Nevada, the Southwest and south through Mexico to Nicaragua.

WESTERN SCRUB-JAY, *Aphelocoma californica*   L: 11–13 in (28–33 cm). A slender jay with blue head, wings, and tail; dark eye patch with white eyebrow, dark bill. Gray-brown back, white throat and underparts. A common call is a rapid *cheek-cheek-cheek*. Stores acorns and pine seeds in the ground. Breeds from late March to mid-August, defending its territory in pairs. Nest is bulky structure of twigs, lined with grass and moss, built low in small tree or shrub. Female lays 2–7 eggs, incubates about 16 days. Inhabits scrub communities on both Sierra slopes. On the east slope prefers Pinyon-Juniper Woodland, Eastern Sierra Riparian Woodlands, Montane Chaparral, and Sagebrush Scrub.   *Distribution*   Resident from southwestern Washington south to southern California and through the Great Basin, east to central Texas, and south through Baja California and Mexico.

*PINYON JAY, *Gymnorhinus cyanocephalus*   L: 9–11 in (23–28 cm). A social, nomadic jay, with dull blue body and white throat. Long, sharp black bill and short tail. On the east slope, depends on Single-leaf Pinyon, Jeffrey, and Ponderosa pines, storing thousands of seeds in the ground on the south side of trees and in bark crevices. When cone crops are poor, flocks wander widely. During these movements, jays from different flocks may mix, but most return to their home flock for breeding. Breeds from March to early July. Builds nest of twigs and lines it with grass. Female lays 3–6 eggs and incubates 17 days. Pinyon Jays nest in colonies of hundreds of birds. Nonbreeding flock members may help breeding pairs. After breeding, birds wander in noisy flock of 50–500 in search of food. High mewing call in flight. Occurs throughout the east slope year-round, preferring Pinyon-Juniper Woodland and low-elevation Montane Forest. May wander to the west slope in fall.   *Distribution*   Range includes central Oregon east to western South Dakota, through the Great Basin and the Southwest. Wanders farther south and east in winter, to northern Mexico, central Texas, western Kansas, and Nebraska.

CLARK'S NUTCRACKER, *Nucifraga columbiana*   L: 12–13 in (30.5–33 cm). Fig. 10.4d, plate 6. Sometimes referred to as the "pine crow"; named for Captain William Clark of the Lewis and Clark expedition. Gray body and head with white around eyes and bill; glossy black wings with white patch on back edge; tail black in center with white edges; long, pointed black bill. Most common call is a grating *kraa-kraa-kraa*. One nutcracker may bury tens of thousands of seeds in small caches of 1–15 or more seeds throughout its home range. Obtains pine seeds by tearing closed cones apart with sharp bill.

Carries 50–100 or more seeds in throat pouch to storage sites. Preferred foods in the Eastern Sierra are the large seeds of Whitebark, Limber, Single-leaf Pinyon, Jeffrey, and Ponderosa pines. May wander widely when cone crops fail. Will attack mice, squirrels, and other small vertebrates. Breeds from March to late July. Builds a platform of twigs lined with grass and pine needles on a horizontal branch. Typically lays 2–5 eggs; both sexes incubate 17–18 days. In summer, travels in family groups; winter and spring, may forage in larger groups. Resident on both Sierra slopes. Prefers Subalpine Forest in summer, Montane Forest and Pinyon-Juniper Woodland in winter. *Distribution* From central British Columbia and southwestern Alberta through the Rocky Mountains, coast ranges, Sierra Nevada, southern California ranges and south to northern Baja California and Nuevo León, Mexico.

*BLACK-BILLED MAGPIE, *Pica pica* L: 17–22 in (43–56 cm). Fig. 10.4h. A conspicuous, large bird of streamside communities. Black head, back, and breast; white belly and shoulders; metallic, greenish blue wings. Long, streaming metallic green tail that is longer than head and body. Sturdy, pointed, black bill. A common call is repeated, high-pitched *kak-kak-kak.* Conspicuous on fence posts next to roads. Feeds on carrion and roadkills; often forages on the ground. May land on the back of a deer or cow to search for ticks and other parasites. Its short flights seem labored. Breeds from mid-March to late June. Small colonies nest in shrubs or trees near stream or river. Female constructs a domed, bulky nest from mud and twigs, usually on a horizontal limb close to trunk, lining it with hair, grass, and other soft materials. Nest is often used by the same pair more than 1 year; old nests are used by other birds for roosting. Female lays 3–9 eggs and incubates 16–18 days. The birds travel in pairs or family groups in summer, in larger groups in winter. Resident along the east slope, rarely wandering to the west slope. Nests in Eastern Sierra Riparian Woodlands and feeds in adjacent Montane Meadows, Sagebrush Scrub, Alkali Meadows, or farmlands. *Distribution* Resident from southern Alaska and southern Yukon south through interior British Columbia; east to western Ontario; south to east-central California, east across the Great Basin to eastern Oklahoma and south to northern New Mexico. Also occurs on other continents.

The **American Crow** *(Corvus brachyrhynchos)* is, like the Common Raven, all black but smaller than the raven. On the east slope it inhabits High Desert Riparian Woodland, farmlands, and towns. Breeds and overwinters locally from Alpine Co. north, and from southern Mono Co. south. Resident at lower elevations on the west slope.

COMMON RAVEN, *Corvus corax* L: 21.5–27 in (54.5–68.5 cm). Ws: 46–56 in (1.2–1.4 m). All black with purple luster. Heavy, long, pointed bill with thick bristles hiding nostrils at base. Wedge-shaped tail. In flight, may be mistaken for a hawk, since it soars, hovers, and circles like a bird of prey. Typical call is a hoarse *croak-croak* or *cur-rock*. Often in pairs or small flocks; occasionally in large, aerial gatherings of hundreds of ravens. Mates for life; spectacular courtship flight with male tumbling in air. Builds a large, bulky nest of branches and twigs lined with bark, hair, and other soft material. Nest sites are on cliffs, in rocky crevices or crotches of conifers. Female lays 2–8 eggs and incubates 18–20 days. Occurs year-round in the Sierra at all elevations, commonly in Subalpine and Montane communities, Pinyon-Juniper Woodland, and desert scrub communities. *Distribution* Resident in Alaska, most of Canada, northern Great Lakes region, New England, Appalachian Mountains, the western United States to the Rocky Mountains and south through Mexico to Nicaragua. Also on other continents.

LARKS: FAMILY ALAUDIDAE

Renowned songsters, male larks sing flutelike or warbling songs during display flights hundreds of feet above their territories. These displays end in steep, spectacular dives. Larks occur in treeless areas, including deserts, grasslands, plains, sandy beaches, heaths, meadows, and farmlands. Their terrestrial lifestyle is reflected in their sturdy legs and feet with a long hind claw; they walk instead of hop. Pointed wings enable them to manuever and fly rapidly. Nest sites are natural depressions in the ground or excavated hollows that are lined with feathers, plant down, and grass. Despite their grouping under the order of "perching birds," this family of songbirds does not perch in trees but dwells on the ground.

HORNED LARK, *Eremophila alpestris* L: 7–8 in (18–20 cm). Fig. 10.4f. Common inhabitant of alpine and High Desert grasslands. Bold facial pattern: black forehead ending in small, black feather "horns"; black across bill and cheek, black collar, yellow face. Brown back, white belly, sides of tail edged with white. Flight slightly undulating, with wings folded to side briefly after each beat. Song, a weak and high-pitched *pit-wit, wee-pit, pit-wee, wee-pit,* much repeated in flight, as male circles high above the territory. Forages on the ground for seeds and grain in winter and for insects, spiders, and small mollusks in summer. Nests early March to July, laying 2–5 eggs. Female builds nest and incubates 11 or 12 days, fed by male. Resident at lower elevations on both Sierra slopes, but also nests in Alpine Meadow and Alkali Meadow communities, and farmlands The high-elevation birds are considered a different subspecies (*E. a. sierrae*) than those at lower elevations (*E. a.*

ammophila). Most high-elevation birds migrate to lower elevations to over-winter. *Distribution* Breeding range includes most of North America, south to Colombia. Winter range includes southern Canada, the United States, and south to Colombia. Also occurs on other continents.

SWALLOWS: FAMILY HIRUNDINIDAE

The swallows and martins of the Hirundinidae share many characteristics with the swifts and are easily confused with them (see Apodiformes, Apodidae). However, they spend less time in flight than the swifts, fly closer to the ground, and have a prominent bend in the wing. The wings of swallows are long, narrow, and pointed, the ideal shape for fast flight; some species have a forked tail, allowing rapid maneuvers. Although the bill is small, swallows catch insect prey on the wing, sweeping them up with their wide mouth. They commonly perch on power lines, fences, and rooftops. Reflecting an aerial lifestyle, swallows have short legs and weak feet, and walk with much difficulty.

Swallows make twittering or buzzy sounds. Some species are extremely social, nesting in colonies of hundreds of birds and feeding together away from the colony in insect-rich locations. In breeding colonies pairs of swallows defend their own nesting territory, which consists of the nest and a small space around it. Several species of colonial swallows build nests of mud pellets, attaching them where a vertical support meets an overhead surface. These swallows take advantage of man-made structures for nesting sites, particularly the underside of bridges and eaves. Swallows also migrate in flocks, roosting at night together in marshes—each bird perched on a tall stem. In spring northward progress during migration depends on weather; the birds follow the warming trend north, dependent on a good supply of insects.

The **Tree Swallow** *(Tachycineta bicolor)* is similar to the Violet-green Swallow but less common. Dark green-blue above and no white rump patches. Nests in tree cavities on both Sierra slopes; prefers Riparian Woodlands and Montane Forest.

VIOLET-GREEN SWALLOW, *Tachycineta thalassina* L: 5–5.5 in (12.5–14 cm). Fig. 10.4b. Primarily a swallow of the mountains. Dark above with green-and-purple gloss, white underparts extending above the eye, and 2 prominent white rump patches. Tail slightly forked. Forages for flying insects over streams, lakes, and meadows, vocalizing *tsip tseet tsip*. Male sings a courtship song in flight before sunrise. May nest in colonies up to 20 pairs in tree cavities, rocks, and crevices. Constructs nest of stems, grass, and feathers; lays 4–6 eggs and incubates 14–15 days. Breeds from early May to early August on both Sierra slopes. Nests in Subalpine and Montane forest and in High Desert scrub communities. *Distribution* Breeding range includes central Alaska and central Yukon south through western Canada and east to southwestern Saskatchewan, through the western United States and as far east as Nebraska, south through Baja California to Oaxaca and Veracruz. Winters from Baja California south through Mexico to Nicaragua.

NORTHERN ROUGH-WINGED SWALLOW, *Stelgidopteryx serripennis* L: 5– 6 in (12.5–15 cm). Named for small hooks on outer primary wing feathers. Brown upperparts and throat, white belly. Slightly forked tail, rounded rather than angular wing. Rarely vocalizes; call is *quiz-z-zeep*. Forages for insects alone or with other swallow species; tends to fly low over ground. Courting male chases female. Usually nests in burrow made by other birds or mammals in stream- or riverbank, in natural rock crevices or man-made structures. Sometimes male and female will dig burrow with their feet. Builds nest of twigs, grass, and stems. Occasionally, 2 or 3 other pairs will nest nearby. Lays 4–8 eggs, incubates 16 days. Breeds from May to mid-July on both slopes; on the east slope from Placer Co. south in Montane Forest, Eastern Sierra Riparian Woodlands, and High Desert scrub communities where suitable nest sites are available. *Distribution* Breeding range includes southern Canada, the United States, and Mexico south to Costa Rica. Wintering range includes extreme southern United States, and Mexico south to Panama.

The **Bank Swallow** *(Riparia riparia)* requires a steep streambank of sand, dirt, or fine gravel for nest burrows; nests in colonies. Brown above, white below with brown band across chest, and slightly forked tail—resembling Northern Rough-winged Swallow. Occurs locally at lower elevations on both Sierra slopes; known to nest in Inyo Co.

CLIFF SWALLOW, *Petrochelidon pyrrhonota* L: 5–6 in (12.5–15 cm). Fig. 10.4a. A locally common swallow that nests in large colonies. Dark blue crown, back, and tail; cream forehead; white streaks on back; reddish throat, face, and rump; gray flanks and white belly. Tail is square at tip. Long song consists of low grating sounds. Forages over meadows and grasslands, catching insects on the wing; occasionally eats spiders and juniper berries. Nests in colonies of a few to more than a thousand pairs on cliffs, barns, large dams, and under bridges. Builds nest of mud or clay pellets over a period of 5–14 days. Nest is jug-shaped, with protruding entrance on side and a round nest chamber lined with grass. Incubates 1–6 eggs for 11–16 days. Other species of birds may roost in old Cliff Swallow nests. Breeds from early May to mid-July on both Sierra slopes. Nests near a mud source at appropriate sites in Montane Meadows and Forest, Eastern Sierra Riparian Woodlands, Desert Marshlands, High Desert communities adjacent to Alkali Meadows, and farmlands. *Distribution* Breeds in most of Alaska, Canada, the United States, to south-central Mexico. Winters from Paraguay and central Brazil to central Argentina.

BARN SWALLOW, *Hirundo rustica* L: 5.8–7.8 in (14.5–20 cm). A graceful, colorful swallow, with deeply forked tail. Metallic dark blue above; forehead and underparts brick red. Long, twittering song. Forages in flight for insects; drinks and bathes in flight as well. Will follow farmers in the field, taking grasshoppers, crickets, and other insects stirred up. Male chases female in long courtship flight. Breeds from early April to late August, both sexes building a nest of clay pellets mixed with dry grass and lined with feathers. Nest is constructed against a vertical surface or on a horizontal surface; sites include cliffs and caves. Lays 4–7 eggs, sometimes 2 broods; incubates 13–17 days. Breeds along the west slope, and on the east slope from Alpine Co. north and southern Mono Co. south. Occurs near streams, lakes, ponds, and other wetlands; found in Montane Forest and, especially, Montane Meadows, Montane Riparian Woodland, Alkali Meadows, and farmlands. *Distribution* Breeding range includes southeastern Alaska, Canada except the northernmost regions, the United States south to central Mexico. Winters from Panama south through South America. Also occurs on other continents.

CHICKADEES, TITMICE, AND TITS: FAMILY PARIDAE, FAMILY REMIZIDAE, AND FAMILY AEGITHALIDAE

These small, tame birds are acrobatic foragers, often hanging upside-down from twigs as they search for insects. The names *tit* and *titmouse* come in part from an Old Icelandic word *titr,* meaning "small." Most species are shorter than 6 inches (15 cm), with rounded wings, ten primary wing feathers, small but sturdy pointed bills, and soft, thick plumage. The chickadees have black-and-white facial patterns; the tits and titmice of North America are gray-brown, some with pointed crests. The songs are pleasing and easily recognized; the calls may be hoarse or squeaky. These birds are not migratory; their omnivorous food habits and thick plumage help them withstand cold winters. Some species form communal roosts during cold weather, a behavior that conserves body heat.

All members of these families forage for insects and vegetable matter. Several species, including the Black-capped and Mountain chickadees, also eat seeds. These birds perch on a branch, hold a seed with their feet, and hammer it open. Many of the seed-eaters will store excess for winter. A single bird may store hundreds of seeds per day. Mountain Chickadees hide seeds under bark and in holes in conifers. Studies of chickadees demonstrate that, like the corvids, these birds use spatial memory to return to the precise locations of their food stores. During the winter, chickadees and titmice may forage with mixed flocks of Brown Creepers, kinglets, nuthatches, and woodpeckers. These mixed flocks move through the forest, with each species searching tree trunks and branches for food in its own manner. During spring, these birds become territorial and and form pairs. Males and females share the incubation and parental duties.

FAMILY PARIDAE

MOUNTAIN CHICKADEE, *Poecile gambeli* L: 5–5.8 in (13–15 cm). Fig. 10.5d. A chickadee of high elevations. Black cap and throat, white face, thin black stripe through eye. Brown-gray back and sides with darker wings and tail. Calls include *chick-a-dee-a-dee-a-dee* and whistled *fee-bee-bay*. Gleans insects from tree canopy and plant material; takes conifer seeds from opened cones. Breeds from mid-April to mid-August. Nests in natural cavities or woodpecker holes, laying 5–9 eggs and incubating 14 days. Common on both Sierra slopes at high elevations; on the east slope inhabits Subalpine and Montane forest and Pinyon-Juniper Woodland. *Distribution.* Resident from southeastern Alaska and southern Yukon to northwestern and central British Columbia east to southwestern Alberta and south through most of the western montane United States (except for the coastal ranges), the Southwest, and northern Baja California.

The *Juniper Titmouse *(Baeolophus griseus)* is gray above with a pointed crest and whitish below. Eastern Sierra slope. Inhabits lower Montane Forest with oak, Montane Riparian Woodland, and Pinyon-Juniper Woodland on the east slope, from about Mono Lake south. Previously know as the Plain Titmouse.

FAMILY REMIZIDAE

The * Verdin *(Auriparus flaviceps)* is the size and color of the bushtit but has a yellow head and throat, small red wing epaulet, and shorter tail. A desert species, it frequents High Desert Riparian Woodland, Shadscale and Alkali Sink scrub from Inyo Co. south.

FAMILY AEGITHALIDAE

BUSHTIT, *Psaltriparus minimus* L: 3.8–4.5 in (9.5–11.5 cm). A small, active tit of scrub communities. Brown head, gray back, long tail, and white

Fig. 10.5.   Birds. (a) House Wren, *Troglodytes aedon.* (b) White-breasted Nuthatch, *Sitta carolinensis.* (c) Brown Creeper, *Certhia americana.* (d) Mountain Chickadee, *Poecile gambeli.* (e) American Robin, *Turdus migratorius.* (f) Hermit Thrush, *Catharus guttatus.* (g) American Dipper, *Cinclus mexicanus*

underparts. Males have dark eyes, females have light eyes. Travels in flocks of 6–50 birds, except during nesting. Constant call is high *tsit-tsit-tsit,* which probably keeps flock members together. Gleans insects and spiders; eats fruit, flying short distances from perch to perch. Breeds from early March to late July. The nest is a pendulous bag, with entrance hole on one side, woven of twigs, leaves, and stems bound with spider silk; it requires 34–50 days to build. The Bushtit lays 5–7 eggs, incubates 12 days. Nonbreeding birds may help pairs feed young. Occurs along the west slope and on the east slope in Plumas Co. and from Alpine Co. south. Found in Pinyon-Juniper Woodland, High Desert Riparian Woodland, and Sagebrush Scrub. *Distribution* Resident from southwestern British Columbia south along the coast ranges, through the Great Basin and southern Rocky Mountains east to western Oklahoma and central Texas, and south to southern Baja California and from Mexico to Guatemala.

NUTHATCHES: FAMILY SITTIDAE

Nuthatches forage as they walk up or down tree trunks, clinging to the bark with their short legs and long, clawed toes. One foot may be turned slightly to secure a hold. They may walk down the undersides of branches as well. After walking down to the base of a tree, they often fly to the top of a nearby tree and start down again, searching for insects and their eggs, larvae, pupae, and spiders. May also search leaf litter on the ground. Pygmy nuthatches also forage through twigs, cones, and conifer needles. With their thin, straight bill that has an upward slant at the tip, nuthatches can pry food from the crevices of bark. Some species may use a small bark scale to help probe crevices or pry up other bark scales. Nuthatches also feed on seeds. The name *nuthatch* may derive from their habit of placing seeds in bark crevices and whacking them open with the bill. Usually nonmigratory, nuthatches in winter months join mixed foraging flocks of chickadees, titmice, Brown Creepers, kinglets, and woodpeckers. These small, plump birds of the forest have large heads, thick necks, stubby tails, and ten primary wing feathers. All are gray above with black crowns and white below. The sexes are similar.

RED-BREASTED NUTHATCH, *Sitta canadensis*   L: 4.5–4.8 in (11.5–12 cm). The nuthatch of coniferous forest. Gray above with black cap and eyestripe; white face, throat, and corners of tail; red-brown breast and belly. Call is a high-pitched, nasal *ank-ank-ank.* Pries open cone scales to obtain seeds. Breeds from early May to early August. May dig nesting hole in dead tree or use old woodpecker hole; builds nest of grass, roots, moss, and bark. Spreads tree sap around nest hole, probably to discourage predators. Female lays 4–7 eggs and incubates 12 days. Resident on both Sierra slopes, on the east slope in dense Montane and Subalpine forest. Local populations may migrate downslope in fall. *Distribution* Southeastern Alaska and southern Canada to south of the Great Lakes region, through the Appalachian Mountains and the Northeast; the western United States east to the western Dakotas.

WHITE-BREASTED NUTHATCH, *Sitta carolinensis* L: 5–6 in (13–15 cm). Fig. 10.5b. The largest North American nuthatch. Gray above with black cap and black on back of neck (both gray in female), black edging to wings. White face, chest, and corners of tail; light underparts. One call is a loud *quank-quank-quank-quank*. Males sing a song with 8 or 10 *what* or *hah* sounds. Caches excess seeds and acorns in trees for winter food. Breeds from mid-March to late July. Nests in natural cavities in dead trees or in old woodpecker holes; lines nest with bark, feathers, fur, and hair. Female lays 5–9 eggs and incubates 13–14 days. Resident on both Sierra slopes. On the east slope inhabits Subalpine and Montane forest, especially lower-elevation forests with oaks, Montane Riparian Woodland, and Pinyon-Juniper Woodland. Some populations move downslope in fall. *Distribution* Resident from southern interior British Columbia east through most of southern Canada, most of the United States except the Great Plains, and south to south-central Mexico.

PYGMY NUTHATCH, *Sitta pygmaea* L: 3.8–4.5 in (9.5–11.5 cm). The smallest nuthatch of the Sierra. Gray-brown cap that extends to eye and black stripe through eye. Gray upperparts, white underparts and tail corners. Call is a soft, constant twitter. A social nuthatch; travels in family groups and in large flocks during winter. Flock members keep warm in cold weather by roosting together in tree cavities. The Pygmy Nuthatch harvests conifer seeds and occasionally fly-catches. Stores seeds for winter in bark crevices. Breeds from mid-April to early August, laying 4–9 eggs. Pair excavates nesting hole in rotted stump or uses old woodpecker hole; lines hole with soft material. Female mostly incubates, 15 or 16 days. Resident on both Sierra slopes; on the east slope in Montane Forest and Pinyon-Juniper Woodland. *Distribution* Resident in south-central British Columbia south to northern Baja California, in the Rocky Mountains, and from the Southwest to south-central Mexico.

CREEPERS: FAMILY CERTHIIDAE

Creepers, true to their name, forage by spiraling up tree trunks and searching bark crevices for insects, larvae, pupae, and eggs. When they reach the top, they fly to the bottom of another tree and begin again. These small birds probe for food with their thin, downward-curving bills. They prop themselves against tree trunks using their long tail feathers, stiffened and pointed at the tip, much as woodpeckers do. Strong feet with long, sharp claws help secure a hold. Brown, black, and white above, creepers are well camouflaged against tree bark and difficult to see from a distance; the sexes are similar. Creepers are related to both the nuthatches and the titmice. Except during the breeding season, creepers may join mixed flocks of titmice, woodpeckers, nuthatches, and chickadees to forage.

BROWN CREEPER, *Certhia americana*   L: 5 – 5.8 in (12.5 – 14.5 cm). Fig. 10.5c. Brown with white streaks and speckles above, white underparts, and white streak over eye; red-brown rump and tail. Call is a faint, hissing *zi-i-i-it*. Eats insects and other invertebrates. Breeds from mid-April to mid-August. Places nest against tree trunk behind loose strip of bark; sometimes uses a tree cavity or woodpecker hole. Builds nest of bark, moss, twigs; lines it with feathers. Lays 4 – 8 eggs, incubates 14 – 15 days. Resident on both Sierra slopes. Prefers large trees with loose or hanging bark in dense Subalpine and Montane forest and Montane Riparian Woodland. *Distribution* Ranges from southern Alaska to interior British Columbia and across southern Canada, south through the West and Southwest, and south through Mexico to Nicaragua.

WRENS: FAMILY TROGLODYTIDAE

Wrens, small active birds, are cryptically colored brown or brown-gray with subtle black-and-white markings; the plumage color varies with the habitat of the species. Sexes are similar. The bill is slender, pointed, and slightly curved downward, a useful tool for searching crevices and holes for insects. Wrens have short, rounded wings, and their flight is weak. The tail is often cocked upright, giving these birds a characteristic profile. Each species has a loud, distinct song. Some male wrens are polygynous and build several nests on their territory. The family name refers to cave dwelling, possibly an allusion to the crevice- and hole-nesting tendencies of the wrens.

ROCK WREN, *Salpinctes obsoletus*   L: 5 – 6.3 in (12.5 – 16 cm). A wren of talus slopes and rocky cliffs. Gray-brown above, white stripe over eye, streaked white breast, buffy rump and flanks. Song consists of repeated phrases, similar to mockingbird. Forages on ground and around rocks for insects. Breeds mid-March to late July, laying 4 – 10 eggs. Nests in cavities and crevices in cliffs and among rocks; builds nest of grass and stems, lined with fur and feathers. Places small stones near entrance to nest. Resident throughout both Sierra slopes at all elevations and in all communities with suitable talus slopes or rocky outcrops; tolerates aridity. *Distribution* Breeds from south-central British Columbia east to Saskatchewan and south through the western United States, except east of the coast ranges; east to western Dakotas, south through Baja California to Costa Rica. Winters in the Southwest to central Texas and south to Costa Rica.

The **Canyon Wren** *(Catherpes mexicanus)* is chestnut brown, except for a white throat and breast. More common on the west slope than on the east, it inhabits rocky, shaded canyons.

BEWICK'S WREN, *Thryomanes bewickii*   L: 5–5.5 in (12.5–14 cm). Reddish brown above with white stripe over eye; wings and tail with black barring. White underparts, and white spots along edge of the long, rounded tail. Male sings with head thrown back and tail hanging down; song consists of several notes followed by a trill. Forages for insects in trees and shrubs, sometimes on ground. Breeds from early April to early August, with 2 broods. Nests in tree cavities, woodpecker holes, or in piles of brush. Female lays 4– 11 eggs and incubates 14 days. Both Sierra slopes, except for much of Mono Co. Prefers Eastern Sierra Riparian Woodlands and shrubby, lower Montane Forest. *Distribution* Resident from southwestern British Columbia south through the Pacific states and Great Basin; east to the Great Lakes region to Pennsylvania and New York, and south through Baja California and south-central Mexico.

HOUSE WREN, *Troglodytes aedon*   L: 4.5–5.3 in (11.5–13.5 cm). Fig. 10.5a. Gray-brown above with small black bars on wings and tail; pale gray-brown below. Loud, bubbling song. Forages on the ground, on bark and foliage of trees, and in piles of debris, seeking insects, spiders, and other invertebrates. Breeds from mid-April to late August, laying 5–9 eggs; incubates 13–15 days. There are often 2 broods. Male builds several nests of twigs in woodpecker holes or natural tree cavities and may have more than 1 mate. Female lines nest with catkins and cocoon silk. A common wren of the mountains, resident along both Sierra slopes. Inhabits Eastern Sierra Riparian Woodlands and Montane Forest. *Distribution* Breeds from southern British Columbia across southern Canada, south through most of the United States, and south through South America. Winters in the Southeast, Southwest, and south through South America.

The **Winter Wren** *(Troglodytes troglodytes)* is the smallest of North American wrens. Uncommon in the Sierra. Occurs on the east slope in Montane and Subalpine forest and Eastern Sierra Riparian Woodlands. Breeds from Sierra Co. north and winters from Mono Co. north. Resident on the west slope.

The **Marsh Wren** *(Cistothorus palustris)* requires marsh vegetation for nesting. It has a broad white stripe over the eye and a short tail. Breeds on both Sierra slopes.

DIPPERS: FAMILY CINCLIDAE

Dippers seldom fly over land. Instead, they live close to fast-flowing, permanent, rocky streams and small lakes, even above treeline. They pursue their prey—aquatic insects, larvae, mollusks, and small vertebrates—by wading into the water, by swimming on the surface (with unwebbed feet), by diving to the bottom and walking on the streambed, and by swimming underwater. They swim by "flying"—that is, by flapping their wings for propulsion. Unusual characteristics facilitate this active, aquatic lifestyle. For example, the plumage has a thick undercoat of down for warmth. At the base of the tail the preen gland, which is several times the size of other perching-bird preen glands, provides oil for grooming and waterproofing the plumage. A flap over the nostrils keeps out water during underwater swimming and diving, a transparent membrane protects the eyes against spray and may be closed during a dive, and large, sharp-clawed feet permit walking underwater. Their territory includes about 0.5 mile (800 m) of stream, which they cover in the course of foraging each day. They travel between foraging locations by flying low, over the water.

These aquatic songbirds have plump bodies, stubby upward-pointing tails, short wings, and long, strong legs with large feet; their bills are short and thin. While perched on a rock, they bob up and down constantly, "dipping." Their song is loud, long, and warbling—audible above rushing water. Several dipper traits, including the song, suggest a kinship with the wrens and thrushes. Dippers are solitary, except during the nesting season. The sexes are similar. The female builds a large, soft nest in a safe site, such as behind a waterfall, on rocks, or under a bridge. During winter, dippers move downstream as the water begins to freeze at higher elevations.

AMERICAN DIPPER, *Cinclus mexicanus*   L: 7–8.5 in (18–21.5 cm). Fig. 10.5g. The only aquatic songbird in North America, also known as the Water Ouzel. Sooty gray all over, with a narrow, pointed bill and sturdy, yellow legs. Forages for insects, mollusks, and small vertebrates both streamside and in deeper water. Breeds from April to August; some males have more than 1 mate. Female builds domed nest of moss and grass with side entrance, lays 4–5 eggs, and incubates 15–17 days; may raise 2 broods. Occurs throughout both Sierra slopes along cold, fast-running streams and sometimes at small lakes, in Subalpine and Montane forest, Montane Riparian Woodland, and

Alpine Rock and Meadow Communities. During winter, moves to lower elevations. *Distribution* Resident from western and central Alaska and north-central Yukon south through western Canada and the western United States, east to southwestern North Dakota, and south to Panama.

KINGLETS: FAMILY REGULIDAE

The small, energetic kinglets move about a tree rapidly, foraging through foliage. They join winter flocks of foraging chickadees, nuthatches, Brown Creepers, and woodpeckers.

GOLDEN-CROWNED KINGLET, *Regulus satrapa* L: 3.5−4.5 in (9−11.5 cm). Fig. 10.6a. A small, active bird of higher elevations. Green-brown above, pale below; dark wings and tail, 2 white wing bars; small, thin bill. Male has distinctive striped head: orange crown bordered by yellow, black, and light stripes, with black stripe through eye. Female has yellow crown. Call consists of 3 high, soft notes; song similar to chickadee. Gleans insects, spiders, and insect eggs from foliage in branch tips; also fly-catches, hovers near foliage, and feeds at sapsucker holes. Breeds from early April to early August; lays 5−11 eggs, incubates 14−15 days. Attaches round, cuplike nest of moss and lichens to end of drooping conifer limb. Resident of both Sierra slopes in dense Subalpine and Montane forest. *Distribution* Breeding range includes southern Alaska and southern Yukon, south to British Columbia and east across Canada, the western United States, east to the Great Lakes region and the Northeast, and south to Guatemala. Wintering range includes southern Alaska, across southern Canada, south to Baja California and to Guatemala.

RUBY-CROWNED KINGLET, *Regulus calendula* L: 3.8−4.5 in (9.5−11.5 cm). Small, active forest bird, similar to Golden-crowned Kinglet. Green-brown above, paler below, dark wings and tail, 2 white wing bars, white eye ring. Male has central red crown, often concealed. Song has several loud, ascending phrases. Forages for small insects and spiders on foliage and twigs, fly-catches, and also catches insects while hovering; eats some fruit and seeds. Breeds mid-May to mid-August, laying 4−9 eggs; incubates 13 or 14 days. Nest similar to that of Golden-crowned. Inhabits both Sierra slopes in Subalpine and Montane forest and Montane Riparian Woodland. After breeding, moves down from higher elevations. *Distribution* Breeds in most of Alaska and Canada except the far north, the West, the Great Lakes region, and the Northeast. Winters in southern British Columbia, the West, Southwest, parts of the Midwest and Southeast, and south to Guatemala.

Fig. 10.6. Birds. (a) Golden-crowned Kinglet, *Regulus satrapa*. (b) Yellow Warbler, *Dendroica petechia*. (c) Yellow-rumped Warbler, *Dendroica coronata*. (d) Wilson's Warbler, *Wilsonia pusilla*. (e) Western Tanager, *Piranga ludoviciana*. (f) Red-winged Blackbird, *Agelaius phoeniceus*. (g) Western Meadowlark, *Sturnella neglecta*

The small, slender gnatcatchers have long, thin bills. They move their tails up and down, and from side to side. Within trees and shrubs they move about energetically in search of insect and spider prey.

BLUE-GRAY GNATCATCHER, *Polioptila caerulea*   L: 4–5 in (10–12.5 cm). A small, drab, energetic bird of lower elevations. Often holds tail upright and flicks it from side to side. Slender and long-tailed. Blue-gray above, white below; black tail with white outer tail feathers. White eye ring, black eye stripe, and thin, slightly decurved bill. Twangy, nasal call. Gleans insects and spiders from twigs and foliage; fly-catches. Breeds from mid-March to late August. Cup-shaped nest of plant fibers, covered with lichens and lined with grass, hair, and feathers. Anchors nest to shrub branch with spider webs. Female lays 3–5 eggs and incubates 11–15 days. Nests on both slopes; uncommon on the east slope from central Mono Co. south. Inhabits Pinyon-Juniper Woodland and Eastern Sierra Riparian Woodlands. *Distribution* Breeds from northern California east to southern Wyoming, the southern Great Lakes region, and the Northeast; and south to Honduras. Winters in southern California, western Arizona, east to central Texas and the Gulf states, and south to Honduras.

THRUSHES: FAMILY TURDIDAE

Of all the Passerines, the thrushes have the most melodic songs, particularly the Hermit Thrush of the United States and the nightingale of Europe and Africa. The thrushes feed on insects, and fruits and berries in season. Several plants depend on fruit-eating birds to disseminate their seeds. In the Sierra, thrushes consume the fruits of cherry, Twinberry, currant, gooseberry, blueberry, serviceberry, elderberry, and mistletoe. Some seeds pass through their digestive tracts unharmed and are inadvertently dispersed by the birds as they fly and perch.

The **Western Bluebird** *(Sialia mexicana)* is dark blue with a red-brown upper back and chest band. In the Eastern Sierra, breeds from Alpine Co. north and Inyo Co. south. Inhabits lower Montane Forest, Eastern Sierra Riparian Woodlands, and Pinyon-Juniper Woodland. Year-round resident on the west slope.

MOUNTAIN BLUEBIRD, *Sialia currucoides*   L: 6.5–7.8 in (16.5–20 cm). A bluebird of high elevations. Male is intense blue above, gray-blue below with white belly. Female brown except for blue on rump, tail, and wings, and gray head and back. Immature like female but with spotted breast. Male sings a soft, warbling song. Flies from a low perch to catch insects on the ground; also dives for insects from a low hover. Eats berries. Breeds from early April to late August; lays 4–8 eggs, incubates 13 days. Builds nest of stems, roots, grasses, and bark in natural tree cavity or woodpecker hole. Resident on both Sierra slopes, in Subalpine and Montane forest, Montane Riparian Woodland, often in Mountain Meadows, and Pinyon-Juniper Woodland. *Distribution* Breeds in central Alaska and the southern Yukon, east to central Manitoba, south through the montane western United States, east to western South Dakota. Winters from Oregon south through California, southern Nevada east to Oklahoma, south to northern Baja California and central Mexico.

TOWNSEND'S SOLITAIRE, *Myadestes townsendi* L: 8–9.5 in (20.5–24 cm). An uncommon, slender thrush. Gray, with buffy wing bars, white outer tail feathers, and white eye ring. Immature birds have black-and-white spots over breast and back. Sings a long, warbling song from high perch. Forages for insects in trees and on the ground; fly-catches from perch. Eats fruits and berries, particularly juniper berries, during winter months. Breeds from early May to late August. Places nest on ground or on cliffs, under overhang, or under downed trees. Constructs nest of twigs, sticks, and grass. Lays 3–5 eggs, sometimes 2 broods. Resident along both Sierra slopes. In the Eastern Sierra nests in Subalpine and Montane forest; winters in lower Montane Forest, Pinyon-Juniper Woodland, High Desert Riparian Woodland, and Sagebrush Scrub. *Distribution* Breeds from east-central Alaska to southern Yukon and west-central Mackenzie, east to southwestern Alberta, south through the western United States to northern Mexico. Winters from southwestern Canada south through the western United States, east to Missouri, Oklahoma, and Texas to northern Mexico.

SWAINSON'S THRUSH, *Catharus ustulatus* L: 6.5–7.8 in (16.5–20 cm). A thrush of dense thickets and forests. Similar to Hermit Thrush. Uniform brown above, white below with dark-spotted breast; white eye ring. Sings a melodic series of notes that rise in pitch. Forages on ground and in trees for insects; also takes fruit and berries. Breeds from mid-April to mid-August. Constructs nest cup of twigs, sedges, leaves, ferns, lichens, and moss in crotch of tree or shrub. Female lays 3–5 eggs and incubates 11–14 days. Nests on both Sierra slopes, on the east slope from Mono Co. north. Inhabits Montane Riparian Woodland and Montane Forest near Montane Meadows. *Distribution* Breeding range includes Alaska, Canada except northernmost regions, western United States, Great Lakes area, the Northeast. Winters from northern Mexico to central South America.

HERMIT THRUSH, *Catharus guttatus* L: 6.5–7.8 in (16.5–20 cm). Fig. 10.5f. A mountain bird noted for its flutelike song, consisting of a clear tone followed by rapidly ascending and descending notes, often alternating; song repeated at different pitches. Brown above, white below with dark-spotted breast. Tail and rump red-brown. White eye ring and slender bill. Forages on the ground for insects; in fall and winter eats berries and fruit from shrubs. Breeds from early May to mid-August, laying 3–5 eggs; incubates 12 days. Builds nest of twigs, bark, moss, grass, and ferns in small tree. Both Sierra slopes during the breeding season in dense Subalpine and Montane forest and Montane Riparian Woodland. *Distribution* Nests in Alaska, Canada except northernmost regions, and the United States except the Southeast. Winters in southern British Columbia, northern United States, western United States to southern Baja California, the Southeast and south to Guatemala.

AMERICAN ROBIN, *Turdus migratorius* L: 9–11 in (23–28 cm). Fig. 10.5e. The familiar backyard thrush of North America. Dark gray above. Male has nearly black head, wings, and tail; white eye ring; breast and belly rusty. Female with lighter breast, juvenile with speckled breast. Song is a series of whistled phrases of several notes each. Forages on the ground for insects, particularly worms, and in trees and shrubs for fruit. Male arrives on breeding ground ahead of female and establishes territory; highly aggressive and reacts to the color red. Robins breed from early April to late August, laying 3–7 eggs per clutch and often raising 2 broods; the female incubates 12–14 days. Nest is a cup of twigs and grass, usually in fork or crotch of tree. Year-round resident of both Sierra slopes in all communities, except higher elevations in winter, but nests only in forests and woodlands. *Distribution* Breeds up to treeline in Alaska and Canada, south through most of the United States and Mexico to Oaxaca. Winters from southwestern British Columbia south to Guatemala.

The **Varied Thrush** *(Ixoreus naevius)* is an uncommon migrant and fall and winter resident on both Sierra slopes. Resembles the American Robin but has orange wing bars and a black band across the breast. Prefers shady, low-elevation canyons in Montane Forest and Sagebrush Scrub.

MOCKINGBIRDS AND THRASHERS: FAMILY MIMIDAE

Named for the mimicking songs of its members—the mockingbirds, catbirds, and thrashers—the Mimidae incorporate the calls of other species and the sounds of their environment into individual, territorial songs. The Northern Mockingbird, for example, has an especially large repertoire of sounds and uses them in various combinations. The mimids build large, cuplike nests on the ground, in trees, and in shrubs. They have slender bodies, short, round wings with ten primary wing feathers, and long tails. The sexes are similar. Members of the Mimidae tend to forage on the ground for insects and seeds; they also glean insects from foliage and eat berries. Except during the breeding season, they are solitary.

The **Northern Mockingbird** *(Mimus polyglottos)* is gray above with white wing bars and white outer tail feathers. Occurs on both Sierra slopes. On the east slope from southern Mono Co. south in residential areas, on farmlands, and in High Desert Riparian Woodland.

*SAGE THRASHER, *Oreoscoptes montanus* L: 8–9 in (20.5–23 cm). A shy, short-billed thrasher of sagebrush communities. Brown-gray above, light breast streaked with heavy, black spots; 2 white wing bars and white spots on outer tail feathers. Eyes are yellow, bill slender and straight. Male sings sweet, warbling song from top of tall shrub in the morning. Runs on the ground while foraging for insects, such as grasshoppers and crickets; gleans insects from foliage; eats some berries. Breeds from early April to mid-August; lays 4–7 eggs, incubates 13–17 days. Nest is constructed of twigs, stems, and sage leaves, and lined with hair, fine plant material, and feathers; it is concealed in low shrubs or on the ground. The Sage Thrasher breeds only on the east slope from Kern Co. north. Prefers Sagebrush Scrub but may occur in other High Desert scrub communities. *Distribution* Breeds from southern British Columbia south to Kern Co., through the interior western United States east to western Oklahoma, and the Southwest. Winters from central California south, the Southwest east to central Texas, and south through Baja California and northern Mexico.

The **Le Conte's Thrasher** *(Toxostoma lecontei)* is an uncommon resident of High Desert communities. Light brown with long, black bill. Year-round resident in Sagebrush, Alkali Sink, and Shadscale scrub communities on the east slope, from southern Mono Co. south. Also southern portion of west slope.

PIPITS: FAMILY MOTACILLIDAE

The family name Motacillidae refers to the tendency of its members, the wagtails and pipits, to pump their long tails up and down. They are slender birds with terrestrial habits, foraging for insects by walking or running. They have a long hind claw and a thin, pointed bill. Motacillids' habitat preferences are similar to those of larks: They frequent open, treeless habitat, such as arctic and alpine tundra, seashores, and meadows, and build nests on the ground. Also, they sing their courtship song on the wing during an aerial display.

AMERICAN PIPIT, *Anthus rubescens*   L: 6–7 in (15–18 cm). A winter resident of lakeshores and grasslands. Dark-streaked gray-brown above and buffy below. Thin, sharp bill, white stripe over eye, and white outer tail feathers. Call like *pippit* in flight. Forages in wetlands; wades into small pools or snowbanks; eats some seeds and berries, insects, small invertebrates. Winter resident of both slopes; on the east slope in Alkali Meadows, Desert Marshlands, and farms. A small number nest from June to mid-August in Alpine Meadows. *Distribution* Breeds in Alaska and northern Canada, south through British Columbia and southwestern Alberta through the western United States. Winters from British Columbia south along the west coast and from New York south along the east coast, in the Southwest, Southeast, and south to El Salvador.

WAXWINGS: FAMILY BOMBYCILLIDAE

Winter residents of the east slope include the *Bohemian Waxwing *(Bombycilla garrulus),* which usually does not occur on the west slope, and the smaller **Cedar Waxwing** *(B. cedrorum),* which occurs throughout both slopes. Both species are brown with a black face mask and chin, pointed crest, waxy yellow tail tip, and red-tipped secondary wing feathers. They travel widely in flocks, searching for berries in lower Montane Forest, Pinyon-Juniper Woodland, Eastern Sierra Riparian Woodlands, and High Desert scrub communities.

SILKY-FLYCATCHERS: FAMILY PTILOGONATIDAE

The **Phainopepla** *(Phainopepla nitens)* is a summer resident of the high desert. The male is black with white wing patches; the female is gray. Both have red eyes and a pointed crest. Diet includes insects and berries, especially mistletoe and juniper. Occurs on the west slope and from central Inyo Co. south on the east slope in Pinyon-Juniper Woodland and Shadscale Scrub.

WOOD-WARBLERS: FAMILY PARULIDAE

The small, active wood-warblers, with their thin, pointed bills, eat mostly insects. Of the 56 warblers in North America, most overwinter in Mexico or farther south. Warblers are renowned singers whose songs are usually composed of trills. Many are boldly colored black and white or have bright yellow undersides. During the breeding season the males, which defend territories, are more brightly colored than females. Most warblers forage and nest in trees or shrubs. If nestlings are threatened by a predator, parent warblers may try to distract predators by feigning injury. Cowbirds, which lay their eggs in the nests of other birds, frequently use warblers as hosts. In the 1960s the Brown-headed Cowbird was a contributing factor to the decline of the endangered Kirtland's Warbler, a species restricted to central Michigan. Birds of this family and subsequent families have 9 primary wing feathers.

ORANGE-CROWNED WARBLER, *Vermivora celata* L: 4.5–5.5 in (11.5–14 cm). Drab warbler of lower-elevation shrub communities. Both sexes green-yellow below with subtle streaks, olive green above, with orange-brown cap on head; cap is seldom visible. Song a rapid trill of *chippy* notes rising and falling in pitch. Forages in shrubs and trees for insects. Breeds from late April to mid-July; the female lays 4–5 eggs and incubates 11–13 days. Builds nest on the ground or low in shrub, using grass, bark, leaves, and plant down. Nests on both Sierra slopes; on the east slope in Montane Chaparral, High Desert Riparian Woodland, and shrubby areas with oak in Montane Forest. Moves upslope after breeding. *Distribution* Nests in much of Alaska and Canada, and in the United States throughout the West. Winters in interior and coastal southern California, central Arizona, Texas, southern Florida, and the Gulf states south to El Salvador.

The related **Nashville Warbler** *(V. ruficapilla)* occurs in Montane Chaparral with some trees, shrubby areas in lower Montane Forest, and Pinyon-Juniper Woodland. Breeds on both slopes; on the east slope from Mono Lake north. A common spring and fall migrant in the southern portion of the Eastern Sierra.

YELLOW WARBLER, *Dendroica petechia* L: 4.5–5 in (11.5–12.5 cm). Fig. 10.6b. Bright yellow warbler of riparian communities. Male has yellow underparts with reddish streaks on breast and side; yellow head. Upperparts olive with black on wings and tail. Female duller than male. Song is rapid and musical, like *wee-wee-wee-witita-weet*. Gleans insects and spiders from foliage and twigs; fly-catches. Breeds from early May to early August, laying 3–6 eggs and incubating for 11 days. Female builds nest on tree limb or in branch fork; uses grass, lichens, moss, and plant fibers, wrapped with spider's silk. One of the most common host species of cowbirds. If cowbird lays an egg in nest, female may build a new floor over all the eggs and lay a new set. Both Sierra slopes. Inhabits Eastern Sierra Riparian Woodlands and shrubby areas in wet Mountain Meadows. *Distribution* Nests in most of Alaska and Canada except the northernmost regions, in the United States except parts of the Southeast, northern Baja California, and central Mexico. Winters from the Southwest and Florida south to Bolivia and Brazil.

YELLOW-RUMPED WARBLER, *Dendroica coronata* L: 5–5.5 in (12.5–14 cm). Fig. 10.6c. A common warbler of coniferous forest. The Audubon subspecies, *D. c. auduboni,* occurs in the Sierra. Male is blue-gray with black streaks above; yellow throat, belly, and wing patch; black breast; bright yellow patch on head, rump, and sides. Female resembles male, but colors are dull. Song is a soft trill. Forages energetically in canopy for insects by gleaning from foliage, hovering, or fly-catching; eats some berries and seeds. Breeds from mid-May to mid-August. Builds nest of twigs, bark, and stems on tree limb or in fork. Lays 3–5 eggs, incubates 12–13 days. Both Sierra slopes; on east slope, inhabits Montane and Subalpine forest and Pinyon-Juniper Woodland. *Distribution* Breeds in Alaska and most of Canada, the western United States, Great Lakes region, Appalachians, Southwest, and northern Baja California south to Guatemala. Winters in southwestern British Columbia, the western states, the Southwest, from Kansas east across the central United States, the East, and Southeast, south to Honduras.

These other *Dendroica* warblers are less common on the east slope. The **Black-throated Gray Warbler *(D. nigrescens)*** inhabits lower Montane Forest and Pinyon-Juniper Woodland from Mono Co. south and Sierra Co. north. The **Hermit Warbler *(D. occidentalis)*** prefers Subalpine and Montane forest primarily from El Dorado Co. north; uncommon to the south. A spring and fall migrant throughout both Sierra slopes, the **Townsend's Warbler *(D. townsendi)*** may occur in Montane and Subalpine forest and Eastern Sierra Riparian Woodlands.

MACGILLIVRAY'S WARBLER, *Oporornis tolmiei* L: 4.8–5.5 in (12–14 cm). A shy warbler that rarely leaves cover. Male has yellow belly and sides, olive upperparts, dark gray hood, broken white eye ring. Female has lighter hood. Song consists of *churry* repeated several times, followed by *cheery* repeated several times. Forages for insects in shrub thickets close to ground; will probe and scrape ground litter. Constructs nest of stems and grass on or close to ground in thickets. Lays 2–6 eggs, incubates 11–13 days. Occurs on both slopes; on the east slope, breeds from early May to late July. Prefers humid, shrubby cover in Eastern Sierra Riparian Woodlands and Montane Chaparral near water. *Distribution* Breeds from southern Yukon south through British Columbia and east to western Alberta, and south through the western United States, east to South Dakota, with scattered populations south to northern Mexico. Winters from southern Baja California and northern Mexico south to Panama.

COMMON YELLOWTHROAT, *Geothlypis trichas* L: 4.5–6 in (11.5–15 cm). A striking, widespread warbler of thickets and brush. Male has bright yellow chin, throat, and breast with white belly; olive green upperparts; black mask bordered above by white. Female lacks mask. Song is a loud, distinct *wichity* repeated several times. Diet includes insects and larvae obtained from foliage, cattails, or the ground. Female lays 3–6 eggs and incubates 12 days. Builds nest of grass, sedge, stems, and leaves, lined with grass and hair, often on ground in marsh or streamside thicket. Breeds on both slopes; on the east slope from early April to early August. Occurs at mid- and lower elevations in Eastern Sierra Riparian Woodlands, Montane Meadows, Desert Marshlands, and Alkali Meadows. *Distribution* Breeding range includes southeastern Alaska, much of Canada, the United States, and northern Baja California to south-central Mexico. Winters from northern California, southern Arizona and New Mexico, the Gulf Coast, and southern Florida south to Panama.

WILSON'S WARBLER, *Wilsonia pusilla*   L: 4.5–5 in (11.5–12.5 cm). Fig. 10.6d. A warbler of thickets and shrubs. Male has bright yellow head and underparts, olive green back, glossy black cap. Female usually lacks black cap. Song is a series of staccato *chips;* the final few notes are lower in pitch. Gleans insects near ground, from trees, and in shrubs; will fly-catch near water. Nests on ground under shrubs; builds nest of moss, leaves, and grass, lined with fine grass. Female lays 4–6 eggs and incubates 11–13 days. A male occasionally mates with more than 1 female. Breeds along both Sierra slopes; on the east slope from late April to early August. Prefers shrubby vegetation around lakes, streams, and marshes, particularly in Montane Riparian Woodland and Mountain Meadows. *Distribution* Nests in most of Alaska and Canada, the West, northern Minnesota, and the Northeast. Winters in southern Baja California, the Gulf states, and south to Panama.

The **Yellow-breasted Chat** *(Icteria virens)* looks like a large version of the Common Yellowthroat. Occurs on the east slope, from Mono Co. south; breeds in High Desert Riparian Woodland. Numbers have declined in the Eastern Sierra.

TANAGERS: FAMILY THRAUPIDAE

The tanagers are brilliantly colored, primarily tropical birds, with only 5 species in North America. Their bills are thick but not conical like those of the finches. Small fruits and berries are their dietary staple. In North America, the tanagers are forest dwellers that live in pairs during the breeding season. Females are mostly yellow; the males are red or red and yellow.

WESTERN TANAGER, *Piranga ludoviciana*   L: 6.5–7.5 in (16.5–19 cm). Fig. 10.6e. A colorful bird of montane coniferous and riparian communities.

Male has red head and chin, bright yellow body, black wings and tail, and 2 white wing bars. Female has yellow underparts, green head and back, with wings and tail like male. Pale, thick bill. Male sings hoarse but short robinlike song from high perch. Forages in middle and upper tree canopy for insects; eats fruit. Builds nest cup of twigs, grass, roots, and moss at end of tree limb. Female lays 3–5 eggs and incubates 13 days. Breeds along both Sierra slopes; on the east slope from early May to early August. Prefers Montane and Subalpine forest, Montane Riparian Woodland, and Pinyon-Juniper Woodland. *Distribution* Breeding range includes southeastern Alaska, southern Mackenzie, British Columbia east to Saskatchewan and south through the western United States, east to South Dakota and south to northern Baja California. Wintering range includes southern Oregon, coastal California, southern Arizona, and southern Baja California to Costa Rica.

EMBERIZID FINCHES: FAMILY EMBERIZIDAE

This family consists of sparrows, towhees, juncos, and their allies. All have pointed, conical bills that reflect a seed-eating diet; they also eat insects. All tend to be territorial and monogamous; the sexes are nearly identical. Because many sparrows are similar in appearance, they are difficult to recognize. Fortunately, they can be identified by their pleasing, distinctive songs. In fact, the Mountain subspecies of the White-crowned Sparrow has "song dialects." The birds breed in willows and other shrubs of high-elevation meadows. The males in each area may share a distinctive song pattern, a variant of the basic White-crowned Sparrow song. Because sparrows typically learn song patterns early in life, young males grow up singing the local pattern.

GREEN-TAILED TOWHEE, *Pipilo chlorurus* L: 6.3–7 in (16–18 cm). Fig. 10.7e. A small towhee of dry, brushy communities. Male and female have a red-brown cap; white throat with 2 dark stripes; green-tinged back, wings, and tail; and gray underparts. Underwings are bright yellow; bill is a slender, light-colored cone. Distinct mewing call. Forages under brush by scratching in litter for seeds. Also eats insects and berries. Builds nest of grasses and twigs, lined with fine material, on the ground or in low shrubs. Breeds along both slopes; on the east slope from northern Inyo Co. north from mid-May to late August. Prefers Montane Chaparral and Sagebrush Scrub. *Distribution* Breeding range includes the drier mountain regions of the West, from southeastern Washington east to southeastern Wyoming, south to southern California, and east to western Texas. Winters in the Southwest south through Baja California and central Mexico.

SPOTTED TOWHEE, *Pipilo maculatus* L: 7–8.8 in (18–22.5 cm). A conspicuous, brightly colored towhee. Both sexes have a black hood, bright red eyes, black conical bill, and black upperparts with white wing bars and white streaks on the back. Sides are bright rust-red, belly is white. Song consists of a trill or may end with a buzzy trill. Scratches noisily in forest litter for food; particularly likes to forage under piles of branches and twigs. Gleans insects from foliage; takes seeds and fruits from branches. Female builds nest on or near ground using grass, leaves, and bark. Lays 2–6 eggs, incubates about 12 days. Year-round resident on both Sierra slopes. Breeds on the east slope from late April to late August. Occurs in dense shrub communities, except at subalpine elevations, including Eastern Sierra Riparian Woodlands, Montane Chaparral, and Sagebrush Scrub. *Distribution* Breeds from southern British Columbia east to southern Saskatchewan, south to northern Baja California, and through Mexico to Guatemala. Winters in all but the northernmost breeding range.

The *American Tree Sparrow *(Spizella arborea)* winters in small numbers in fields, Sagebrush Scrub, and other Eastern Sierra scrub communities.

CHIPPING SPARROW, *Spizella passerina* L: 5–5.8 in (12.5–14.5 cm). Named for its call notes and song with rapid, chipping trill. Both sexes have a bright chestnut crown and a white stripe over the eye bordered by a narrow black stripe through the eye. Upperparts streaked brown and black, underparts light gray and white. Forages on ground or low in foliage for seeds and insects. Nest is an open cup made of grasses and lined with hairs, built in shrubs and trees or, rarely, on ground. Female lays 3–5 eggs, incubates 11–14 days. Summer resident along both Sierra slopes; on the east slope, breeds from late March to early August in dry, open Subalpine and Montane forest. *Distribution* Breeds from eastern Alaska through all but northernmost Canada, south through most of the United States to northern Nicaragua. Winters in the southern United States south through Baja California and Mexico to Tehuantepec.

BREWER'S SPARROW, *Spizella breweri* L: 5–5.2 in (12.5–13 cm). A shy, small sparrow of arid scrub communities. Gray-white stripe over eye, brown-streaked upperparts, white belly, and gray breast. Song consists of a long, buzzy trill, much like the sound of a cicada. Forages on the ground in summer mostly for insects; in fall and winter for seeds. Breeds from April to mid-August. Builds nest of grasses, in shrub, lays 3–5 eggs per nest, and incubates 11–13 days. A rare breeder on the west slope; on the east slope prefers Sagebrush Scrub. *Distribution* Breeds from southwestern Yukon, western Canada, and southwestern North Dakota south to central Arizona and northwestern New Mexico. Winters in its southernmost breeding range, east to Texas and south through central Mexico.

BLACK-CHINNED SPARROW, *Spizella atrogularis* L: 5–5.5 in (12.5–14 cm). A juncolike sparrow with gray head, breast, and rump; flesh-colored bill, brown-streaked upperparts, white wing bars, and white belly. Male has a black chin patch in breeding season. Song is a clear, descending trill. Forages through shrubs for insects and seeds. Nests from late April to early August, laying 2–5 eggs and incubating about 13 days. Constructs nest of grasses, in a small, dense shrub. A rare summer resident on the west slope; a common breeder on the east slope in southeastern Mono Co. and throughout Inyo Co., preferring Pinyon-Juniper Woodland and Sagebrush Scrub. *Distribution* Breeding range includes central California and the Southwest to western Texas and south through central Mexico. Winters in the Southwest and south through its breeding range.

VESPER SPARROW, *Pooecetes gramineus* L: 5–6.5 in (12.5–16.5 cm). Named for its habit of singing in the early evening. Dark streaks on brown back, brown streaks on white underparts, white belly, notched tail with white outer tail feathers, bend of wing red-brown, thin bill. Song ends with trill. Walks on ground, foraging for insects and seeds. Builds nest on ground under shrub or in grass. Female lays 3–6 eggs and incubates 11–13 days. Winters on the west slope; breeds on the east slope from late April to mid-August. Prefers dry fields with scattered shrubs, dry Montane Meadows, and open Sagebrush Scrub. *Distribution* Breeds in southern Mackenzie, across southern Canada, the western United States south to central California, central Arizona, and central New Mexico; Missouri, Tennessee, and North Carolina. Winters across the southern United States and north along the east coast to Connecticut; in southern Baja California, and south through interior Mexico.

The **Lark Sparrow** *(Chondestes grammacus)* is an uncommon inhabitant of lower-elevation, open Sagebrush Scrub or Pinyon-Juniper Woodland. It has white underparts, a red-brown patch behind the eye, and red-brown and white stripes on crown. Resident on the west slope; breeds in Inyo Co. south and in eastern Placer, Nevada, and Sierra counties.

BLACK-THROATED SPARROW, *Amphispiza bilineata* L: 4.8–5.2 in (12–13 cm). A small desert sparrow with distinctive black-and-white markings. Dark gray head, black bib, and bright white stripe above the eye and below the ear; gray back, black tail with white outer feathers; light underparts. Song consists of musical notes and trill. Usually forages on the ground or in shrubs for seeds and insects. Nests from mid-April to late June, laying 3–4 eggs. Builds nest cup of grasses lined with hair, placed low in shrubs. A rare breeder on west slope; on east slope, breeds from Mono Lake south. Occurs in most High Desert communities—Pinyon-Juniper Woodland, and Sage-brush, Shadscale, and Creosote Bush scrub. *Distribution* Breeds from southeastern Oregon east to southwestern Wyoming, and south to north-western Oklahoma and most of Texas; in the Great Basin and south through Mexico to Hidalgo. Winters in southeastern California, the Southwest, and south through its breeding range.

SAGE SPARROW, *Amphispiza belli* L: 5–6 in (12.5–15 cm). A shy, common sparrow of scrub communities. Gray head and back with white patch below the ear, brownish wings, and black tail. Light underparts with a single dark spot on breast. Habitually flicks tail. Song consists of several weak, high notes. Eats insects and seeds by foraging on ground or low in foliage. In winter, may travel in flocks. Usually hides under a shrub or on the ground. Nests from early April to mid-August, laying 3–5 eggs and incubating 13–16 days. Builds cuplike nest of twigs and grasses, in depression in ground or in shrub. Common resident on the west slope. On the east slope, breeds in southern Plumas and southern Lassen counties, and from Mono Lake south; winters in Inyo Co. Prefers Sagebrush and Shadscale scrub. *Distribution* Breeding range includes most of the Great Basin, east to southwestern Wyoming and northwestern Colorado, and south to northeastern Arizona, northwestern New Mexico, and central Baja California. Winters from central California east to southwestern Utah, to western Texas, and south to central Baja California and northern Mexico.

SAVANNAH SPARROW, *Passerculus sandwichensis* L: 4.5–6 in (11.5–15 cm). A widely distributed sparrow of fields and grassy areas. Dark brown streaks above; light underparts streaked with brown; a short, notched tail. Faint yellow line over the eye during breeding season. Bill a flesh-colored, slender cone. Song consists of a few sharp notes followed by a trill. Inhabits grassy areas, such as farmlands and Montane Meadows. Forages on the ground for seeds and insects. Female builds nest of grass, lined with finer materials, in a hollow or scrape on ground under cover. She lays 2–6 eggs and incubates 11–13 days. Breeds on the east slope from April to July, often raising 2 broods. Winters on both slopes, but not in Mono Co. *Distribution* Breeds in Alaska, Canada, and most of the northern United States except the East; also breeds locally in the highlands of Mexico and Guatemala. Winters in the eastern and southern United States, south through Baja California and Mexico to Honduras.

FOX SPARROW, *Passerella iliaca* L: 6.2–7.5 in (15.5–19 cm). Fig. 10.7a. A robust, ground-dwelling sparrow named for the reddish color (foxlike) of the eastern populations. Sierra birds are gray-brown above with rusty rump,

Fig. 10.7.   Birds. (a) Fox Sparrow, *Passerella iliaca*. (b) Dark-eyed Junco, *Junco hyemalis*. (c) Evening Grosbeak, *Coccothraustes vespertinus*. (d) Cassin's Finch, *Carpodacus cassinii*. (e) Green-tailed Towhee, *Pipilo chlorurus*. (f) Gray-crowned Rosy Finch, *Leucosticte tephrocotis*. (g) Red Crossbill, *Loxia curvirostra*. (h) White-crowned Sparrow, *Zonotrichia leucophrys*

white underparts streaked with brown, and a central breast spot. The song is loud and flutelike. Forages in ground litter by scratching backward vigorously with both feet. Eats berries, fruit, insects, spiders, seeds, and buds. Nests from mid-May to early August, laying 2–5 eggs. Females usually incubate, 12–14 days. Nest is a cup of twigs, stems, and grasses, lined with feathers or hair, and built on ground or low in shrub or tree. Common year-round on both Sierra slopes. During the breeding season prefers Montane Chaparral, Montane Riparian Woodland, and open Montane Forest. Widespread during the winter. Some winter residents have migrated to the Sierra from more northern latitudes. *Distribution* Breeds from northern Alaska through Canada (except for the northernmost regions) and south along the interior Pacific states to southern California, central Nevada, central Utah, and central Colorado. Winters from southern Alaska and southern British Columbia south along the Pacific Coast to the Southwest and northern Baja California; part of the north-central United States; and from southeastern Canada south along the Atlantic Coast, the Southeast, the Gulf Coast to northern Sonora.

SONG SPARROW, *Melospiza melodia* L: 5–7 in (12.5–18 cm). Common, conspicuous sparrow of wetlands. Brown above with dark streaks on back and wings. Light stripe above eye, light patch below ear, and dark stripe through eye. White underparts streaked with dark brown and a conspicuous central breast spot. Size increases and color darkens at higher latitudes. Song varies among individuals; begins with long notes followed by short notes and a trill. The Song Sparrow forages on the ground, in shrubs, and in trees for insects, seeds, berries, and fruit. Breeds from late March to mid-August; may raise 2 broods. Female lays 3–6 eggs per nest and incubates 12–13 days. Nest is usually on the ground, sometimes in shrub. Occurs year-round on both Sierra slopes; however, breeders move south in fall and are replaced by migrants from the north that overwinter. Prefers wet Montane Meadows, willows and other shrubs around lakes and along creeks, Eastern Sierra Riparian Woodlands, and Desert Marshlands. *Distribution* Breeds from the Aleutian Islands and Alaska through most of Canada except the northernmost parts, south to central Baja, central Mexico, and across the United States from northern New Mexico east to South Carolina. Winters in the Aleutian Islands, southern Alaska, coastal British Columbia, southern Ontario east to Nova Scotia, the northern United States, and south through the breeding range, including the Gulf Coast and southern Florida.

LINCOLN'S SPARROW, *Melospiza lincolnii* L: 5–6 in (12.5–15 cm). A sparrow of high-elevation wetlands. Resembles Song Sparrow, but has shorter tail, gray stripe over eye, light eye ring, and grayer upperparts. Buffy band on breast has fine streaks; belly is white. Bubbling, wrenlike song. Forages on the ground by scratching with both feet; eats seeds, insects, and spiders. Breeds from mid-May to early August. Female lays 3–5 eggs and incubates 10–13 days. Nests on the ground or in small shrubs; nest is a cup constructed of grasses and lined with hair or feathers. Breeds, but does not winter, on the east slope in Mountain Meadows, willow thickets around lakes and creeks, and Montane Riparian Woodland. Year-round resident on the west slope. *Distribution* Breeds in western Alaska, much of Canada south to southern California, western Nevada, eastern Arizona, and northern New Mexico; the Great Lakes states and east through New England. Winters from northern California, the Southwest, Oklahoma, Kansas east to northern Georgia, the Gulf Coast, and south to Honduras.

WHITE-CROWNED SPARROW, *Zonotrichia leucophrys* L: 5.5–7 in (14–17.5 cm). Fig. 10.7h. Recognized by a striking white cap with a broad black stripe high on each side and a narrower black stripe through the eye. Dark brown stripes and light brown cap on first-year birds. Bill flesh-colored; back, rump, wings, and tail brown; 2 white wing bars; underparts gray with white belly. Song consists of clear whistles and a trill. The species, which consists of 5 subspecies, is known for great geographic variation in song. Not only do the subspecies differ in song, but the various populations differ within 4 of the subspecies. Breeds from early March to late August. Nest, constructed of twigs and grasses, and lined with hairs, is placed on ground or in shrub. Raises 2 broods with 3–7 eggs per nest; female incubates about 12 days. *Distribution* Breeds throughout Alaska, western and northern Canada, the U.S. Pacific coast, and high mountain ranges of the West. Winters in southern British Columbia and the western, central, and southern United States through Mexico to Michoacan.

The Mountain subspecies *(Zonotrichia leucophrys oriantha)* breeds on both Sierra slopes and other western mountain ranges, and in September migrates to Mexico. Occurs in wet Mountain Meadows and moist foothill meadows, around lakes, and near creeks, preferring dwarf willows and other shrubs for nesting. The Gambel's subspecies *(Z. l. gambelii)*, which breeds in Alaska and Canada, overwinters on both slopes. Wintering Gambel's Sparrows occur in open, shrubby lower Montane Forest, Sagebrush Scrub, and farmlands. They forage on the ground and in shrubs for seeds, insects, and vegetable material, including leaves, catkins, and blossoms.

Two relatives of the White-crowned Sparrow, *Harris' Sparrow *(Zonotrichia querula)* and the **White-throated Sparrow** *(Z. albicollis)*, are uncommon winter residents on the east slope. Both may occur in moist foothill meadows, High Desert Riparian Woodland, and scrub communities. The Harris' Sparrow occurs throughout the east slope, the White-throated Sparrow only from Inyo Co. south.

GOLDEN-CROWNED SPARROW, *Zonotrichia atricapilla*   L: 6–7 in (15–18 cm). A winter-flocking associate of the White-crowned Sparrow, which it resembles with the exception of a bold yellow crown bordered on each side by a single broad, black or brown stripe. Song consists of 3 descending whistles, *Oh dear me*. Forages on the ground and in low shrubs for insects, seeds, buds, and flowers. Winter resident on both Sierra slopes. Prefers lower Montane Forest, Sagebrush Scrub, High Desert Riparian Woodland, and Pinyon-Juniper Woodland.   *Distribution*   Breeding range extends from western and southern Alaska south through British Columbia, northernmost Washington, and southwestern Alberta. Winters from southern Alaska south to southern Arizona and northern Baja California.

DARK-EYED JUNCO, *Junco hyemalis*   L: 5–6 in (12.5–15 cm). Fig. 10.7b. The distinctive Oregon subspecies, *J. h. oreganus,* is common in the Sierra Nevada. Male has a black or dark gray hood; female's hood is lighter. In both sexes the back is reddish brown, sides are light brown or cinnamon, and the tail has white outer feathers. Bill is distinctly conical and flesh-colored. Song is a slow, musical trill, a common forest sound in summer. Juncos forage on the ground or low in shrubs and trees for seeds and insects. Breeds from early April to late August. Nest is an open cup made of grasses, on the ground. Female usually lays 3–5 eggs and incubates 11 or 12 days. Travels in flocks after breeding. Year-round resident on both Sierra slopes. Prefers Subalpine and Montane forest adjacent to meadows. Migrates to lower elevations during winter.   *Distribution*   The Oregon subspecies breeds from southeastern Alaska south through British Columbia, east to southwestern Saskatchewan, south along the west coast to northern Baja California, and east to northwestern Wyoming. Winters from southeastern Alaska, southern British Columbia, and the Northwest east to the Midwest, and south to northern Mexico.

BUNTINGS AND ALLIES: FAMILY CARDINALIDAE

This family consists of buntings, cardinals, and some grosbeaks (see also Fringillidae).

BLACK-HEADED GROSBEAK, *Pheucticus melanocephalus*   L: 7–7.5 in (18–19 cm). A common, distinctive bird of streamside communities. Male has black head, black wings with white wing bars, and a black tail with white edges; underparts and back are brown-orange. Female has light stripes on top and sides of head, bordered by dark areas; brown wings and tail; and orange underparts streaked with black. Underwings of both sexes yellow; bill heavy, conical, light-colored. Song is a series of ascending and descending notes, with brief pauses. Diet includes insects, conifer seeds, buds, and fruits. Usually nests in deciduous tree or shrub. Female builds nest cup of roots and twigs; lays 2–5 eggs; both sexes incubate 12–14 days. Breeds on both slopes; on the east slope from late April to mid-August in Eastern Sierra Riparian Woodlands, and usually near water in Montane Forest and Pinyon-Juniper Woodland. *Distribution* Breeds in western Canada, south through the western United States, east to western Kansas; northern Baja California and south through Mexico to Guerrera and Oaxaca. Winters in southern Baja California and Mexico south to Oaxaca and Veracruz.

The **Blue Grosbeak** *(Guiraca caerulea)* is a beautiful but uncommon species of High Desert Riparian Woodland. Males are almost entirely deep blue; females are brown. Breeds on the west slope; rare, breeding resident on the east slope in Owens Valley.

LAZULI BUNTING, *Passerina amoena* L: 5–5.5 in (12.5–14 cm). A colorful finch of shrubby communities. Male has light blue head, throat, and upperparts with rusty brown wash on breast and sides; white belly; dark tail and wings, with 2 white wing bars. Female is brown with light belly, faintly blue on wings and tail. Short, conical bill. Sings a rapid song of varied phrases, ending with a *buzz* or *beep*. Forages on ground, in shrubs, and in trees for insects and seeds. Breeds from mid-April to early August. Builds nest cup of dried grasses in dense thicket of shrubs near water. Female lays 1–6 eggs and incubates 11–14 days. Both Sierra slopes. On the east slope, breeds in Eastern Sierra Riparian Woodlands and, to some extent, Sagebrush Scrub and other scrub communities. May migrate to higher elevations after breeding. *Distribution* Breeding range includes southwestern Canada, the western United States east to western Oklahoma. Wintering range includes southern Arizona south through Mexico to Guerrero.

BLACKBIRDS: FAMILY ICTERIDAE

The 22 species of blackbirds, grackles, meadowlarks, orioles, and cowbirds of North America do not look or act alike but belong to the same subfamily. They vary in length from about 6.5 to 21 inches (16.5–53 cm); plumage colors and patterns also vary from all black to startling combinations of black and red, yellow, or orange. Females are usually less brightly colored than males. Bills vary in shape from long and pointed to short and conical; diets consist mostly of insects and seeds. Several blackbirds are marsh dwellers and compete with each other. For example, the Yellow-headed Blackbird prevents Red-winged Blackbirds from nesting in the best areas. Meadowlarks and Bobolinks inhabit grasslands and fields; orioles occupy forests or shrubland. Several members of this subfamily sing distinctive or, in the case of the meadowlarks, melodic songs. Several species are polygynous, and the females assume the major burden of raising young. The 2 cowbird species (plus a recent invader) that occur in North America are *brood parasites*—the females lay their eggs in the nests of host birds and play no role in raising young. In the Sierra Nevada, the Brown-headed Cowbird is a recent arrival. Not until the 1950s and 1960s were there records of large numbers and sightings at higher elevations. The populations of some songbirds—including the Yellow and Wilson's warblers; Solitary, Warbling, and Bell's vireos; Song Sparrow; and Willow Flycatcher—have declined since cowbird populations have increased. They are known hosts of the cowbird, but other factors may also be involved in their decline.

RED-WINGED BLACKBIRD, *Agelaius phoeniceus* L: 7–9.5 in (18–24 cm). Fig. 10.6f. A common bird of marshy wetlands. Adult male is glossy black with bright red wing epaulets bordered at lower edge by yellow; year-old male has red patches but resembles female. Female has light underparts heavily streaked with brown, brown upperparts with light streak over eye. Bill is conical, but sharply pointed. Male sings a loud, squeaky *kon-ka-ree* while clinging to a cattail. Diet includes insects and grain. In winter, Red-winged Blackbirds may forage in large flocks with cowbirds, starlings, and other blackbird species. They nest from late March to late July. Males arrive first and establish territories. More than 1 female may settle in a male's territory; thus, some males are polygynous. Female builds nest in cattails, shrub, tree, or on ground, constructing it of leaves and sedges, lined with fine grass. She lays 2–6 eggs and incubates 11–13 days. Breeds at all elevations on both Sierra slopes, but overwinters at lower elevations. *Distribution* Breeds in southern Alaska, Canada except the northernmost regions, the United States south to Costa Rica and northern Baja California. Winters from southern British Columbia south through most of the United States to Costa Rica.

WESTERN MEADOWLARK, *Sturnella neglecta* L: 8–11 in (20.5–28 cm). Fig. 10.6g. A bird of grasslands and farmlands known for its musical song. Both sexes are yellow below with dark-streaked white sides and undertail. A large black V on breast; mostly mottled and streaked black and brown above;

white stripe over eye. Outer feathers of short tail mostly white; bill long and pointed. Song is loud and melodious, frequently repeated; usually delivered from perch. The Western Meadowlark forages on ground, sometimes in flocks in winter, for insects, spiders, snails, and grain; occasionally feeds on carrion. Breeds from March to late August, raising 2 broods. Female lays 3 – 6 eggs per brood and incubates 13 –14 days. Builds nest on ground, concealed by grass. Nest of stems and grass has a dome and side entrance. Year-round Sierra resident. On east slope, occurs in Montane and Alkali meadows, grassy areas of forest and shrub communities, and farmlands. *Distribution* Breeding range includes southwestern Canada, the Great Lakes region, across the West and Midwest to western Ohio and south to central Mexico. Winter range includes Washington and Oregon south through the West and Midwest, and the Gulf states south to central Mexico.

YELLOW-HEADED BLACKBIRD, *Xanthocephalus xanthocephalus* L: 8 – 11 in (20.5 – 28 cm). A striking, robust blackbird of marshes. Upper body of male—head, neck, breast, and shoulders—is bright yellow; area around eyes and rest of body is black, except for 2 white patches on each wing. Female is smaller than male and brown, with yellow on throat and breast. Bill is stout at base and pointed. Song sounds like a raspy *klee-klee-klee-ko-kow-w-w.* The Yellow-headed Blackbird establishes breeding territories in marshy areas around lakes and ponds at montane and lower elevations. Feeds in meadows, including Desert Marshlands and Alkali Meadows. Walks along shorelines and in fields searching for insects and snails; also forages in marshes by gleaning and fly-catching. Males arrive on marsh before females and establish territories, which are fiercely contested. A male may acquire more than 1 mate. Localized breeding on both slopes. On the east slope, breeds from late April to late July. Female builds baglike nest of grass and cattails, suspended from supporting stalks above water; lays 2 – 5 eggs and incubates 12 – 13 days. *Distribution* Nests from interior British Columbia east to western Ontario, Minnesota, Michigan, and south through the West and Midwest. Winters from eastern Arizona to western Texas, and south to south-central Mexico.

BREWER'S BLACKBIRD, *Euphagus cyanocephalus* L: 8 – 10 in (20.5 – 25.5 cm). Widespread, year-round Sierra resident. Males are all black with purple gloss on head, green gloss on body, and yellow eyes. Female is mostly gray-brown, lighter below, with brown eyes. Long, pointed bill. Song is a hoarse whistle. Makes a variety of other squeaking and whistling sounds in flocks. Diet includes insects, spiders, snails, seeds, and grain. Forages on ground; jerks head forward slightly with each step. Will travel as far as 6 miles

(10 km) from nesting to foraging areas. Breeds from early April to late July, often in colonies of up to 20 pairs. Male may have more than 1 mate. Builds nest of twigs, conifer needles, grass, and mud, on ground, in shrub, or in conifer. Female lays 3–7 eggs, incubates 12–14 days. Breeds and forages in Mountain Meadows, Montane Forest near water, Eastern Sierra Riparian Woodlands, and in marshy and grassy areas. Moves downslope in winter. *Distribution* Breeds from British Columbia through southern Canada to Ontario; from the western United States east to Oklahoma, and the Great Lakes region. Winters from southern British Columbia and Alberta through the western United States, east to Kansas and Oklahoma, the Gulf states and southern Florida, and southern Baja California to south-central Mexico.

BROWN-HEADED COWBIRD, *Molothrus ater*   L: 6–8 in (15–20.5 cm). A small blackbird that frequents fields and pack stations. Male cowbird is black with brown head; female is gray-brown. Bill conical, like that of finch. Song resembles *Glug-glug-glee;* male also makes thin, shrill whistle in flight. Breeds from late April to mid-July. Male defends large breeding territories in Eastern Sierra Riparian Woodlands and adjacent Montane Forest. Female cowbirds will search for the nests of other birds within these territories, usually laying 1 egg per nest. One female may lay up to 40 eggs per season. After laying egg in host nest, female cowbird may return and remove 1 or 2 host eggs. Cowbird eggs hatch after 10–12 days; young are raised by host species. Most host birds are usually the same size or smaller than cowbirds. Young cowbird may force nestlings of host species out of the nest. Some species, such as the American Robin, can distinguish cowbird eggs from their own and throw them out. Some hosts build a new nest floor over all the eggs and lay again. Cowbirds will travel as far as 7 mi (11 km) to feeding areas, which include pack stations, meadows, grasslands, and farmlands, where they forage for insects, grass, seeds, and grain. Will take insects flushed by livestock; also take insects from the backs of horses and cows. In the Eastern Sierra, numbers increased before 1980, attributed to disturbances from livestock grazing, logging, irrigation, and urbanization. Recent population estimates, however, suggest a decline in numbers. The decline of several songbird species may be partly the consequence of increased cowbird parasitism, but there is no clear evidence to date. Resident on both Sierra slopes, overwinters at lower elevations. *Distribution* Breeding range includes southern Canada, most of the United States, northern Baja California to central Mexico. Wintering range includes western California, the Southwest, southern Midwest and southern Great Lakes region, New England, the Gulf Coast, the East and Southeast, Baja California, and Mexico south to Oaxaca and Veracruz.

BULLOCK'S ORIOLE, *Icterus bullockii* L: 7–8 in (18–20.5 cm). A brightly colored inhabitant of riparian communities. Male has orange underparts, face, and rump; black cap, back, wings, central tail, streak through eye, and narrow streak on throat and chest; broad white wing patch. Female is olive gray above, buffy white below, with darker wings. Long, sharp bill. Song consists of musical whistles. Forages in trees and shrubs for caterpillars, other larvae, and insects; eats fruit and takes nectar from flowers. May nest colonially. Nest, constructed by female, is a pendulous bag of plant fibers, hair, and string suspended from tree branch. Female lays 3–6 eggs, incubates 12–14 days. Breeds locally on both slopes; on the east slope from late April to mid-July in Eastern Sierra Riparian Woodlands. *Distribution* Breeds from interior British Columbia east to southwestern Saskatchewan and south to northern Baja California and northern Mexico. Winters along coastal California, along the Gulf Coast, and from central Mexico to Guatemala.

* **Scott's Oriole** *(I. parisorum),* a yellow-and-black species, breeds in Pinyon-Juniper Woodland and Joshua Tree Woodland from southern Mono Co. south.

CARDUELINE FINCHES: FAMILY FRINGILLIDAE

Of the Fringillidae's three subfamilies, only the cardueline finches are represented in the Sierra Nevada. They too, as discussed in the family Emberizidae, have a conical bill that is specialized for cracking seeds. The size and shape of the bill in fringillid finches varies with diet. The most extreme bills are nearly parrotlike and belong to the grosbeaks. Their very large bills can handle large fruits and seeds, such as crabapples and acorns. Finches with smaller, slender bills consume small seeds and insects. Crossbills (genus *Loxia*), with their oddly crossed upper and lower beak, can pry apart cone scales (of pine, hemlock, fir, spruce, larch) and lift out a seed with the tongue. Other foraging adaptations include mouth pouches in rosy finches and the Pine Grosbeak that enable them to carry food to the nest. Several other finches that winter in cold climates have a crop in which they store seeds before they roost. These seeds are digested at night and provide the energy to maintain body temperature. Most finches supplement their seed diet with insects, spiders, fruits, and sometimes vegetable matter.

Many cardueline finches inhabit northern coniferous forests in North America and depend on conifer seeds for food. Periodically, the coniferous species in a region fail to produce cones; thousands of finches then wander widely in search of food. In these unusual years, cardueline finches occur well outside of their normal range.

GRAY-CROWNED ROSY FINCH, *Leucosticte tephrocotis* L: 5.8–6.8 in (14.5–17.5 cm). Fig. 10.7f. A finch of rugged, alpine habitats that occurs in the high mountains of the Pacific states. The colorful males are brown with deep pink on the wings and rump, a black patch on the forehead, and a gray crown. Female is duller with a gray crown. Vocalizations consist of call notes but no true song. Often forages on snowbanks for frozen insects; also fly-catches, gleans, and forages on the ground for insects; consumes seeds as well. Breeds from June to early August. Female lays 3–6 eggs and incubates 12–14 days. During nesting, both sexes develop a pair of upper throat pouches that enable them to carry food to their young. Builds nest of grass in rock crevices, openings in cliffs, or among boulders. Resident year-round on the east slope; on the west slope from Tulare Co. north. Occurs in Alpine Rock and Meadow Communities near snow fields, talus slopes, and rocky outcrops. Some rosy finches remain at high elevations in winter; others move eastward to lower elevations, traveling in large flocks after nesting.   *Distribution*   Breeding range includes the Aleutian Islands, parts of Alaska, northwestern Canada south through the Cascades, Sierra Nevada, north-eastern Oregon, northwestern Montana, and central Idaho. Winters in the Aleutian Islands, southwestern Canada south to eastern California, central Nevada, central Utah, northern New Mexico, and northwestern Nebraska.

PINE GROSBEAK, *Pinicola enucleator* L: 8–10 in (20.5–25.5 cm). Large red finch. Numbers vary from year to year. Male is red to purple-red; female is gray with yellow-brown tinge on upperparts. Large, conical bill; slightly forked, dark tail; and dark wings with white wing bars. Song is a low warble. Forages in trees and shrubs, and on the ground. Eats conifer seeds in fall; also eats seeds of deciduous trees, large fruits, berries, and insects. Breeds from late May to early August. Female lays 2–6 eggs and incubates about 13 days. Nest of twigs, lined with grass and hair, is built on conifer crotch or branch or in shrub. During breeding, both sexes develop paired, upper throat pouches for carrying food. Resident on the east slope from Mono Co. to southern Plumas Co.; on the west slope from Tulare Co. north. May wander outside typical range in winter. Prefers open Subalpine and Montane forests near meadows and Montane Riparian Woodland.   *Distribution*   Breed-ing range extends from Alaska through Canada, except northernmost areas, south to central California, central Nevada, Arizona, New Mexico, northern Great Lakes, and New England. Winters throughout the breeding range and in the Southeast.

PURPLE FINCH, *Carpodacus purpureus* L: 5.5–6.2 in (14–15.5 cm). Easily confused with the other *Carpodacus* finches. Mature male is purple-red over head, upperparts, breast, and upper belly, with white lower belly. Female and first-year male are brown above with broad white stripe over eye, underparts with large brown streaks. Both have brown wings, brown notched tail, and heavy, conical bill. Songs consist of rapid notes and warbling. Forages on the ground and in shrubs and trees for seeds, insects, berries, and buds. Nests from mid-April to mid-August, sometimes raising 2 broods. Female lays 2–7 eggs and incubates 12–13 days. Builds nest—constructed of twigs and grass, and lined with hair—on tree branches. Breeds from Mono Co. north on the east slope; year-round resident on the west slope. Prefers shady Montane Forest and Montane Riparian Woodland. *Distribution* Breeding range includes northwestern Canada east to Newfoundland, south through the West to northern Baja California; the Great Lakes states and the Northeast south to West Virginia. Winters from southwestern British Columbia south to northern Baja, and southeastern Canada south to southeastern Texas, the Gulf Coast, and central Florida.

CASSIN'S FINCH, *Carpodacus cassinii* L: 6–6.5 in (15–16.5 cm). Fig. 10.7d. A colorful finch of high elevations. Male has a bright red crown contrasting with brown neck and back; red also on rump, throat, and breast; white belly. Female is brown above with narrow, light stripe over eye and white underparts with distinct narrow, brown streaks. Both sexes have a robust, conical bill, brown wings, and brown tail, slightly notched. Song is a short, loud warble. Breeds from late April to late July. Female lays 3–6 eggs and incubates about 13 days. Builds nest of twigs, lined with finer materials, on conifer branch. Forages both on the ground and in trees for insects, berries, and buds. In fall, Cassin's Finches typically forage in cones on trees for conifer seeds. Resident on the west slope. Breeds throughout the east slope; overwinters from Mono Co. north. Occurs at edges and in open Subalpine and Montane forest and Montane Riparian Woodland; in winter may descend to lower elevations. *Distribution* Breeds from southwestern Canada, western Montana and Wyoming, south through the interior of the Pacific states to northern Arizona and New Mexico. Winters throughout its breeding range and south to northern Baja California and the highlands of Mexico to Zacatecas.

HOUSE FINCH, *Carpodacus mexicanus* L: 5–5.8 in (12.5–14.5 cm). The common reddish finch of lower elevations. Red on male restricted to crown, breast, and rump. Female has streaked underparts. Unlike the other *Carpodacus* finches, male has streaked sides and female has no eye stripe. Both sexes have large, conical bills and brown upperparts, wings, and tail; tail is slightly notched. Loud, warbling song, often repeated. Forages on the ground for seeds; also takes fruits, berries, blossoms, and buds. Breeds from March to August; lays 3–6 eggs, incubates 12–17 days. Feeds young by regurgitating food from crop. Builds nest in trees, under eaves of buildings, or on ground. Nest constructed of grasses, stems, and leaves, lined with hair. Several birds may nest in same vicinity. Breeds but does not winter on the east slope; resident on the west slope. Prefers open Montane Forest, Pinyon-Juniper Woodland, Sagebrush Scrub and other scrub communities, and urban and agricultural areas. *Distribution* Breeds from southern British Columbia through most of the western United States, east to central Texas and south through Baja California and Sonora to Tehuantepec. Winters throughout the breeding range. Introduced to the Northeast in 1950; now occurs in southern Canada, parts of the Midwest, New England, and the Southeast.

RED CROSSBILL, *Loxia curvirostra*　L: 5.5–6.5 in (14–16.5 cm). Fig. 10.7g. A conifer-seed-eating finch that is nomadic and unpredictable in occurrence. The breeding male is orange-red above and below, and the female is yellow-brown. Both have dark wings and tail; tail is notched. The long, conical bill is crossed at the tips. Song consists of whistles and warbles. Forages in conifer cones, including fir, hemlock, and pine, prying open scales and extracting seeds; also takes the seeds of deciduous trees and shrubs, buds, foliage, and insects. Actively seeks salt—will eat it off highways. Nests anytime from January to August. Female lays 2–5 eggs and incubates 12–18 days. Builds nest of twigs and rootlets, lined with grasses, fur, or feathers, in cluster of conifer foliage on branch. Resident and nomadic on the east slope and on the west slope from Tulare Co. north. Breeds in open Subalpine and upper Montane forest. Wanders in flocks through other communities in winter searching for food. *Distribution* Breeding range extends from southeastern Alaska and northwestern Canada east to Newfoundland and south to southern California and Baja California, east to western Texas, the upper Great Lakes states, the Northeast, and south to eastern Tennessee and western North Carolina, and south through the highlands of Mexico to Nicaragua. Winters in the breeding range but wanders elsewhere erratically. Also occurs on other continents.

PINE SISKIN, *Carduelis pinus*　L: 4.5–5.2 in (11.5–13 cm). A small, nomadic finch that travels in flocks. Heavy, dark streaks on gray-brown back and on light underparts; dark wings with yellow patches; dark, notched tail with yellow base; slender bill. Long, buzzy call: *shreeee*. Forages for seeds and insects in conifers, deciduous trees, shrubs, weeds, and wildflowers, and on the ground. Breeds from early April to late July. Female lays 2–5 eggs and incubates 13 days. Nest is a platform of twigs and grass, lined with hair and feathers, built in foliage of trees. Resident on the west slope and on the east slope from Mono Co. north. Breeds in open Subalpine and Montane forests. In winter wanders widely, in Pinyon-Juniper Woodland, Montane Chaparral, Montane Riparian Woodland, and Sagebrush Scrub. *Distribution* Breeding and wintering range includes Alaska, parts of Canada east to Newfoundland; south to southern California and northern Baja California, the Southwest east to western Texas, western Oklahoma, Kansas, Missouri, and through the southern Great Lakes states; south through the Mexican highlands to Michoacan and central Veracruz.

LESSER GOLDFINCH, *Carduelis psaltria* L: 3.8–4.2 in (9.5–10.5 cm). A small finch with bright yellow underparts. Male and female are bright yellow below and greenish on the back and rump; both have black wings with white patches, and black tail with white at base. Male has a black cap. Song is canarylike with distinct, scratchy notes. Forages on the ground and low in foliage for seeds, insects, leaves, and buds; particularly likes the seeds of weeds and wildflowers. Nests from early April to mid-August. Female lays 3–6 eggs and incubates 12 days. Nest of grasses and bark, lined with softer material, is built on tree limbs or in dense shrubs. In winter, pairs form large flocks. The Lesser Goldfinch breeds on the east slope from Mono Lake south and Placer Co. north; winters from Mono Co. south. Resident on the west slope. Prefers dry, open forest and woodland, including Montane Forest, Pinyon-Juniper Woodland, Eastern Sierra Riparian Woodlands, and Montane Chaparral. *Distribution* Breeds and winters from southwestern Washington and western Oregon to northern California and northern Nevada, northern Utah, northern Colorado, northwestern Oklahoma, and southern Texas south to South America.

The **American Goldfinch** *(Carduelis tristis)* in spring and summer has yellow on underparts and upperparts. Resident on the west slope but only winters in the Eastern Sierra from Mono Co. south at lower elevations.

EVENING GROSBEAK, *Coccothraustes vespertinus* L: 7–8.5 in (18–21.5 cm). Fig. 10.7c. A robust, large-billed finch of coniferous forest. Male has yellow patch along bill and eyes, yellow belly and back, brown neck. Wings are black with a large, white patch; tail is short, black, and notched. Female is gray with yellow tinge and has 2 white wing patches. Undulating flight exposes white wing patches. Commonly travels in flocks. Common call is *peet peet kreeck.* Forages in trees and shrubs, sometimes on ground, for fruit, berries, conifer seeds, other seeds, buds, and insects. Nests from June to late August, laying 2–5 eggs per clutch. Female incubates 12–14 days. Nest is a shallow cup built of twigs and roots, placed in foliage on a branch. Breeds and overwinters along both Sierra slopes. Prefers Montane and Subalpine forest for breeding; may wander to lower elevations in winter. *Distribution* Breeds from northern British Columbia and northern Alberta east to New Brunswick and Nova Scotia, south to central California, Nevada, southeastern Arizona, southern New Mexico, the Midwest, parts of the Northeast, and south of the border to central Mexico. Winters throughout the breeding range and irregularly south to southern California, southern Arizona, Texas, the northern Gulf States, central Florida, and south through Mexico to Oaxaca.

MOVING ON

Hundreds of different birds inhabit the Eastern Sierra at one season or another. A few, such as the chickadees and nutcrackers, stay year-round, but many are migrants. Many come in summer, hatching their young just when insects are most abundant. After the young are fledged, in late summer, they leave for lower elevations or warmer climates. Some that nest farther north, such as the Rough-legged Hawk, come south in winter to Eastern Sierra valleys where they find plenty of mice and ground squirrels. Others merely stop briefly to feed, en route to their northern nesting grounds in spring or their southern wintering grounds in fall. Of all the animals, the birds are the most mobile. When it becomes too hot or too cold, when food becomes scarce, when water sources dry up, some birds remedy the problem by flying somewhere else. At dawn they may fly to the top of a tree to catch the warmth of the rising sun, or they may seek shade in dense vegetation at high noon. As the seasons change, some species, such as the nutcrackers, fly up and down the mountain slopes—nesting in the Pinyon or Jeffrey pine forests in spring, flying up to the Whitebark Pine to feed in summer and fall, then returning to the lower elevations for the winter. Others avoid Eastern Sierra winters by flying hundreds or even thousands of miles south, to lower elevations and warmer climates, wherever food is abundant.

You may note that we mention that many bird species are declining, sometimes without being specific about where and why. This reflects the limits of present knowledge. Yet solid evidence shows that when habitat is destroyed—forests clear-cut, brush cleared, marshes drained, water diverted, or domestic animals introduced—many animals that fed, bred, and found cover in that habitat can no longer survive there.

RECOMMENDED READING

Ehrlich, Paul R., D. S. Dobkin, and D. Wheye. 1988. *The Birder's Handbook.* New York: Simon and Schuster.

Gaines, David. 1992. *Birds of Yosemite and the East Slope*. Mammoth Lakes, Calif.: Artemisia Press.

Peterson, Roger T. 1990. *Field Guide to Western Birds*. New York: Houghton Mifflin.

Ryser, Fred A., Jr. 1985. *Birds of the Great Basin*. Reno: University of Nevada Press.

Small, Arnold. 1994. *California Birds: Their Status and Distribution*. Vista, Calif.: Ibis Publishing.

CHAPTER ELEVEN

MAMMALS

DIANA F. TOMBACK

Mammals are wonderfully diverse in body form, behavior, and lifestyle. Many species are aquatic, ranging from the amphibious otters and seals to the completely aquatic porpoises and whales. Mammals have even conquered the air—the gliding flying squirrels, flying opossums, and those aerial performers, the bats. All ways of life we can imagine—burrowing underground or living in trees, meat-eating or plant-eating, solitary or social—are practiced by mammals. Yet, because a large number of the 4000 or so mammals in the world are nocturnal and secretive, most people are not familiar with even the most common species that live near them.

Mammals differ from other vertebrates in many ways. The most obvious difference is a covering of hair or fur. By reducing the loss of body heat, this insulation helps mammals maintain a high, constant body temperature; this enables them to survive in cold climates and to be active at night. In addition, females have mammary glands for nursing young. Young mammals remain with their mothers for some period after birth; they are well developed and experienced before striking out on their own. Mammals also have specialized teeth that reflect their diet. For example, carnivores have pointed, sharp teeth for grasping and tearing, while herbivores have molars with wear-resistant grinding surfaces. In contrast, birds have no teeth at all, and the unspecialized teeth of amphibians and reptiles are used only to grasp prey.

Unlike birds, which may migrate with the seasons to more favorable climates, most terrestrial mammals remain year-round in the same habitat. Mammals are able to endure extreme heat, cold, and aridity because of specialized behavior, physiology, and anatomical features. For example, thick fur enables land mammals to tolerate arctic cold, and thick layers of fat or blubber enable marine mammals to tolerate cold water. Most mammals prevent heat stress by sweating or panting, which cool body surfaces by evaporating body water. However, in arid environments where water is limited, larger mammals such as antelopes conserve body water by increasing their body temperatures each day and unloading the heat at night.

A number of mammals, and many bird species as well, are able to conserve energy by reducing their body temperature and their metabolism when food and water are scarce. Small mammals, such as bats and mice, may become torpid each day. Torpor is a deep, comatose condition in which body temperature drops nearly to air temperature, breathing rate and heart rate slow to less than one breath or beat per minute, and blood flow to the periphery of the body is greatly reduced. Individuals of some species enter torpor only if their energy intake was insufficient for the day, or if the weather is extremely cold. Larger mammals such as bears hibernate through the winter in a state of dormancy. Although dormancy involves a drop in body temperature of only a few degrees and a 50 percent decrease in metabolic rate, the energy savings are great.

Because the Eastern Sierra encompasses a wide array of habitats and conditions, the mammals are also diverse and interesting. The greatest challenges to mammals in the Eastern Sierra are the summer heat and aridity associated with the High Desert plant communities and the cold, snow, and scarce food of mountain winters.

EXPLANATION OF TERMS

Species descriptions that follow usually provide two measurements: *body* refers to the length from the tip of the nose to the rump; and *tail* refers to the length from the base to the tip of the tail. *Total length,* used for some species, includes head, body, and tail. Metric measurements are usually rounded to the nearest 0.5 cm. The *breeding period* usually includes the time of birth through lactation. Mammals that occur only in the Eastern Sierra (and not on the west slope) are noted with an asterisk (*).

SHREWS AND MOLES: ORDER INSECTIVORA

The *insectivores* (animals that feed on insects) are direct descendants of the oldest true mammals (the first placental mammals), which evolved about 200 million years ago. In fact, the shrews are the oldest group of living insectivores, some virtually unchanged over the course of 30 million years. Shrews and moles are small, secretive burrowing animals with poor eyesight but a keen sense of smell.

Small and mouselike, shrews weigh less than 1.2 oz (35 g). Their body size is at the theoretical lower limits for mammals, which presents a major challenge. Since shrews have a large surface area in relation to body volume, their bodies lose heat rapidly. To compensate, they have a high metabolic rate and must eat constantly. In fact, they are active around the clock. For example, the Northern Water Shrew has a ninety-minute activity cycle: It hunts for thirty minutes and rests for sixty. As a result of this frenzied lifestyle, shrews live only about fourteen to eighteen months.

Vicious carnivores, shrews will attack another shrew or mouse twice their size. They search for food with quick erratic movements, changing direction unpredictably. Some have a toxin in their saliva that immobilizes prey. In turn, Sierra shrews are eaten by owls, Steller's Jays, and trout. All shrews have five-toed feet, a scaly tail, and incisor teeth that protrude forward and are used like pincers for handling food. California shrews' teeth have red pigment on them. The nose—long and tapered with well-developed whiskers—is constantly in motion, poking here and there in ground litter. Because shrews have poor eyesight, they use smell, whiskers, and ultrasound twitters (similar to bat radar) bounced off objects to guide their movements. Bats are close relatives.

NORTHERN WATER SHREW, *Sorex palustris*   Body: 3.2–3.5 in (8–9 cm). Tail: 2.5–3.3 in (6.5–8 cm). Fig. 11.1b. Black-gray above with white, gray, or brownish underparts. Long, bicolored tail and naked feet. The large hind feet are fringed with stiff hairs along the sides and toes. Requires cold streams with vegetation cover on the banks for foraging and protection from predators. Swims, walks on the bottom of streams, and "walks on water" using air bubbles trapped in the stiff hairs of its feet and in its fur. Stiff, thick hairs on the hind feet make them into broad paddles. Diet includes insect larvae, salamanders, mice, carrion, and other shrews (including its own species!). Breeds from late February through midsummer; nests in crevices near water. Gives birth to 2 or 3 litters per year, 4–8 young per litter. Resident on the western Sierra slope from Fresno Co. north; on the east slope from northern Inyo Co. north. Prefers high-elevation Montane Riparian Woodland and moist Montane Meadows. *Distribution*   Range includes southern Alaska, southern Canada, mountainous regions of the United States, and the Great Lakes region.

VAGRANT SHREW, *Sorex vagrans*   Body: 2.3–2.8 in (6–7 cm). Tail: 1.5–1.8 in (4–4.5 cm). Reddish brown in summer and black to light gray in winter. Forages in moist areas under ground litter. Diet includes insects, spiders,

Fig. 11.1.   Mammals. (a) Pika, *Ochotona princeps*. (b) Northern Water Shrew, *Sorex palustris*. (c) California Myotis, *Myotis californicus*. (d) Mountain Beaver, *Aplodontia rufa*. (e) American Beaver, *Castor canadensis*. (f) White-tailed Jack Rabbit, *Lepus townsendii*. (g) Yellow-bellied Marmot, *Marmota flaviventris*

earthworms, slugs, some plant material, and other shrews. Breeds from February through November; nests in stumps, logs, or dense vegetation. Gives birth to 1 or 2 litters, 2–9 young each. Inhabits both Sierra slopes. On the east slope from northern Inyo Co. north in Mountain Meadows, Montane Riparian Woodland, and in wet sites in open coniferous forest at all elevations. *Distribution* Ranges from Alaska and western Canada south through the Pacific Northwest, Rocky Mountains, Sierra Nevada, and coastal ranges.

*INYO SHREW, *Sorex tenellus* Body: 2.4 in (6 cm). Tail: 1.6 in (4 cm). Grayish brown with pale underparts and bicolored tail. A little-known species. Absent from the western Sierra. In the Eastern Sierra, in Inyo Co., primarily in lower-elevation communities, such as High Desert Riparian Woodland and aspen stands; also in Sagebrush Scrub and Pinyon-Juniper Woodland.

MT. LYELL SHREW, *Sorex lyelli* Body: 2.4 in (6 cm). Tail: 1.5 in (4 cm). A rare and little-known species. Brownish color above and gray below with bicolored tail. Prefers moist habitats and Montane Riparian Woodland above 6900 ft (2100 m). Restricted to the general vicinity of Mount Lyell, which includes a small part of both Mono Co. and southeast Yosemite National Park.

*MERRIAM'S SHREW, *Sorex merriami* Body: 2.2–2.6 in (5.5–6.5 cm). Tail: 1.3–1.6 in (3.5–4 cm). Upperparts light brownish gray, whitish underparts and feet, bicolored tail. The only *Sorex* species that often lives far from streams and meadows. Eats spiders, grasshoppers, caterpillars, and other insect larvae. Forages in rodent burrows and runways. Breeds from March to June; 5–7 young per litter. Absent from the west slope; on the east slope in Mono Co. Prefers arid communities with grass understory, such as Sagebrush Scrub and Pinyon-Juniper Woodland. *Distribution* The Great Basin and Rocky Mountains, from eastern Washington east through Montana, and south to Arizona and New Mexico.

TROWBRIDGE'S SHREW, *Sorex trowbridgii* Body: 2.4–2.7 in (6–7 cm). Tail: 1.8–2.5 in (5–6.5 cm). Large shrew, uniformly dark gray, black, or sometimes brown, with bicolored tail. Forages in forest litter for insects, centipedes, and earthworms, plus much vegetable matter, including conifer seeds. Breeds from April to August; nests under logs or in shallow holes; 2–3 litters per year, 3–6 young per litter. Occurs on both Sierra slopes; on the east slope from Alpine Co. north. Common in dense Montane Forest, particularly in those areas with much litter accumulation, and often some distance from water in damp areas. *Distribution* Ranges from southern British Columbia south along coastal regions to southern California; inland through northern California and south along the Sierra.

Burrowing animals that are rarely observed aboveground, moles possess many unusual features for their underground lifestyle. Their football-shaped bodies, pointed at each end, enable them to move forward or backward in their long burrows; in addition, their short, soft fur can lie flat in either direction. Although their eyesight is poor, their tactile sense — using stiff hairs that are distributed over the body — is well developed. Moles have no outer ears, but can literally "hear" with their entire body, picking up the sound vibrations of prey or predator.

Moles dig through soil by alternately pushing each broad front foot sideways. A specialized, fused skeleton provides a sturdy frame for attachment of powerful muscles. Excavated dirt is pushed up into mole hills. A single mole may construct 200–500 hills in six months; a mole commonly digs 10–15 ft (3–4.5 m) in an hour. Actually, a mole home range features two kinds of burrows: a network of temporary, shallow surface burrows and several permanent burrows 6–8 inches (15–20 cm) deep. Most moles are territorial and solitary, except during the breeding season when males enter the burrow systems of females.

BROAD-FOOTED OR CALIFORNIA MOLE, *Scapanus latimanus*   Body: 4.4–6 in (11–15 cm). Tail: 0.8–1.8 in (2–4.5 cm). Dark gray, blackish, or coppery brown with a hairy tail. Long, naked snout; broad front feet with palms facing out. Forages year-round belowground by burrowing in search of insects, earthworms, small mammals, small reptiles and amphibians, plant material, and seeds. Young are born in March and April, 1 litter per year, 2–5 young per litter. Inhabits both Sierra slopes; the east slope from northern Inyo Co. north. Occupies moist, crumbly soil over a wide elevation range. Prefers areas with grasses and forbs in Montane Riparian Woodland, lower Montane Forest, wet Montane Meadows, and Pinyon-Juniper Woodland.   *Distribution*   All but the most arid regions of California, southern Oregon, northern Baja California.

BATS: ORDER CHIROPTERA

Bats are the only true flying mammals. The *wings* of bats are large, hairless membranes between the elongated digits of the forefeet and the hind legs and part or most of the tail. Powerful flight muscles are attached to the breastbone. Most bats are insect-eaters. Since bats are nocturnal, they locate, track, and capture insects and other prey by echolocation (like radar) rather than by vision. The larynx produces ultrasonic clicks that are emitted through the mouth or nose at a pulse rate of 10–200 clicks per second. The sound bounces off prey or other objects, and the ears receive the faint echoes of the emitted clicks. As a bat zeroes in on a target, the click echoes become more and more rapid. The bat then speeds up its own clicking rate and lowers the frequency of the clicking sounds. Rather than disrupt its concentration by moving its head, the bat captures prey using a complex maneuver: It nets the insect in a wing or tail membrane and passes it to the mouth.

Because of their small body size, active lifestyle, and high metabolic rate, particularly during flight, bats have high energy requirements. This places a premium on hunting efficiency and energy conservation, particularly in cold climates. During the day and throughout the coldest months, bats re-treat to roosting sites in caves, trees, and buildings that provide shelter from the cold as well as protection from predators. Roosting or hibernating bats hang upside-down by their hind feet, with wings folded forward around the body to minimize heat loss. For temperate-zone bats, winter is a difficult time because insects are not available. Bats hibernate through this period in a state of torpor. In the weeks before hibernation, they may accumulate two or three times their normal amount of fat. As they enter torpor, their metabolic activity decreases, permitting their body to cool to air temperature, but never below freezing.

Mating usually occurs when both sexes of bats gather at roosting sites before hibernation. Later, males may periodically become active and continue to inseminate hibernating females. Females store the sperm during hibernation and do not conceive until spring. The young are born after about two to three months of gestation. Pregnant females and those with young form small nursery colonies in caves, mines, or trees, under bridges, or in rock crevices. The males usually roost alone.

Although it is widely believed that bats should never be handled because they carry the rabies virus, Dr. Rick Adams, University of Wisconsin, insists that rabies infects less than 1 percent of bats and is less of a problem than in many other mammal groups. However, the wisest strategy is to not handle any wild animals.

VESPER BATS: FAMILY VESPERTILIONIDAE

MYOTIS BATS   Total length: 2.9–4 in (7.5–10 cm). Seven species of small *Myotis* bats occur throughout both Sierra slopes. Color ranges from buffy brown to golden brown to dark brown. No accessory skin structures on the nose. Pointed flap of skin in front of each ear. Unlike other bats, no *Myotis* bat has furred wings or tail membranes. Ear, wing, and tail membranes are contrasting dark brown or black.

All *Myotis* species hunt flying insects, particularly moths and beetles. The

Fringed Myotis hunts insects in the forest canopy; the other *Myotis* bats hunt over ponds and streams and over land. Females give birth to a single offspring between early May and late August. The **Little Brown Bat** *(Myotis lucifugus)*, **Yuma Myotis** *(M. yumanensis)*, **Long-legged Myotis** *(M. volans)*, **Long-eared Myotis** *(M. evotis)*, and **California Myotis** *(M. californicus)* (fig. 11.1c) occur at all elevations, except alpine, in forest, riparian, and scrub habitats. The **Fringed Myotis** *(M. thysanodes)* and **Western Small-footed Myotis** *(M. ciliolabrum)* occur at all elevations and habitats except for Subalpine Forest and Alpine Rock and Meadow Communities. *Distribution* The Western Small-footed Myotis ranges from central Mexico through the western United States to southern Canada; also part of the Midwest, and the Northeast. The Little Brown Bat is widely distributed—in northern Mexico, the United States, Alaska, and Canada (but not most of the Northwest Territories). The other *Myotis* bats range from central Mexico through the western United States to Canada.

SILVER-HAIRED BAT, *Lasionycteris noctivagans*   Total length: 3.6–4.2 in (9–10.5 cm). Dark fur tipped with silver. The membrane connecting both legs is partly furred. Forages over both water and land, earlier in the evening than most bats. Gives birth and nurses 1–2 young between late May and late August. Inhabits both Sierra slopes. Prefers Sagebrush Scrub and other shrub communities with trees or buildings for roost sites, Pinyon-Juniper Woodland, Riparian Woodlands, and lower Montane Forest. *Distribution* Ranges from northeastern Mexico through most of the United States, southern Canada, coastal British Columbia, and southern Alaska.

WESTERN PIPISTRELLE, *Pipistrellus hesperus*   Total length: 2.4–3.4 in (6–8.5 cm). Small bat with distinctive buffy or smoke gray color; contrasting black nose, ears, feet, and wing membranes. Prefers to roost in rock crevices and boulder piles. Gives birth to 1 or 2 young in late June and nurses until early August. Found on both Sierra slopes; on the east slope from Alpine Co. south. Prefers High Desert communities, such as Sagebrush, Alkali Sink, and Shadscale scrub; also Riparian Woodlands and Pinyon-Juniper Woodland. *Distribution* Ranges from southern Washington through the western Great Basin and from northern California to central Mexico; the Southwest to western Texas and Oklahoma.

BIG BROWN BAT, *Eptesicus fuscus*   Total length: 4.2–4.8 in (10.5–12 cm). Large bat with dark brown, glossy fur and black ears, feet, and wing membranes. Diet includes many flying beetles and also nonflying insects; forages above meadows and among trees. Females give birth to 1 young between late May and early July. Both Sierra slopes at all elevations. On the east slope prefers Montane Forest and Pinyon-Juniper Woodland. *Distribution* Ranges from southern Canada south through the United States to Colombia and Venezuela.

WESTERN RED BAT, *Lasiurus blossevillii* Total length: 3.6–4.4 in (9–11 cm). Brick or rufous red fur tipped with white; rounded ears. Fur-covered membrane connects legs. Forages over wet Montane Meadows for moths, beetles, crickets, cicadas, and other flying insects. Roosts and hibernates in trees. Females give birth to 1–5 young in late May and June. Breeds on both Sierra slopes. Prefers Eastern Sierra Riparian Woodlands and aspen stands. Migrates in August and September, wintering at lower elevations west of the mountains. *Distribution* Ranges from southern Canada through the United States (except the Great Basin and Rocky Mountains) and south to Chile and Argentina.

The **Hoary Bat** *(L. cinereus),* a relative of the Western Red Bat, breeds on the west slope and only migrates through the east slope in spring and summer. Prefers Montane Forest, Montane Riparian Woodland, and Pinyon-Juniper Woodland.

TOWNSEND'S BIG-EARED BAT, *Plecotus townsendii* Total length: 3.5–4.3 in (9–11 cm). Brown bat with long ears joined at the base; 2 large bumps between the eyes. Forages during the night for moths in the air and from foliage. Females give birth to 1 young from mid-May through June. All communities on both Sierra slopes except Subalpine Forest and alpine communities. *Distribution* Ranges through the West to southern British Columbia, east to Oklahoma and Kansas, and south through Mexico; also occurs in the central Appalachians.

PALLID BAT, *Antrozous pallidus* Total length: 4.5–5.3 in (11.5–13.5 cm). Gray to buffy with long, winglike ears; ears and wing membranes dark brown. A bat of High Desert communities. Forages on or close to the ground, eating ground-dwelling or low-flying insects, such as scorpions, crickets, and beetles, and also lizards. Females bear 1–2 young from late April to late June. Both Sierra slopes; on the east slope from mid–Mono Co. south. Common in Pinyon-Juniper Woodland, High Desert Riparian Woodland, and Sagebrush, Creosote, Shadscale, and Alkali Sink scrub communities. *Distribution* Ranges from southern British Columbia south through the Pacific states, from Nevada east to Kansas, and south to central Mexico.

The **Spotted Bat** *(Euderma maculatum)* is a poorly known, rare species that occurs in the southern portion of the west slope and on the east slope from

Alpine Co. south. It is thought to inhabit Pinyon-Juniper Woodland and High Desert scrub communities.

FREE-TAILED BATS: FAMILY MOLOSSIDAE

BRAZILIAN FREE-TAILED BAT, *Tadarida brasiliensis*   Total length: 3.5–4.4 in (9–11 cm). Chocolate brown or red-brown; half the tail extends beyond the membrane connecting both legs. Hunts moths over meadows and in trees. Gives birth late May to early July usually to 1 young. Nursery colonies may consist of thousands of bats; one of the largest nursery colonies is in Lava Beds National Monument and has more than 100,000 bats. Females in these colonies do not discriminate; they nurse any young from their immediate group. Inhabits both Sierra slopes. Occurs at all but the highest elevations of the east slope. Breeds in open Pinyon-Juniper Woodland and in High Desert scrub communities.   *Distribution*   Ranges from California and the Southwest east to Missouri, the Southeast, and south to northern South America.

PIKAS, RABBITS, AND HARES: ORDER LAGOMORPHA

Lagomorphs are often confused with rodents because of their similar food habits and prominent front incisors. Rodents, however, have two upper incisors while lagomorphs have four: two front teeth and two smaller "peglike" teeth directly behind. Like the incisors of rodents, lagomorph incisors grow continuously. Nearly worldwide in distribution, lagomorphs occur naturally on every continent except Australia (where they have been introduced). Rabbits and hares inhabit tropical forest, grassland, desert, deciduous forest, coniferous forest, and arctic and alpine tundra. Some species, such as the Black-tailed Jack Rabbit, have even adapted to the suburbs. All lagomorphs are grazers and browsers, eating grasses, twigs, bark, and foliage. Despite their diverse habitats, lagomorphs' body form differs little.

All lagomorphs are *coprophagous*—that is, they produce a special kind of fecal pellet that they eat directly from the anus. These soft, green pellets are rich in protein and other nutrients; they are the product of bacterial digestion in the *cecum,* an intestinal pocket found in many herbivores. The intestine itself produces hard, brown pellets, which are not eaten. Coprophagy usually occurs when individuals are resting. If deprived of coprophagy, they die in a few weeks.

PIKAS: FAMILY OCHOTONIDAE

The family Ochotonidae consists only of the little animals known as Pikas, which look more like guinea pigs than rabbits. Their hind legs are barely longer than their front legs, and they have no tail. They require talus slopes adjacent to meadows, shrubs, or forest. Pikas are well-adapted to cold, but their thick fur coat makes them vulnerable to heat stress. They are active all winter, foraging aboveground on rocks for lichens or in nearby meadows for grass, or tunneling beneath the snow. During summer they take refuge in cool spaces under the large rocks of talus slopes. During winter Pikas supplement scarce food by feeding on the "haystacks" they constructed the previous summer and fall. Pikas work hard from July through November gathering extra grasses and forbs, and piling them neatly on rocks to dry in the sun. Recent studies indicate that Pikas may select plants with natural "preservatives" (chemicals that inhibit decay) for their hay piles. Each Pika defends a territory, protecting its haystacks from neighbors. While sitting on their favorite boulders, Pikas often give territorial calls.

Pikas are always alert for predators and readily voice an alarm squeak at the approach of any large animal, including humans. They easily escape Coyotes, Martens, and hawks by ducking into crevices and retreating through complex, narrow passages among the rocks. However, weasels are small and agile enough to follow a Pika; these predators are so threatening that Pikas do not emit alarm calls at their approach, perhaps hoping to be inconspicuous.

Pikas may not be very social, but they have several vocal signals for predator and territorial defense, and also for courtship. Aggression among Pikas is lowest in spring during snowmelt, when males expand their territories to overlap with those of neighboring females. Courtship ensues, and the first litter is born shortly thereafter. Males do not participate in parental care. Before the second litter is born, female aggression increases, and the males retreat to their previous territories. Juveniles begin to build hay piles their first summer but will not seek a territory until the following spring. Because a talus slope is usually filled with Pika territories, this is a difficult time in a Pika's life. Young males are usually forced into the poorer talus territories, and females may end up dispersing to other areas.

PIKA, *Ochotona princeps*   Body: 6.4–8.5 in (16–21.5 cm). Fig. 11.1a. Small and plump with no visible tail; short, broad, rounded ears edged with white. Grayish buff to dark brown above and buffy underparts. Does not hibernate but stores "haystacks" of grasses and forbs under rocks for winter. Lives in crevices among the rocks. Eats grasses, herbs, lichens, conifer twigs, and the foliage of shrubs and trees. Breeding period from April through late September. Two litters per year, 1–5 young per litter. Both Sierra slopes. Prefers talus slopes and rock slides near alpine communities, Mountain Meadows, Subalpine and Montane forest, and Montane Chaparral; occasionally at lower elevations adjacent to Pinyon-Juniper Woodland and Sagebrush

Scrub. *Distribution* Range includes the higher mountainous regions of the western United States and southwestern Canada.

RABBITS, COTTONTAILS, AND HARES: FAMILY LEPORIDAE

Rabbits and cottontails are plump and have relatively short hind legs, ears, and necks. Because they cannot run fast, they need brushy habitats to hide from predators. Their nests are usually shallow depressions in the ground lined with grass and hair pulled from the female's body. The young are born naked with closed eyes. In contrast to rabbits and cottontails, hares are larger and more gangly, and have long necks, hind legs, and ears. (Bugs Bunny is a hare and not a rabbit.) They run rapidly and need only sparse shelter for protection; jack rabbits have been clocked up to 70 mph (117 kph). Many hares do not build nests, only depressions in the soil. The young, born fully furred with open eyes, leave the nest within twenty-four hours.

All leporids run a high risk of becoming dinner for a predator; few individuals live more than two years. Snowshoe Hares are a staple food for Great Horned Owls, Coyotes, weasels, foxes, Bobcats, and Canada Lynx. Rabbits, cottontails, and jack rabbits are food for Gopher Snakes, rattlesnakes, Red-tailed Hawks, Northern Harriers, Golden Eagles, and Barn Owls. Defenses include evasive running, retreating into the shelter of brush, and "freezing" in place. Some species scream shrilly if caught, which may startle the predator into loosening its hold. Hares and jack rabbits, with their long ears, rely more on hearing than sight to detect predators. This well-developed sense, coupled with high-speed running, enables them to wander some distance from cover with a good chance of escaping a predator.

One consequence of the high predation rate is a high reproductive rate. Females give birth to several litters a year. Shortly after giving birth, a female mates again; ovulation is stimulated by copulation, ensuring a high pregnancy rate. Hares' courtship behavior is most unusual. Male hares track receptive females by their scent, but females attack males who approach them hesitantly. If a male is persistent, courtship begins: One of them jumps into the air and the other runs underneath, and then vice versa, and so on. Often, the jumping individual urinates on the other as it runs below. After the jumping sequence the female runs off, zigzagging, leading the male on a courtship chase; after it ends, the hares mate.

*PYGMY RABBIT, *Brachylagus idahoensis* Head and body: 8.5–11 in (21.5–28 cm). Ear: 1.4–2 in (3.5–5 cm). Very small rabbit; short hind legs. Black to dark gray; may have a slightly pink hue. Gray tail with brown above, black-and-white whiskers, small white spot next to each nostril. Active at dusk and dawn, and occasionally during the day. Eats sagebrush, grasses, and forbs. Digs burrow at the base of a shrub with an entrance hole less than 3 in (8 cm) in diameter. A poor runner, rarely ventures far from burrow. Preyed upon by Long-eared Owls and Coyotes. Young are born May through early August, 4–8 per litter. Only on the east Sierra slope and only in Mono Co. Prefers Sagebrush Scrub and Pinyon-Juniper Woodland with a shrub understory. *Distribution* The Great Basin, including small areas of eastern California.

*MOUNTAIN COTTONTAIL, *Sylvilagus nuttallii* Head and body: 12–15.4 in (30.5–39 cm). Ear: 2.1–2.6 in (5.5–6.5 cm). Gray-brown fur above, white underparts and tail, slight red-orange wash to neck, and black whiskers. Requires rocks and shrubs for cover. Eats grass, also foliage of sagebrush and juniper. Breeds from April to late July; 1–8 young per litter. Females have 4–5 litters per year. Nests consist of hollows in the ground lined with dried grass and fur. Absent from the west slope. In the Eastern Sierra from southern Inyo Co. north, in Sagebrush Scrub, Pinyon-Juniper Woodland, and other open woodland with sagebrush understory. *Distribution* Ranges from southwestern Canada south through the Great Basin and the Rocky Mountain states to northern Arizona and New Mexico.

DESERT COTTONTAIL, *Sylvilagus audubonii* Head and body: 12–15 in (30.5–38 cm). Ear: 2.6–3.1 in (6.5–8 cm). Pale gray with yellow wash, white underparts and tail. Black whiskers, white at tips; long ears. Prefers shrubby habitat for breeding but feeds primarily on grasses. In the most arid parts of its range, may obtain water by eating cactus. Breeds from October through June with 2–4 litters per year and 3–4 young per litter. Both Sierra slopes; on the east slope from southern Mono Co. south in Pinyon-Juniper Woodland, Sagebrush Scrub, and other scrub communities. *Distribution* Range includes northern California and the southern Great Basin; the Rocky Mountains east to the western Plains states; the Southwest south to central Mexico.

SNOWSHOE HARE, *Lepus americanus* Head and body: 13–18.7 in (33–47.5 cm). Ear: 3.5–4 in (9–10 cm). Dark brown in summer, white in winter (white only on tips of hairs). Tail always brown below. Notably large hind feet and small ears. Mostly active at night. In summer feeds on green vegetation, such as grasses and willows; in winter on twigs, bark, and buds. In winter the white fur provides camouflage. Young are born from April through August, 1–6 per litter. Females have 2–4 litters per year and build nests lined with fur and grasses, unlike other hares. Both Sierra slopes; on the east slope from Mono Co. north, commonly in higher-elevation communities with a shrubby understory, such as Montane Riparian Woodland, Montane Chaparral, and Montane and Subalpine forest bordering meadows. *Distribu-tion* The higher mountain regions of the West; coniferous forest in the northern and eastern United States, Alaska, and Canada.

WHITE-TAILED JACK RABBIT, *Lepus townsendii* Head and body: 18 –22 in (45.5 –56 cm). Ear: 3.7 – 4.8 in (9.5 –12 cm). Fig. 11.1f. Gray-brown in summer, white in winter; tail always white. Ears longer than head. An agile runner, it has been clocked at 40 mph (67 kph). Active at night. Feeds on green vege-tation in summer, and on bark, twigs, and buds in winter—nibbling high in shrubs. Migrates to lower elevations in winter and tunnels through snow in search of food. Breeds from April through October, 1 litter per year, 1–9 young per litter. Female builds a nest under dense vegetation, lining it with grass, leaves, and fur. Uncommon on the east slope; limited distribution on the west slope. Prefers Mountain Meadows, open grassy Subalpine and Montane forest, and grassy shrub associations, such as Montane Chapar-ral and Sagebrush Scrub. *Distribution* Ranges from northwestern Can-ada south through the interior Northwest, east to western Wisconsin and Illinois, and south to northern New Mexico.

BLACK-TAILED JACK RABBIT, *Lepus californicus* Head and body: 17–21 in (43 –53.5 cm). Ear: 6 –7 in (15–18 cm). Named for the black stripe along top of tail. Large hare with gray, gray-brown, to blackish fur above; white under-parts; and long, black-tipped ears. Important prey species for carnivores and birds of prey, and frequently encountered as road kills. Primarily nocturnal but often observed during the day. Has been clocked up to 35 mph (58 kph). Eats a variety of plants. Thought to be *outcompeting* (displacing) the White-tailed Jack Rabbit where their ranges overlap. Breeds year-round, but par-ticularly April through June. Has 3– 4 litters per year, 1–8 young per litter. Gives birth in dense vegetation with no nest. Occurs on both Sierra slopes. On the east slope, common in Sagebrush Scrub and other High Desert scrub communities. *Distribution* Range includes most of California, Baja California, the Great Basin, the Southwest; the southern Plains states north to South Dakota and south to central Mexico.

RODENTS: ORDER RODENTIA

Roughly 40 percent of all mammal species are rodents. Of all mammal or-ders, the Rodentia has the greatest proliferation of numbers and forms. They exhibit an amazing number of behavioral, physiological, and anatomical adaptations to a great variety of habitats and lifestyles. All rodents have two continually growing front incisors in both the upper and lower jaw, which enables them to gnaw on plants or trees; the *cheek teeth,* premolars and molars, grow continually as well. While rodents are gnawing, they can close off the incisors from the rest of the mouth with folds of skin, so that they do not swallow the material. Like the lagomorphs, rodents produce moist, green fecal pellets in the cecum, aided by bacterial action; they eat these pellets to further extract protein and other nutrients.

MOUNTAIN BEAVER: FAMILY APLODONTIIDAE

There is only one species of Mountain Beaver in the world; it is the oldest and most primitive living rodent, and is not closely related to the true Beaver. The species has survived virtually unchanged for about 55 million years. Its cheek teeth are different from those of modern rodents, having cup-like surfaces and no roots. The Mountain Beaver has five toes on the front and back feet, whereas most rodents have four toes on their front feet.

Mountain Beavers dig complex burrow systems in dense thickets near creeks or streams, sometimes positioning one opening so that water trickles in. They are active year-round. Preparing for winter, they will store piles of vegetation near or in burrow entrances. They often procure fresh vegetation by climbing into conifers and clipping off branches; they descend by climbing down headfirst or by dropping directly onto the ground below. Many foresters consider Mountain Beavers pests because they damage trees. Indeed, *Aplodontia* do eat seedlings, remove bark (girdling the tree), and cut branches. But the effects are minimal, and surely there is much yet to be learned from this species about the ancestral rodent stock.

MOUNTAIN BEAVER, *Aplodontia rufa*   Body: 12–17 in (30.5–43 cm). Tail: 0.8–1.6 in (2–4 cm). Fig. 11.1d. Uncommon. Dark brown overall, except for white spot below each small, round ear. Small eyes, tiny, chubby tail, toes sharply clawed. Primarily nocturnal. Each Mountain Beaver digs an extensive tunnel system, perhaps 900 ft (275 m) long and about 1.5–5 ft (0.5–1.5 m) below the surface. Wanders only a short distance from its burrow system. Herbivorous; forages underground and aboveground, on snow and under snow, and also climbs trees and shrubs. Favors dogwood, blackberry, thimbleberry, lupine, and ferns but will eat bark and twigs. Breeds February through June; 1–4 young per litter, 1 litter per year. Found on both Sierra slopes; on the east slope from northern Inyo Co. north. Inhabits areas with soft soil adjacent to streams in higher-elevation Montane Riparian Woodland, and Montane and Subalpine forest with understory thickets of willow, alder, or dogwood.   *Distribution*   Ranges along the coast from southern British Columbia to northern California, and through the Cascades and Sierra Nevada.

MARMOTS, CHIPMUNKS, AND SQUIRRELS:
FAMILY SCIURIDAE

Features common to the Sciuridae include dense body fur, a fully furred, often bushy tail, large ears, large eyes, and good color vision. All except the flying squirrels are active by day. Some species are highly social and live in colonies; other are solitary and territorial. Several of the sciurids, notably the marmots, ground squirrels, and chipmunks, escape extreme cold as well as heat and drought by hibernating for part of the year. Sierra sciurids range in size from the small Least Chipmunk to the large Yellow-bellied Marmot. Marmots, chipmunks, and ground squirrels are burrowers; the flying squirrel and tree squirrels spend most of their time in trees. Sciurids eat seeds, vegetation, insects, and occasionally small vertebrates. They in turn are preyed upon by birds of prey, several carnivores, and snakes.

Yellow-bellied Marmots are large ground squirrels. Their social system is different from that of most squirrels: Males have one to three mates that share their territories. Because good-quality territories are in short supply, females prefer to share rather than to inhabit an inferior territory. Males benefit from this arrangement by having more offspring; females that share a territory, however, raise fewer offspring. Marmots breed about two weeks after emerging from winter hibernation (prolonged torpor). After a thirty-day gestation period, the young are born helpless—naked, blind, and toothless. They remain in the burrow for twenty to thirty days before joining their parents aboveground. The following year, they disperse to find territories of their own. Marmots communicate by means of whistle signals, upraised tails, and tail-flicking. Sharp alarm whistles warn against predators—most large carnivores and, particularly, Golden Eagles.

Chipmunks also hibernate, usually becoming torpid in October or November and emerging in March. But instead of accumulating fat in preparation for the long winter, in late summer and fall they store seeds of trees and flowers in cache sites or in their burrows. They carry food in cheek pouches, which open into their mouths. During brief periods of arousal during winter, chipmunks snack on these food hoards. Unretrieved seed caches may produce clusters of plant seedlings the following spring. During summer and fall chipmunks have a diverse diet, including foliage, flowers, bulbs, roots, fruits, seeds, insects, fungi, mushrooms, bird eggs, and carrion. The bold striping pattern of chipmunks may be a form of camouflage that enables them to blend in with areas of sunlight and shadow on the forest floor. Those species with the most distinct patterns seem to occur in habitats with highly contrasting open sunlit areas and dark shadows.

The White-tailed Antelope Squirrel lives in a harsh desert environment, with summer temperatures commonly over 115°F (46°C) and little food for much of the year. Unlike other ground squirrels, it does not become torpid during extreme temperatures. Instead, using behavioral and physiological adaptations, it conserves energy and water, and remains active year-round. When air temperatures soar, most mammals increase their metabolic rate, pant, and sweat in order to avoid an increase in body temperature. But this ground squirrel maintains a constant metabolic rate, which may increase its body temperature up to 110°F (43°C) without any damage. The white underside of the tail serves as a reflective shield against the back, slowing heating somewhat. When threatened with overheating, the squirrel retreats to its cool burrow, where it unloads heat. In unusually hot weather, the squirrel may smear saliva over its head, cooling itself by evaporation. In addition, its kidneys are so effective in conserving water that the squirrel can drink saltwater and lose very little to urine.

The *Spermophilus* ground squirrels become torpid during stressful periods—hibernating during winter and *estivating* (a hot-weather version of hibernating) during summer. Consequently, they may be active as few as four or five months of the year. For example, to avoid heat and drought some populations of the California Ground Squirrel begin estivating in late May. With this ability to escape periods of extreme weather and scarce food, *Spermophilus* ground squirrels are able to inhabit desert, mountain, and even arctic regions. Ground squirrels build up their body fat before becoming torpid; they also store food in their burrows to eat during brief periods of arousal. Like chipmunks, they transport food in internal cheek pouches to their burrows. Hibernation usually begins in September or early October; final arousal occurs from mid-March to mid-April, and the squirrels breed shortly after. In short, the annual cycle of most ground squirrels consists of fattening, torpor, and reproduction. Young are born naked, their eyes and ears closed, and without teeth. They remain in their mother's burrow for 22 to 48 days after birth. Most do not disperse far from their birthplace—usually remaining within 0.5 mile (about 1 km). Alarm chirps warn of predators, which include hawks, Golden Eagles, Coyotes, foxes, Mountain Lions, Bobcats, and Badgers.

The tree squirrels differ from other squirrels in several ways: Their fur lacks any distinct spotting or striping patterns, they do not have internal cheek pouches, and they do not hibernate. The ecological requirements of the three Sierra species are similar, but they prefer different foods and spend different amounts of time on the ground. For example, during winter the Douglas Squirrel eats conifer seeds, the Gray Squirrel eats acorns, and the Northern Flying Squirrel eats lichens and mosses. Important predators, which usually take squirrels that are foraging on the ground, include Coyotes, foxes, Bobcats, hawks, owls, and Golden Eagles. Martens can capture squirrels in trees.

YELLOW-BELLIED MARMOT, *Marmota flaviventris*   Body: 13.8−18.9 in (35−48 cm). Tail: 5.1−8.7 in (13−22 cm). Fig. 11.1g. Chubby. Yellow-brown with white-tipped hairs, yellow belly. White between eyes, and buffy side patches on neck. Active day or night. Forages on grasses, foliage, herbs, and insects. In late summer, begins accumulating fat; enters hibernation in September and arouses in March or April. Breeds from mid-March to July; 1 litter with 3−8 young. Found on both Sierra slopes. Inhabits talus slopes and boulder fields up to 12,000 ft (3660 m) in Alpine Rock and Meadow Communities or adjacent to Mountain Meadows; and open, grassy areas in High Desert communities.   *Distribution*   Ranges through the mountain regions of the West, and southern British Columbia and Alberta.

Tamias Chipmunks   Nine chipmunk species inhabit the Eastern Sierra. Identification can be difficult because their size, behavior, and color patterns are similar. All are various intensities of red-brown with gray patches; all have prominent, alternating light-and-dark stripes on the head and sides, and a central dark back stripe. In a few cases their geographic range, altitudinal occurrence, and habitat preference may overlap. Chipmunks build nests in burrows or holes in snags, lining them with leaves, lichens, or feathers. Most species bear young in July, 3−8 in 1 litter per year.

ALPINE CHIPMUNK, *T. alpinus*   Body: 4.3−4.5 in (11−11.5 cm). Tail: 2.8−3.3 in (7−8.5 cm). Small chipmunk with long, silky fur, much gray on head and body, with nearly black back stripe. Head stripes are dark brown or reddish. Tail has light orange-yellow underside, with black tip. On the east slope from southern Mono Co. through northern Inyo Co. Inhabits Alpine Rock and Meadow Communities, Subalpine Meadows, and grassy areas in open Subalpine Forest.   *Distribution*   Limited to the Sierra Nevada, occurring at the highest elevations, from about 9000 ft (2740 m) up.

LEAST CHIPMUNK, *T. minimus*   Body: 3.7−4.5 in (9.5−11.5 cm). Tail: 3−3.5 in (7.6−8.9 cm). Fig. 11.2a. Fur is grayish, short, and rough. Middle back stripe is black with bright white side stripes bordered by black stripes; black stripes extend to base of tail. Tail is long and narrow with yellow underside; it is held straight up when chipmunk runs. Limited on the west slope. In the Eastern Sierra, inhabits open, dry habitats with brushy areas, rocks, and logs, such as Pinyon-Juniper Woodland, Sagebrush Scrub, and open Montane Forest.   *Distribution*   Range includes Canada from northern British Columbia

Fig. 11.2.   Mammals. (a) Least Chipmunk, *Tamias minimus*. (b) Douglas' Squirrel, *Tamiasciurus douglasii*. (c) Golden-mantled Ground Squirrel, *Spermophilus lateralis*. (d) Belding's Ground Squirrel, *Spermophilus beldingi*. (e) Mountain Pocket Gopher, *Thomomys monticola*. (f) Ord's Kangaroo Rat, *Dipodomys ordii*. (g) Brush Mouse, *Peromyscus boylii*

east through Ontario; the Great Basin, mountain regions of the West, and Lake Superior region.

YELLOW-PINE CHIPMUNK, *T. amoenus* Body: 4.5–5.2 in (11.5–13 cm). Tail: 2.9–4.3 in (7.5–11 cm). Distinct black-and-white stripes with black in front of ears and white behind. Ears have light rims; head and shoulders reddish; underside of tail is brown-yellow. Both Sierra slopes; on the east slope from northern Inyo Co. north. Inhabits many plant communities, from Subalpine Forest to High Desert scrub communities. *Distribution* Ranges from the montane regions of British Columbia and the Northwest, south through the northern Sierra.

ALLEN'S CHIPMUNK, *T. senex* Body: 5.3–6.5 in (13.5–16.5 cm). Tail: 3.5–5 in (9–12.5 cm). Large, dark brown chipmunk with indistinct, dull yellowish or dull gray light stripes and black stripes on back and sides. Ears gray behind and black in front, facial stripes brown, undertail red-brown or yellow-brown. Both Sierra slopes; on the east slope from Mono Co. north. Inhabits Subalpine and Montane forest, Pinyon-Juniper Woodland, aspen stands, and Montane Riparian Woodland, particularly areas with shrub understory. *Distribution* Ranges from central Oregon through the Cascade Range to the Sierra Nevada.

LONG-EARED CHIPMUNK, *T. quadrimaculatus* Body: 5–6 in (12.5–15 cm). Tail: 3.3–4.6 in (8.5–11.5 cm). Large, reddish chipmunk with distinct stripes. Large white patch behind ears, black stripe below ear. Tail has white-tipped hairs near end and is red-brown below. On the east slope, from Alpine Co. north in habitats with shrub understory, including Montane Forest, Montane Chaparral, and Montane Riparian Woodland. *Distribution* An exclusively Sierran species, both slopes.

LODGEPOLE CHIPMUNK, *T. speciosus* Body: 4.6–5.3 in (11.5–13.5 cm). Tail: 2.8–4 in (7–10 cm). Brightly colored chipmunk. Top of head and shoulders brown to gray, with contrasting side stripes of white and dark brown. Outer white stripe is broad with faint lower dark side stripe; black mid-back stripe. Tail has buffy edge and dark tip. On the east slope, from northern Inyo Co. north, in Subalpine and Montane forest, aspen stands, Montane Riparian Woodland, and Montane Chaparral. *Distribution* Restricted to both slopes of the Sierra Nevada, western Nevada adjacent to Lake Tahoe, and isolated areas in the southern California mountains.

MERRIAM'S CHIPMUNK, *T. merriami* Body: 4.7–6.5 in (12–16.5 cm). Tail: 3.5–5.6 in (9–14 cm). Large, gray-brown chipmunk with indistinct stripes. All light stripes grayish, all dark stripes brown. Long, fluffy tail, nearly as long as body, with prominent white or buffy edging. Uncommon on the east slope from Inyo Co. south. Prefers shrubby understory in Montane Forest and in other mid- to low-elevation communities. *Distribution* Primarily a southwestern Sierra slope species; also found in the Coast Range south of San Francisco to Santa Barbara.

*PANAMINT CHIPMUNK, *T. panamintinus* Body: 4.5–4.7 in (11.5–12 cm). Tail: 3.1–4 in (8–10 cm). A chipmunk of arid Eastern Sierra habitats. Brightly colored with indistinct stripes. Gray crown and rump, reddish shoulders and back. Lower dark stripe on cheeks and sides is faint. Other dark stripes are red-brown or gray-brown; light stripes are white. Red patch on underside of tail. Prefers cliffs and rocky areas in Pinyon-Juniper Woodland and Montane Forest. *Distribution* Restricted to southern Mono Co., Inyo Co., and adjacent areas in Nevada.

*UINTA CHIPMUNK, *T. umbrinus* Body: 4.5–5 in (11.5–12.5 cm). Tail: 3.4–4.5 in (8.5–11.5 cm). Back of neck and crown of head gray. Lower dark stripe on face and sides is indistinct. Lower white stripe on sides distinct, other dark stripes almost black. Tail edged with buff, has black tip. Absent on the west slope; in the Eastern Sierra in southern Mono Co. and northern Inyo Co., in Subalpine and Montane forest, aspen stands, and Montane Riparian Woodland. *Distribution* Ranges from western Wyoming south through the Great Basin, east to northern Colorado, and south to northern Arizona.

*WHITE-TAILED ANTELOPE SQUIRREL, *Ammospermophilus leucurus* Body: 5.5–6.5 in (14–16.5 cm). Tail: 2.1–2.8 in (5.5–7 cm). A squirrel of desert communities. Gray-brown body, white line on each side of back. Undertail white bordered with black or gray-brown. Runs with tail over back, exposing white undersurface. Eats seeds, plant material, insects, and carrion, obtaining much of its water from food. Stores food but is active year-round. Breeds March through May; a single litter of 5–14 usually in late April. Occurs in the Eastern Sierra from northern Mono Co. south in Alkali Sink, Shadscale, Creosote Bush, and Sagebrush scrub; also Pinyon-Juniper Woodland. *Distribution* Range includes Baja California, southeastern California, the Great Basin, northern Arizona north through Utah, and east to northern New Mexico and western Colorado.

BELDING'S GROUND SQUIRREL, *Spermophilus beldingi* Body: 8–9 in (20–23 cm). Tail: 2.2–3 in (5.5–7.5 cm). Fig. 11.2d. Also known as the "picket pin," because it typically sits upright near burrow entrances. Upperparts gray washed with reddish brown or cinnamon. Broad brown-gray streak down back, contrasting with buff or white underparts and sides. Short tail with black tip. Feeds on grass, herbs, and seeds. Hibernates from September until May, living on accumulated fat. Breeds from mid-May to mid-July with 1 litter of 3–12 young. The social system is unusual: Kin groups of squirrels defend territories. Males leave their birthplace, while related females remain together. One unrelated male will associate with a kin group of females until the young are born. Important predators include Coyotes and Badgers, which dig out burrows and devour entire litters of young squirrels. Both Sierra slopes; on the east slope from northern Inyo Co. north in Mountain Meadows and grassy areas of Montane Forest and Sagebrush Scrub. *Distribution* Range includes eastern Oregon and southern Idaho, southeastern California, western and north-central Nevada.

CALIFORNIA GROUND SQUIRREL, *Spermophilus beecheyi* Body: 9–11 in (23–28 cm). Tail: 5.4–7.9 in (13.5–20 cm). Because they may carry fleas that are vectors for bubonic plague, these ground squirrels should never be handled. Head brown and body gray-brown dappled with buff or white. Sides of head, neck, and shoulders light gray. Dark cape between shoulders to mid-back, belly buff, bushy tail edged with buff. Eats twigs, stems, and leaves of herbs, roots and bulbs, seeds (including acorns), mushrooms, fruits and berries, bird eggs, insects, and carrion. Stores food in burrows. At lower elevations that are hot in summer, may estivate in July and August; hibernates in winter. Lives either in colonies or solitarily. Breeds April to July, with 1 litter of 3–15. The squirrels are considered a pest because they eat crops and because they invade overgrazed pastureland. Common along both Sierra slopes in almost all habitat types; on east slope, prefers grasslands and grassy areas in Montane Forest and Pinyon-Juniper Woodland. *Distribution* Range includes western Oregon, most of California, and northern Baja California.

*TOWNSEND'S GROUND SQUIRREL, *Spermophilus townsendii* Body: 5.5–7 in (14–18 cm). Tail: 1.3–2.8 in (3.5–7 cm). Squirrels often stand like "picket pins" at burrow entrance. Smoke gray body with pink-buff tones and buffy underparts. Undersurface of small, flattened tail is cinnamon with white edges. Requires soils easy to dig; burrow has a characteristic rim 4–6 in (10–15 cm) high around opening. Eats green vegetation, seeds, and carrion; does not store food. When plants dry up in May or July, Townsend's Ground Squirrels estivate, emerging in January or February. Young are born in March and April, 5–15 per litter. Limited to Mono Co. in the Eastern Sierra, chiefly in arid scrub habitats; also in farmlands. *Distribution* Ranges through the Great Basin, north to southern Washington and southern Idaho; in California only in eastern Lassen and Mono counties.

GOLDEN-MANTLED GROUND SQUIRREL, *Spermophilus lateralis* Body: 6–8 in (15–20 cm). Tail: 2.5–4.8 in (6.5–12 cm). Fig. 11.2c. This chipmunk-like ground squirrel is also known as the Copperhead, for its head can be various shades of copper gold. White stripe bordered with black on each side of back, but no stripes on face; back buff, brown, or gray. A common, conspicuous mountain inhabitant that fearlessly takes handouts in campgrounds and picnic areas. Burrows under rocks and logs, and lines nest with grass or leaves. Eats seeds, fruits, underground fungi, mushrooms, flowers, bulbs, insects, bird eggs, and carrion. Caches food in its burrow. Hibernates from October through April. Breeds from April to early September; 1 litter with 2–8 young. After emerging from hibernation, a male will share a burrow with a female until the young are born, then move into a nearby burrow. Both Sierra slopes in open coniferous Montane and Subalpine forest; on the east slope also in Pinyon-Juniper Woodland, Montane Chaparral, and Sagebrush Scrub. *Distribution* Range includes the mountain regions of the West from southern British Columbia and Alberta south.

DOUGLAS' SQUIRREL OR CHICKAREE, *Tamiasciurus douglasii* Body: 6–7 in (15–18 cm). Tail: 4.8–5 in (12–12.5 cm). Fig. 11.2b. Dark olive brown to brown-gray above, with buff, yellow, or rusty belly. Ears tipped with long hairs in winter. Black line along each side between back and belly. White edges to tail, white ring around eye. Commonly scolds and chatters from a tree limb, and often assumes a threatening posture when defending its territory. Eats conifer seeds, flowers, fruits, fungi, insects, and occasionally bird nestlings. Climbs high into conifers and cuts down cones, then stores them under soil and forest litter in moist locations that keep cones from opening. Active all winter; feeds on seeds from stores. Males and females defend separate territories, but when a female is ready to breed, she allows a male to enter her territory for a day. Breeds May to October; 1 litter of 1–9 young. Nests in tree and rock cavities or builds a nest in branches; lines nest with grass and shredded bark. Both Sierra slopes, in mature coniferous Montane and Subalpine forest; also in Pinyon-Juniper Woodland. *Distribution* Ranges from southern British Columbia south through the Coast Range and Cascades to the Sierra Nevada and San Francisco.

WESTERN GRAY SQUIRREL, *Sciurus griseus* Body: 9–12 in (23–30.5 cm). Tail: 10–11.4 in (25.5–29 cm). Large squirrel with dark gray to silver gray upperparts, white belly, and dark feet. Long, bushy tail edged with white. Feeds on acorns, conifer seeds, leaves, fruit, and mushrooms. In fall stores acorns singly in holes in the soil 3–4 in (8–10 cm) deep, for winter food. Probably relocates acorns using acute sense of smell and some spatial memory. Nests either in tree cavity or in a nest of sticks in tree branches, usually at least 20 ft

(6 m) above ground. Males and females defend separate territories. Breeds January to September; 1 or 2 litters of 2–5 young each. Both Sierra slopes; prefers mature Montane Forest, including pine-oak forest. On the east slope from northern Mono Co. north. *Distribution* Range includes Washington, western Oregon, and most of California.

NORTHERN FLYING SQUIRREL, *Glaucomys sabrinus* Body: 5.5–6.4 in (14–16 cm). Tail: 4.5–5.8 in (11.5–14.5 cm). Small squirrel with glossy, thick, soft fur. Lead gray above with cream-white belly. Long whiskers. Folded loose, furred skin along each hind leg; when outstretched, it forms a membrane for gliding. Squirrels glide from tree to tree, as far as 150 ft (46 m). Active at night. Male and female will share a home range in summer but nest separately. In winter many squirrels may den together. Diet includes seeds, insects, meat, fungi, fruit, lichens, and also mosses (particularly in winter). Stores cones and seeds in the ground or in rock crevices. Nests in woodpecker hole or builds platform of twigs, leaves, and bark on branches. Breeds from March to September; 1 or 2 litters, each with 2–6 young. Both Sierra slopes; on the east slope from Alpine Co. north, with a small population near June Lake in Mono Co. Prefers mature, dense Subalpine and Montane forest, and Montane Riparian Woodland. *Distribution* Range includes Alaska, Canada, the coniferous forest of the Great Lakes, and the mountain regions of the western and eastern United States.

POCKET GOPHERS: FAMILY GEOMYIDAE

Pocket gophers are widely distributed, from mountain meadows to lowland plains throughout much of North and Central America. Their local occurrence is limited by their need for light, dry, porous soils at least 4 inches (10 cm) deep. Like moles, pocket gophers are adapted to a digging lifestyle. They dig with their forefeet, whose long, sharp claws grow very quickly. While digging, they can close their eyes and tiny ears with valves, so that these organs are not clogged with dirt. Sensitive facial whiskers enable them to "feel" their way along. Their pelvis is small, allowing them to maneuver in narrow spaces. The upper and lower incisors are very long and continuously growing—each at an astounding rate of about 11.5 inches (29 cm) per year. Furred "lips" that close in back of the incisors keep dirt out of the mouth. Gophers can travel forward or backward in their burrows, because their soft fur bends easily in either direction. Their fur-lined cheek pouches have external cheek openings and can actually be turned inside out. Perhaps because of their solitary lifestyle, pocket gophers have few calls. The most common sound, made by grinding their teeth, probably indicates aggression.

A pocket gopher excavates burrows by digging, but when progress is impeded, it will actually bite off chunks of dirt. It shovels the loosened material backward under the body. After turning around, it will then push the dirt through the burrow system up to the ground surface or into an unused tunnel. The diameter of the burrow corresponds to the size of the gopher and varies from about 2.5 to 3.5 inches (6.5–9 cm). The main tunnel system is usually 4–12 inches (10–30 cm) belowground, and a deep tunnel leads to nesting chambers. Side tunnels may lead to foraging sites aboveground. Discarded food, nesting materials, and fecal matter may be carried into an unused tunnel. Unlike moles or ground squirrels, pocket gophers plug the entrance of their burrow in the center of the mound, which helps prevent flooding from heavy rain.

Although pocket gophers are usually solitary and highly territorial, their aggressive nature decreases briefly during the breeding season, when the home ranges of males and females may overlap. In Botta's Pocket Gopher, and perhaps other species as well, males have several mates, overlapping home ranges with several females. Male pocket gophers are noticeably larger than females; males continue to grow throughout their life, whereas females stop when they become sexually mature. Females of some species have as many as three litters per year. Gestation lasts about 18 or 19 days. Young are born helpless, with closed eyes and ears, and with small slits where cheek pouches will develop. After weaning, young pocket gophers, forced out of the burrow, are on their own.

Pocket gophers have a checkered reputation. Their burrowing habits mix and aerate soil. However, in overgrazed land where they occur in high densities, the resulting soil disturbance leads to the growth of undesirable weeds and annuals.

MOUNTAIN POCKET GOPHER, *Thomomys monticola*   Body: 5.6–6 in (14–15 cm). Tail: 2–3.8 in (5–9.5 cm). Fig. 11.2e. Gray-brown above in winter, russet or yellow-brown in summer. Underparts vary from gray to buffy. Black snout, large black patches behind small, pointed ears. White feet and tail; tail nearly nude. Requires deep, soft soil and abundant green vegetation. Active in winter, Mountain Pocket Gophers construct tunnels in the snow just above the ground surface. The excavated soil forms long, solid cores that remain intact after snowmelt. This gopher eats tubers, bulbs, roots, seedlings, grasses, sedges, leaves, and bark. Will feed on the surface but also chews on underground plant parts. Young are born in May and June; 1 litter of 3–4 young. Found on both Sierra slopes; on the east slope from northern Mono Co. north in Alpine Rock and Meadow Communities, Mountain Meadows, and grassy areas in Subalpine and Montane forest. *Distribution*   Restricted to the Sierra Nevada.

BOTTA'S POCKET GOPHER, *Thomomys bottae*   Body: 4.8–7 in (12–18 cm). Tail: 2.2–3.8 in (5.5–9.5 cm). Buffy brown to light gray above, with gray to white underparts. Small, rounded ears, each with an ear-sized dark patch behind; tail nearly hairless. Diet includes stems, roots, bulbs, tubers, seeds, and grasses. Forages both on top of the ground and from tunnels. May breed from February through October; 1–3 litters of 2–11 young per litter. Remains in the same burrow system for life. Will be displaced (outcompeted) in any area by the presence of another species of pocket gopher. Present on both Sierra slopes. In the Eastern Sierra, in Plumas Co. and from southern Mono Co. south in arid, lower-elevation areas with a grass-forb understory, including Sagebrush Scrub, Pinyon-Juniper Woodland, lower Montane Forest, Montane Chaparral, and Montane Meadows. *Distribution*   Ranges through most of California and the Southwest, east to Colorado, and south through Baja California and northern Mexico.

*NORTHERN POCKET GOPHER, *Thomomys talpoides*   Body: 5–6.5 in (12.5–16.5 cm). Tail: 1.8–2.8 in (4.5–7 cm). Yellow-brown or gray-brown above, gray below. White on throat and under chin. Dark gray or black nose; rounded ears with large, dark patch behind each ear; tail nearly nude. Feeds on roots, grass, wildflowers, aspen, and cactus. Breeds from late March to August; 3–10 per litter, 1 or 2 litters per year. Absent from the west slope. In the Eastern Sierra, in eastern Plumas, eastern Sierra, and Mono counties in almost all communities at lower elevations featuring soft soil and good grass and herb cover. *Distribution*   Range includes southwestern Canada, northern Great Basin, and the Rocky Mountain region east to the Dakotas, and south to northern Arizona and New Mexico.

POCKET MICE, KANGAROO MICE, AND KANGAROO RATS: FAMILY HETEROMYIDAE

Pocket mice, kangaroo mice, and kangaroo rats share many unusual features: external, fur-lined cheek pouches; small, weak front feet; tails longer than the combined head and body length; large, strong hind feet; and grooved upper incisors. Well-adapted to arid environments, they are nocturnal, store food, and derive all their required water from the metabolism of their food, supplemented in some species by succulent vegetation. They are territorial and aggressive, leading solitary lives except for brief periods during the breeding season. Like all small, nocturnal rodents, they are vulnerable to predation by weasels, Badgers, foxes, and owls.

Pocket mice are among the smallest rodents in the world. They have high metabolic rates and lose heat rapidly. Their ability to enter torpor during periods of food scarcity is an important survival strategy. Their 3-foot-deep (1-m) burrows have both nesting and food-storage cavities. During the day while resting, pocket mice plug the burrow entrance, which helps maintain lower interior temperatures.

Kangaroo mice characteristically have silky fur and lack distinct striping on their tails. These otherwise small versions of kangaroo rats also enter torpor—either winter hibernation or summer estivation—during times of food scarcity. The enlarged middle portion of the tail serves as a fat-storage depot. Like pocket mice, kangaroo mice close their burrows during the day; burrows are about 1 foot (0.3 m) below the surface.

Kangaroo rats have very long, strong hind legs. Facial markings are distinct, with various patches and whisker marks; tails are boldly striped, dark on bottom and top, white on the sides, a furry tuft at the end. The peaks of pregnancy generally correspond to the appearance of new vegetation in spring and fall. Life spans in the wild are usually less than one year. Several special physiological and behavioral adaptations enable kangaroo rats to live in hot, dry regions without drinking water. For example, they are active only at night and do not sweat or pant when overheated. Efficient kidneys that concentrate urine to a remarkable 24 percent (versus the usual 6 percent) conserve water; relatively high humidity in the burrow further reduces water losses from respiration.

GREAT BASIN POCKET MOUSE, *Perognathus parvus* Body: 2.5–3 in (6.5–7.5 cm). Tail: 3–4 in (7.5–10 cm). Buffy yellow-orange above with some blackish brown hairs; below, white with reddish tint; buffy line along side. Tail light below with tuft at tip. Diet includes the seeds, leaves, and stems of forbs, grasses, and shrubs, and also some insects. Hibernates December through March or April. Breeds from May to August; 2 litters of 3–8 young per litter. In the Eastern Sierra, from Plumas to Nevada Co., and from Alpine Co. south; limited occurrence on the west slope. Prefers grassy areas in Sagebrush Scrub and Pinyon-Juniper Woodland. *Distribution* Ranges from southeastern British Columbia south through the Great Basin to northern Arizona.

LITTLE POCKET MOUSE, *Perognathus longimembris* Body: 2.2–2.6 in (5.5–6.5 cm). Tail: 2–3.4 in (5–8.5 cm). Smallest member of the Heteromyidae. Soft fur, buffy to gray buff above, buffy or white belly. Small light patches at base of ears. Bicolored tail, light on bottom, tuft at tip. Eats seeds of grasses and forbs. Breeds from March to June; 1–2 litters per year, each with 2–8 young. In the Eastern Sierra from Mono Co. south; restricted to southeastern part of west slope. Prefers gravelly or sandy soil and open areas in arid communities, such as Sagebrush, Shadscale, Alkali Sink, and Creosote Bush scrub, and Pinyon-Juniper Woodland. *Distribution* Ranges from southeastern and central California south to northern Baja California, north to Oregon, and east to Utah.

*LONG-TAILED POCKET MOUSE, *Chaetopidus formosus* Body: 3.2–3.8 in (8–9.5 cm). Tail: 3.8–4.6 in (9.5–11.5 cm). Upperparts gray or brown with white belly. Long, bicolored tail, white below, crested with dark hairs along tip. Eats seeds of shrubs, grasses, and forbs; also insects. Hibernates from November through March. Breeds from March to July; 2–6 per litter, 1 or 2 litters per year. Absent from the western slope; in the Eastern Sierra, from Mono Co. south. Inhabits arid, shrubby communities, such as Sagebrush, Alkali Sink, and Shadscale scrub. Requires rocky soil. *Distribution* Range includes northeastern and southeastern California, and south through northern Baja California, southern and western Nevada, northern Arizona, and eastern Utah.

*DARK KANGAROO MOUSE, *Microdipodops megacephalus* Body: 2.8–3 in (7–7.5 cm). Tail: 2.9–4 in (7.5–10 cm). Gray-brown or blackish, sometimes with red cast. Fur of underparts white-tipped but dark gray at base. Tail black-tipped and thickest in the middle, no tuft or stripes. Hibernates from November through February. Eats seeds and insects. Breeds from late April to late September, 2–7 young per litter. Absent from the west slope; in the Eastern Sierra restricted to eastern Plumas Co. and southern Mono Co. Occurs in fine, sandy, or gravelly soil in Alkali Sink, Shadscale, and Sagebrush scrub communities. *Distribution* Ranges through the Great Basin north to Oregon and east to Utah.

*ORD'S KANGAROO RAT, *Dipodomys ordii* Body: 4–5 in (10–12.5 cm). Tail: 5–6.4 in (12.5–16 cm). Fig. 11.2f. Brown-yellow on top, white below; small, dark whisker marks near nose. White stripes on side of tail wider than dark top and bottom stripes; bottom stripe does not reach tip of tail. Last 40 percent of tail bears a small crest. Eats grass seeds and insects. Breeds nearly year-round; 1–6 young per litter. Absent from the west slope; in the Eastern Sierra in eastern part of Plumas and Sierra counties. Inhabits sandy soils in arid communities, including Shadscale, Alkali Sink, and Sagebrush scrub; also Pinyon-Juniper Woodland. *Distribution* Has the widest geographic range of all kangaroo rats: the Great Basin, the Great Plains north to Alberta and Saskatchewan, the Southwest east to Texas, and much of the arid interior of Mexico.

*CHISEL-TOOTHED KANGAROO RAT, *Dipodomys microps* Body: 4–5 in (10–12.5 cm). Tail: 5.5–6.9 in (14–17.5 cm). Yellow-brown above, whitish below. White stripes on side of tail narrower than dark top and bottom stripes;

tail has short crest and dusky tip. Dark whisker marks near nose. Diet includes grass and wildflower seeds, and fresh vegetation such as saltbush (*Atriplex*). Births peak in February and September; 1–2 litters, each with 1–4 young. Found in the Eastern Sierra from northern Inyo Co. south, in sandy or gravelly soil in moderate to dense Shadscale, Alkali Sink, and Sagebrush scrub; sometimes in Pinyon-Juniper Woodland. *Distribution* Range includes the Great Basin north to Oregon and east to Utah.

*PANAMINT KANGAROO RAT, *Dipodomys panamintinus* Body: 5 in (12.5 cm). Tail: 6.1–7.5 in (15.5–19 cm). Yellow-buff or brownish above, white underparts. Light cheek patches, white spot behind ear, dark whisker patches. Dark stripe on underside of tail tapers to a point; short, light-colored crest at tip. Eats seeds, leaves, and juniper berries. Breeds from March through July; 1 or 2 litters, each with 3–4 young. Absent from the west slope; in the Eastern Sierra, eastern Sierra Co. and from Mono Co. south. Inhabits gravelly or coarse soils in Pinyon-Juniper Woodland and Sagebrush, Alkali Sink, and Shadscale scrub. Ranges up to higher elevations than other Eastern Sierra kangaroo rats. *Distribution* Range includes eastern California, western Nevada adjacent to Lake Tahoe, and arid mountains of southern California.

*MERRIAM'S KANGAROO RAT, *Dipodomys merriami* Body: 4 in (10 cm). Tail: 5–6.4 in (12.5–16 cm). A small kangaroo rat. Light yellow to dark brown above, white below. Whisker patch not connected across nose. White tail stripes wider than dark stripes; long, dark tail crest. This kangaroo rat digs burrow down to moist soil for cooler temperatures. Diet includes seeds of grass and forbs, green vegetation, and insects. Breeding peaks from March through July; 2 litters per year, 1–5 young per litter. Absent from west slope. In the Eastern Sierra, from Inyo Co. south, in sandy to rocky soils in lower-elevation scrub communities, including Alkali Sink, Shadscale, and Sagebrush scrub. *Distribution* Range includes western Nevada, southern California, the Southwest east to Texas, Baja California, and northern Mexico.

BEAVERS: FAMILY CASTORIDAE

Although Beavers are native to most of the United States, they do not naturally occur in the Eastern Sierra. Beginning in the 1940s Beavers were planted at several locations to further water conservation, fishing, and trapping. A few of the colonies have been successful; most have not. Because there is not much aspen along Eastern Sierra streams, the Beavers eventually destroyed most of the aspen groves within their range and killed other trees by flooding. Abandoning these areas, they left a tangle of standing dead trees and downed debris. Transplants, in general, are seldom successful; at worst, they can cause considerable damage.

Beavers are more specialized for swimming than any other rodent. They are capable of remaining underwater for up to 15 minutes because their heart rate slows, conserving oxygen. Body heat is also conserved during dives in cold water: In the base of the tail the warmth of arterial blood is passed to the returning venous blood. The tail also serves to dissipate excess body heat in summer. Other swimming adaptations include nostrils and ears that close underwater, eyes protected by membranes, and lips that, like those of pocket gophers, close behind their large incisors. Their remarkable long, strong incisor teeth grow continuously. The incisors are used to fell trees, cut off branches, and strip off bark; they are supported by a massive skull and jaw, which resist the strain of the gnawing action.

Beavers live in family groups that center on male-female pairs. The maximum colony size is about twelve, with an average of about five. Females usually bear their first young at three years of age; the kits are born fully furred with open eyes. Two-year-olds disperse during ice-free periods. Common predators of Beavers include bears, Coyotes, lynx, Bobcats, and Mink.

AMERICAN BEAVER, *Castor canadensis*   Body: 25–30 in (63.5–76 cm). Tail (bare part): 9–12.7 in (23–32 cm). Fig. 11.1e. Largest North American rodent, weighing up to 74 lb (33.5 kg). Golden brown to rich, dark brown. Most of tail is naked, paddle-shaped, and covered with dark, leathery scales and sparsely scattered coarse hairs. Forelegs shorter than hind legs, but with short, heavy claws for digging; large hind feet have webbed toes. Large, dark orange incisors. Active at dusk and at night. Fells trees to construct dams as well as to provide food. Forages on woody and herbaceous plants, eating leaves, bark, sprouts, buds, fruits, roots, tules, and cattails in summer. Depends on cottonwood, aspen, and willow for winter food supply; stores twigs, branches, and bark, weighted down with mud, on pond bottoms. Breeds from April to July; with 1–8 kits per litter, 1 litter per year. Historically native to Modoc and Lassen counties; introduced throughout both Sierra slopes. Inhabits permanent water with aspen, cottonwoods, or willows nearby, particularly Eastern Sierra Riparian Woodlands; also marshy areas. *Distribution* Ranges through the United States and Canada. Original populations extinct or reduced in most of range; restored by reintroductions and management.

MICE, RATS, AND VOLES: FAMILY MURIDAE

One of the largest rodent families, the Muridae consists of species from nearly every region of the world. North American murids include white-footed mice, grasshopper mice, woodrats, voles, and the Muskrat. In addition, three common species have been introduced from the Old World.

Mice of the genus *Peromyscus,* usually referred to as white-footed mice, have large ears and thin, bicolored tails longer than 75 percent of the head-and-body length. They are nocturnal and active all winter; they store seeds and forage above snow. Wherever they occur, their densities are high—in the Sierra Nevada, as many as five to ten mice per acre (12–25 per hectare). Males and females occupy different home ranges, but may overlap during the breeding season. After a gestation period of 22 to 25 days, the young are born naked, pink, deaf, and blind. They are weaned at five weeks, and young females may breed at eight weeks. These prolific rodents are an important food for Coyotes, foxes, owls, hawks, snakes, weasels, Martens, and Bobcats. Most mice have life expectancies of only six months to a year.

Grasshopper mice are similar to white-footed mice, except for a shorter tail and more robust body. Unlike many other rodents, grasshopper mice are carnivorous and eat mostly insects; grasshoppers are a favorite food. While hunting for prey, they will climb shrubs and enter the burrows of other rodents. They will fiercely attack other rodents, even those their own size, and subdue them by biting the back of their neck or head.

Woodrats may be mistaken for the Old World Norway Rat and Black Rat (see below). However, woodrats have white throats and breasts, and their tails are hairy so the scales do not show. They build houses out of sticks, rocks, and assorted debris, such as bones, paper, cow pies, tin cans, and foil. Their tendency to accumulate junk has earned them the common name "packrats." When they enter campgrounds at night, probably in search of food, they take whatever small items strike their fancy. Woodrats are nocturnal and active year-round. Gestation lasts from 23 to 38 days, depending on the species. The young, born blind and deaf, are weaned after about three weeks. Woodrat predators include owls, Coyotes, foxes, and snakes—particularly Gopher Snakes and kingsnakes.

The voles have a blunt snout; stocky, round body; small, dark eyes; short whiskers; inconspicuous, rounded ears; short legs; a short, sparsely haired tail; and coarse fur. Special adaptations for their herbivorous lifestyle include teeth that grow continuously and have wear-resistant grinding surfaces, and a digestive tract that extracts nutrients from cellulose with the help of microorganisms. Voles eat about 30–35 percent of their body weight in food daily and also practice coprophagy (see Order Lagomorpha). They do not hibernate or enter torpor; family groups cope with cold by huddling together in insulated, underground nests. The short ears, legs, and tail, and the round body, conserve body heat by reducing surface area; fur acts as insulation. Voles are an important food for hawks, owls, shrews, and snakes.

Voles of the genus *Microtus* are widespread in North America, wherever there are grassy fields. Within their home ranges, individuals build runways through vegetation and forage along these paths. Voles also burrow, constructing subterranean food-storage, resting, and nesting areas. Occasionally they build nests aboveground, concealed in clumps of grass, and lined with dried plant material. Voles communicate by high-pitched squeaks, but social contact is minimal since they deliberately avoid one another. Mating occurs when a female is receptive and allows a male to approach. This is risky business for a male; females may attack and wound several males before finally accepting one. After mating, a copulatory plug forms in the female's vagina, preventing other males from mating with her. Young are born naked, their eyes and ears closed. Females may become sexually mature at three weeks; males do not mature until six to eight weeks. Recent studies indicate that certain substances in the grasses that voles eat are similar to mammalian hormones and stimulate reproduction. Reproduction is thus timed with the availability of fresh food.

Some *Microtus* species have population cycles (as do lemmings), in which populations regularly build up and then crash. Each cycle is specific to the species and the habitat—for example, it may last three or four years or eight to ten years. Cycles can be predicted mathematically, although their causes are not clear. Population downswings have been attributed primarily to food scarcity; vegetation recovery time may dictate the length of the cycles.

The Muskrat is an amphibious relative of the voles, valued for its soft, waterproof fur. Although awkward and slow-moving on land, Muskrats are skillful swimmers and divers. They swim by kicking their large hind feet; tail undulations propel them during dives. The name *Muskrat* derives from the long-lasting musky secretions produced by males and females during the breeding season. Males mix the musky secretions with urine to mark territories. Within its home range, each Muskrat constructs a house or den, several feeding huts or platforms, and resting "push-ups." The den, containing the nest, may be either a burrow or a structure of aquatic plants. Its entrance is underwater, but the nest is above water and remains dry. Feeding huts and platforms provide resting areas above water as Muskrats forage in the surrounding vegetation. "Push-ups" are holes in ice, protected on the surface by vegetation, that provide cover and relief from icy water. Active all winter, Muskrats swim in the chilly water beneath ice, but in winter they make only short excursions, afterward warming themselves in sheltered areas.

Some Muskrats defend territories as mated pairs, and several may den together in winter for warmth. Young Muskrats are born blind, nearly naked, and helpless; in about two weeks, the kits are climbing, swimming, and diving. They are nearly grown at three and a half months and reach sexual maturity before they are a year old. Females are very protective of their young, responding quickly to distress calls and carrying the young from place to place by the belly skin. Predators of Muskrats include Coyotes, owls, Raccoons, snakes, and Mink.

Introduced Species   Three Old World murids—the Norway Rat, Black Rat, and House Mouse—were inadvertently introduced to the United States by ship. The Norway Rat arrived in 1775, and the others probably earlier; they have spread throughout North America. Whereas rats inhabit towns and cities, the House Mouse has established populations in many plant communities. Success of these species is due to their prolific reproduction, tolerance for different climates and environments, and omnivorous and opportunistic food habits. They are major pests in cities, invading homes and warehouses, and consuming and contaminating food stores. The Norway Rat is a particular menace, because it may carry fleas that are vectors for bubonic plague. These three introduced species are widespread and may be confused with native species.

WESTERN HARVEST MOUSE, *Reithrodontomys megalotis*   Body: 2.8–3 in (7–7.5 cm). Tail: 2.2–3.8 in (5.5–9.5 cm). Small mouse, dark brown or brown-black above; mixed with buffy color on sides, cheeks, shoulders, and legs, and buffy-gray below. The long tail is darker on top. Upper incisors have deep grooves down the front. This mouse eats fruits, seeds, insects, and green shoots; will forage in shrubs. Its kidneys are so efficient that it can consume seawater with no ill effects. Active year-round and nocturnal. Breeds throughout the year at lower elevations, from spring to fall at higher elevations. Females are sexually mature at 4 months of age and may bear 5 litters per year of 1–9 young. Nest is built of woven dried grasses and weeds on the ground or under cover. Found on both Sierra slopes at all elevations, except above treeline. In Mountain Meadows, and grass and forb understory of Pinyon-Juniper Woodland, Montane Riparian Woodland, Montane Chaparral, and High Desert scrub communities.   *Distribution*   Ranges through the central and western United States, except the northern Rocky Mountains and western Washington, and south through the interior to southern Mexico.

DEER MOUSE, *Peromyscus maniculatus*   Body: 2.8–4 in (7–10 cm). Tail: 2–3.5 in (5–9 cm). Most abundant mammal in North America; widely distributed. Upperparts range from gray-buff to yellow-brown; a light patch in front of ear. White below with white feet. Tail furred with short hairs, narrowing toward the tip, and bicolored (dark above, light below). The Deer Mouse forages on seeds, fungi, insects, fruits, and vegetation, on the ground and in shrubs and trees. Builds grass-lined nests in rotting logs, among rocks, and in burrows. May breed year-round at lower elevations, but probably breeds from mid-March to September at higher elevations. Bears 3 or 4 litters of 3–7 young per year. Inhabits both Sierra slopes in all plant communities. *Distribution* Range includes eastern Alaska, most of Canada, the United States except the Southeast, and interior Mexico. **Note:** This species carries hantavirus. Avoid places contaminated by mouse feces and urine.

*CANYON MOUSE, *Peromyscus crinitus*   Body: 3–3.4 in (7.5–8.5 cm). Tail: 3.1–4.6 in (8–11.5 cm). Soft, long fur, yellow-buff above, white below; white feet. Tail longer than head and body together, faintly bicolored with light brown stripe on top, slight tuft at tip. The Canyon Mouse eats seeds of grasses, forbs, and shrubs. May breed year-round; more than 1 litter per year of 1–5 young. Occurs on the east slope from southern Mono Co. south. Frequents rock canyons, ledges, and cliffs in many arid communities, including lower Montane Forest, Pinyon-Juniper Woodland, Montane Chaparral, and scrub communities.   *Distribution*   The arid regions of southern California, the Great Basin north to Oregon and Idaho, east to western Colorado, and south to northern Arizona and northwestern New Mexico.

BRUSH MOUSE, *Peromyscus boylii*   Body: 3.6–4.2 in (9–10.5 cm). Tail: 3.6–4.4 in (9–11 cm). Fig. 11.2g. Golden brown or dark brown above, white feet and underparts. Distinctly bicolored hairy tail, dark above, longer than head and body together. Varied diet of seeds, leaves, insects, fruits, and fungi. Breeds from April through October; 4 or more litters per year of 2–6 young. Requires shrubs, rocks, and forest litter for cover. Builds grass nest in burrow or log, among rocks, or in crevice. Occurs on both Sierra slopes. In the Eastern Sierra, from Alpine Co. north in open Montane Forest, Pinyon-Juniper Woodland, oak-pine associations, Montane Riparian Woodland, Montane Chaparral, and High Desert scrub communities.   *Distribution* Inhabits most of California, north through Utah and Colorado, the Southwest east to western Texas, and south to Honduras.

PIÑON MOUSE, *Peromyscus truei*   Body: 3.6–4.2 in (9–10.5 cm). Tail: 3.4–4.8 in (8.5–12 cm). Long, silky fur, gray-brown to dark brown above; cream-white underparts; white feet. Bicolored tail, usually about 90 percent of head-and-body length. This mouse feeds on leaves, insects, fungi, and seeds; will climb trees and shrubs for food. Breeds from April through September, as many as 9 litters per year of 1–6 young. Forms nest of bark and grass. Occurs widely on both Sierra slopes. Prefers open habitats, including Pinyon-Juniper Woodland, Montane Chaparral, and High Desert scrub communities. *Distribution* Most of California south through northern Baja California; Nevada and the Southwest north through Utah, east through Colorado, and south through interior Mexico.

*NORTHERN GRASSHOPPER MOUSE, *Onychomys leucogaster* Body: 4–5 in (10–12.5 cm). Tail: 1.1–2.4 in (3–6 cm). Stocky, heavy-bodied mouse with gray or pale sandy brown upperparts. White underparts, light patch in front of ears, and white tip to short tail. Forages in ground litter, in shrubs, and in burrows of other animals; eats insects—beetles, caterpillars, grasshoppers—small mice, lizards, and seeds. Nocturnal and active year-round. Breeds from February to August; up to 10 litters per year of 1–6 young. Pairs build grass-lined nests in abandoned burrows or dig their own. In the Sierra, only on the east slope, in eastern parts of Plumas, Sierra, Nevada, and Mono counties in arid communities with moderate cover, such as Sagebrush, Alkali Sink, and Shadscale scrub, Montane Chaparral, High Desert Riparian Woodland, and Pinyon-Juniper Woodland. *Distribution* Range includes the Great Basin north to Washington, the arid regions of the northern Rocky Mountains, the Great Plains north to southern Canada, the Southwest east to Texas and south to northeastern Mexico.

SOUTHERN GRASSHOPPER MOUSE, *Onychomys torridus* Body: 3.5–4 in (9–10 cm). Tail: 1.6–2.2 in (4–5.5 cm). Stocky mouse, gray or light cinnamon brown above, white underparts, and light patch at base of ear. White-tipped tail about half the head-and-body length. Eats large insects, but sometimes takes mice, reptiles, and seeds. Nocturnal and active year-round. Breeds from May to July; as many as 6 litters of 2–6 young per year. Occurs on the southwestern Sierra slope; in the Eastern Sierra, from northern Inyo Co. south in arid communities with sparse to intermediate cover: Alkali Sink, Shadscale, and Sagebrush scrub. *Distribution* Range includes southern California, the southern Great Basin, the Southwest east to western Texas and south to central Mexico.

BUSHY-TAILED WOODRAT, *Neotoma cinerea* Body: 7–9.7 in (18–24.5 cm). Tail: 4.7–8.8 in (12–22.5). Brown, gray, buffy, and sometimes black above; white or gray below; white legs and feet. Has long, bushy tail like that of ground squirrel. Nocturnal and active year-round. Eats seeds, fungi, berries, bark, leaves, insects, and lizards; stores dried vegetation. Breeds from February to September; 1 litter of 1–6 young. Builds stick house under rocks, in cave, or in hollow log. Constructs nest of leaves, bark, grass, and moss. Inhabits both Sierra slopes. On the east slope, from southern Inyo Co. north, in talus slopes and rock outcrops in Subalpine and Montane forest, Pinyon-Juniper Woodland, Eastern Sierra Riparian Woodlands, and to a lesser extent in Montane Chaparral and High Desert scrub communities. *Distribution* From southern Yukon south through British Columbia, western Alberta, and the West, east to the western Dakotas, south to northern Arizona and New Mexico.

*DESERT WOODRAT, *Neotoma lepida* Body: 5.8–7 in (14.5–17 cm). Tail: 4.3–6.4 in (11–16 cm). Buffy gray to dark gray, with yellow-brown wash. Underparts and feet gray, buffy, or yellow-brown. Short-haired tail, dark above and white below. Diet includes acorns, seeds, fruits, foliage, twigs, bark, and cacti. Will climb up into shrubs or cactus for food. May breed year-round; 1 or 2 litters of 1–5 young each per year. In the Eastern Sierra, only from northern Inyo Co. south. Prefers moderate to dense cover in Pinyon-Juniper Woodland, and Sagebrush, Alkali Sink, and Shadscale scrub. *Distribution* Range includes the Great Basin to southeastern Oregon and southwestern Idaho, east to Colorado, southern California and northern Baja California, and western Arizona.

DUSKY-FOOTED WOODRAT, *Neotoma fuscipes* A large, dark woodrat, gray-brown to black-brown above, and gray to white below with white feet; its long tail is bicolored. Occurs along the west Sierra slope, but on the east slope only in Plumas Co., most commonly in Montane Forest and Pinyon-Juniper Woodland. *Distribution* Ranges from western Oregon south through northern Baja California.

*SAGEBRUSH VOLE, *Lemmiscus curtatus* Body: 3.8–4.5 in (9.5–11.5 cm). Tail: 0.6–1.12 in (1.5–3 cm). Pale buffy gray to ash gray above, white underparts, white feet; long, silky fur. Ears and nose washed with buffy color. Very short tail. Active day or night. Eats primarily green vegetation, bark, and cambium wood. Forages on the ground and in shrubs. Births peak from March to May and October to December, 1–11 young per litter. Females with young live in colonies. Nests are shredded bark or dried grass. In the Eastern Sierra, only in Mono and northern Inyo counties. Prefers Sagebrush Scrub but also occurs in grassy and weedy areas of Pinyon-Juniper Woodland. *Distribution* Range includes the Great Basin north to Washington, and the northern Plains into southern Canada and east to North Dakota and northern Colorado.

WESTERN HEATHER VOLE, *Phenacomys intermedius* Body: 3.5–4.6 in (9–11.5 cm). Tail: 1–1.8 in (2.5–4.5 cm). An uncommon vole of higher elevations. Upperparts ash gray washed with brown, underparts gray-buff; white feet. Bicolored tail about half of head-and-body length. Active at dusk and at night, foraging on ground and in shrubs. Caches food at night to eat next day. Breeds from mid-June to mid-October; 2 litters of 2–8 young. Both Sierra slopes; on the east slope from Nevada Co. south through Mono Co. Lives in grassy and weedy areas or in dense, short vegetation—particularly heather—in Subalpine and Montane forest. *Distribution*　Ranges through the Canadian coniferous forest, the higher mountains of the Northwest and the West.

The **Western Red-backed Vole** (*Clethrionomys californicus*), rarely encountered, has dark chestnut brown upperparts, a buffy or gray-white belly, white feet, bicolored tail. On the Eastern Sierra slope, only in Plumas Co.; on the west slope, only in Shasta and Plumas counties. Optimum habitat is dense, mature Subalpine and Montane forest. Ranges from southwestern British Columbia south to northern California.

MONTANE VOLE, *Microtus montanus*　Body: 4–5.5 in (10–14 cm). Tail: 1.2–2.8 in (3–7 cm). Upperparts buffy gray-brown to black-brown, becoming more buffy on sides, grading into dark gray belly and dusky feet. Tail less than half head-and-body length. Dark, nearly concealed ears. Constructs underground burrows and builds nests under snow. Breeds throughout the year, but births peak in June; 1–10 young per litter. Breeding stimulated by plentiful food. Diet includes grass, sedges, and forbs. Found on both Sierra slopes. On the east slope, from northern Inyo Co. north in moist meadows at all elevations, from alpine meadows to Desert Marshlands. *Distribution*　Range includes southern British Columbia south through the Great Basin, and the Rocky Mountains.

LONG-TAILED VOLE, *Microtus longicaudus*　Body 4.5–5.3 in (11.5–13.5 cm). Tail: 2–3.7 in (5–9.5 cm). Large vole with buffy gray-brown to black-brown upperparts, dark gray belly, dirty-white feet. Bicolored tail, half or more of head-and-body length, tipped with long hairs. The Long-tailed Vole eats grasses, sedges, forbs, seeds, bark, and bulbs. Breeds from April to October; 3–4 litters per year of 2–8 young. Both Sierra slopes; on the east slope, from central Inyo Co. north. Inhabits soft soil in moist grassland at all elevations, particularly in Alpine and Mountain meadows, also in Montane Riparian Woodland. *Distribution*　Ranges from eastern Alaska south through British Columbia and the West, east to South Dakota, and south to northern New Mexico and Arizona.

CALIFORNIA VOLE, *Microtus californicus*　Body: 4.8–5.7 in (12–14.5 cm). Tail: 1.6–2.7 in (4–7 cm). Buffy brown, gray-brown, or dark brown above; red-brown mid-back, lighter on sides grading into gray or white underparts. Prominent ears, bicolored tail, pale feet. Eats grasses, sedges, forbs, and developing seeds. May breed year-round, in response to abundant food and cover; 2–5 litters of 1–9 young per year. Populations cycle every 3 to 4 years with peak densities of 200 per acre (500 per hectare). Inhabits most of the western Sierra slope. On the east slope, from southern Mono Co. through northern Inyo Co., especially in Eastern Sierra Riparian Woodlands; also in moist Montane Meadows and other grassy areas. *Distribution* From southern Oregon south through western California to northern Baja California.

MUSKRAT, *Ondatra zibethicus* Body: 10–14 in (25.5–35.5 cm). Tail: 7.1–11.6 in (18–29.5 cm). Large, aquatic rodent with soft, thick, red-brown to dark brown fur, covered by long guard hairs. Short, round ears. Large, partly webbed hind feet with stiff hairs along sides of toes. Long, scaly, naked tail flattened on the sides. Mostly herbivorous; forages on banks or underwater for grasses, sedges, cattails, and other aquatic plants or small animals, such as tadpoles and snails. Nocturnal and does not hibernate. Births peak from March to August; 1–4 litters per year of 1–12 young. Native to the northeastern Sierra slope south to Mono Co. Introduced to most of the west slope. Requires permanent streams, ponds, or marshes. *Distribution* Range includes most of Alaska, Canada, and the United States, except for western California and the Southeast. Introduced to many parts of the United States, Europe, and Asia.

HOUSE MOUSE, *Mus musculus* Body: 3.2–3.4 in (8–8.5 cm). Tail: 2.8–3.8 in (7–9.5 cm). Gray-brown to buffy brown above with gray or buffy belly. Scaly tail sparsely haired. Nocturnal and active year-round. Breeds from mid-April to early January; 1–5 litters per year of 3–12 young. Introduced from Europe. Widespread on both Sierra slopes in cities and in most plant communities with an herbaceous understory, except at the highest elevations. *Distribution* Range includes most of North, Central, and South America, and other parts of the world.

The **Norway Rat** *(Rattus norvegicus)* may be distinguished from woodrats by its long, scaly tail and nearly naked ears. Introduced from the Old World, it has spread to cities and towns on both Sierra slopes, and may live in Eastern Sierra Riparian Woodlands. Omnivorous. Its range is nearly worldwide.

JUMPING MICE: FAMILY DIPODIDAE

If startled, jumping mice will leap several feet, surprising their predators. Large, strong hind feet and legs, and a long tail—similar to those of the kangaroo rats and mice—are adaptations evolved for jumping. Some species depend on jumping as their principal mode of locomotion. In contrast to the Heteromyidae, the Dipodidae have cheek pouches with internal rather than external openings, and they prefer forest and meadow communities rather than arid environments. They build a nest of grasses and sedges above-ground under a cover of litter. In late summer, preparing for hibernation, they accumulate fat. Jumping mice hibernate (prolonged torpor) in burrows or under logs, beginning as early as September.

WESTERN JUMPING MOUSE, *Zapus princeps*   Body: 3.5 – 4 in (9 –10 cm). Tail: 4.8 – 6.1 in (12 –15.5 cm). Uncommon. Moves by jumping. Dark yellow-gray band down back, graded yellow to white on sides; white belly. Tail longer than combined head-and-body length, brown to gray above and white below. Large hind feet and small, weak forefeet. Small ears bordered with yellow; front incisors are grooved. Active day or night. Diet includes seeds of grasses and forbs, insects, and fruits. Breeds from early June to late August; single litter of 2 – 8 young. Hibernates from about mid-September to mid-May. Found on both Sierra slopes. On the east slope, from northern Inyo Co. north, at higher elevations in Montane Riparian Woodland with an herbaceous understory, and moist Mountain Meadows; also in open Sub-alpine and Montane forest near meadows and streams.   *Distribution* Range includes the higher elevations of southwestern Canada and the West.

PORCUPINES: FAMILY ERETHIZONTIDAE

The Porcupine is the only mammal in the United States and Canada bearing quills. Quills—stiff, enlarged hairs covered with barbules that hook toward the base of the shaft—are a Porcupine's primary defense against predators. With their slow, awkward gait, Porcupines cannot outrun their enemies, although they are good tree-climbers. Porcupines should not be molested. Flicks of their tail will drive the quills, which are densely and loosely attached to the back and tail, into a closely approaching enemy. The barbules help a quill work itself deep into flesh and resist efforts to withdraw it—as many a dog owner has learned. Yet Porcupines are able to pull out the quills inflicted by other Porcupines during fights. Despite their quills and arboreal habits, Great Horned Owls, Coyotes, Bobcats, wolves, and especially Fishers have learned to subdue them. Fishers attack the face and soft, unprotected belly of a Porcupine.

Porcupines are solitary animals most of the year, except for brief breeding encounters and during winter when many individuals may den together to stay warm. Vocal exchanges during these social events consist of whines, screeches, and grunts. Courtship in Porcupines is a complex affair.

As suggested by the old joke—"How do Porcupines do it? Very carefully, of course!"—they do work around the quill problem. The male must first coax a female out of a tree; he then emits low grunts and sprays the female with urine. A series of chases, vocal exchanges, and wrestling bouts ensues. Finally, the receptive female crouches, elevates her hindquarters, and flips her dangerous tail out of the way as the male approaches from the rear. A copulatory plug forms in the vagina of the female shortly after mating, preventing encounters with other males. Male Porcupines sometimes battle fiercely over unmated females. Gestation requires about 210 days. The single young is born with teeth, eyes open, and quills erupted. Porcupines probably live as long as five to seven years in the wild.

Porcupines eat conifer bark, sometimes girdling a tree and killing it. Hence the timber industry considers Porcupines to be forest pests and tries to eliminate them. However, in wild forests, conifers and Porcupines have existed together for thousands of years, and the forests have survived quite well.

COMMON PORCUPINE, *Erethizon dorsatum* Body: 19.9–22 in (50.5–56 cm). Tail: 6.9–11.8 in (17.5–30 cm). Fig. 11.3e. Yellow-brown to black above with stiff, sharp quills on back and short tail. Small face with blunted snout. Heavy-bodied and short-legged; feet have massive claws. Has a clumsy, ambling walk and climbs trees awkwardly. Needs rocks, caves, logs, or burrows for den. Feeds on bark and cambium layer of conifers, twigs, buds, fruits, grasses, forbs, and leaves. Microorganisms in gut break down plant cellulose. Breeds from February to October; 1 young per year. Occurs on both Sierra slopes in Subalpine and Montane forest, Montane Riparian Woodland, and Pinyon-Juniper Woodland. *Distribution* Range includes most of Alaska, Canada, and the United States (except the Southeast and southern Plains states) south to northern Mexico.

CARNIVORES: ORDER CARNIVORA

The meat-eaters or Carnivora of the Eastern Sierra include Coyotes, foxes, Mountain Lions, Bobcats, bears, Raccoons, Martens, weasels, Fishers, Badgers, and River Otters. Terrestrial carnivores are nearly worldwide in distribution, absent only from Australia and Antarctica. All have adaptations for hunting and meat-eating: special teeth (upper premolars and lower molars with shearing edges), powerful jaws, well-developed canine teeth, simple digestive tracts, good sense of smell, and long bodies with powerful legs. Carnivores also have a high ratio of brain to body weight. Hunting requires athletic prowess and complex behavior—the ability to learn and associate the presence of prey with various cues and to lie in wait for opportune moments.

Members of the Canidae, the dog family, have long, bushy tails; pointed, erect ears; long legs; five toes on the forefeet and four on the hind feet. They walk on their toes, a gait that increases speed. Hearing and smell are more acute senses than sight. Eastern Sierra members of this family are the Coyote and foxes. Wolves *(Canis lupus)* are probably extinct in the Sierra, although rare, unsubstantiated sightings have been reported.

Coyotes are most active in early morning and sunset hours, but their habits change with the seasons. The kind of prey available influences the size of their home range and their social organization. When rodents are the mainstay, solitary Coyotes or pairs hunt them most effectively, and the home range tends to be large. When larger mammals and carrion are the principal food, Coyotes form packs, and the home range is usually small. The social organization may shift from pairs and lone animals in summer to packs in winter in response to the available food. Courtship may last for many months before mating occurs, and the pair bond may endure several years. Den sites include caves, crevices, rock ledges, or tunnel systems. After copulation, male and female may be locked together for as long as thirty minutes. Gestation requires about sixty-three days; the young are born blind and helpless. Males begin providing food when the pups are three weeks old and able to handle semisolid regurgitated food. Pups are weaned at five to seven weeks. By nine months the young attain adult weight. Coyotes have traditionally been hunted both for bounty and for their luxurious pelts. There is little information to support the view that Coyotes are primarily responsible for most annual livestock losses, particularly those of sheep. Losses caused by disease and other factors outweigh by far the losses caused by Coyote predation. Despite widespread efforts to control Coyote populations by hunting, trapping, and the use of repellents and lethal chemicals, their numbers remain healthy—a tribute to their adaptability and resilience.

Foxes communicate by means of howls, screeches, and short, yapping barks, particularly during the breeding season. Their diet changes with the season; for example, the Red Fox eats carrion and rabbits in winter, rodents in spring, and plant material and insects in summer. Male and female foxes form pairs and defend an area together. On the upper part of the undertail, foxes have a scent gland that exudes a musky odor; males use the secretions to mark their territories. Within their territory, members of a pair may dig or locate a den; only one litter is raised each year. After the pups are born and until they are old enough to be left alone for several hours, the male hunts for the female. By ten weeks the pups wander by themselves for short distances.

From mid-September to early October, the young foxes disperse. They become sexually mature at one year and may live as long as four or five years.

COYOTE, *Canis latrans*   Body: 32–37 in (81–94 cm). Tail: 11–16 in (28–40.5 cm). Fig. 11.3a. Doglike. Buffy gray and black or gray and reddish above, with white throat and underparts, and orange-brown legs, feet, and ears. Generally darker at high elevations. Plumed, dark-tipped tail, held down or between legs when the animal is running. Diet includes gophers, mice, ground squirrels, rabbits, insects, carrion, and fruit. Breeds from March through July; 1–11 young, 1 litter per year. Inhabits both Sierra slopes; alone or in groups, in all plant communities. *Distribution* Range includes much of Alaska, western and southeastern Canada, the United States (except for parts of the East and Southeast) south to Panama.

RED FOX, *Vulpes vulpes*   Body: 21.6–24.6 in (55–62.5 cm). Tail: 13.8–15.7 in (35–40 cm). Uncommon. Small with full, plumed tail. Color phases include the red phase with rich, yellow-red fur, white-tipped tail, blackish legs and ears, and white belly; cross phase, which is similar to red phase but with a broad line of darker fur down the back and across the shoulders; silver phase; and black phase. Protected by law from hunting or trapping. Eats marmots, ground squirrels, woodrats, mice, rabbits, birds, insects, fruit, and carrion. Occasionally caches food. Needs rock crevices, abandoned burrows, or hollow logs for dens. Breeds from February to October; litter of 1–12 young. Occurs on most of the west slope; on the east slope, from northern Inyo Co. north. Prefers a mix of forest or shrub and meadow; in Subalpine and Montane forest, Montane Chaparral, Montane Riparian Woodland, and aspen stands and shrubby parts of Alpine Rock and Meadow Communities. *Distribution* Ranges through Alaska, Canada, and most of the United States, except for the Great Plains and parts of the arid Southwest and the Southeast. Also occurs on other continents.

*KIT FOX, *Vulpes macrotis*   Body: 15–20 in (38–51 cm). Tail: 8.9–12 in (22.5–30.5 cm). Uncommon, desert-dwelling species. Small, slender fox with large ears. Pale gray or buffy yellow with rust tinge above light belly, black tip on tail. Eats mice, kangaroo rats, squirrels, rabbits, lizards, fruit, and insects. Young are born in February or March; 3–5 per litter. Protected by law from trapping and hunting. Found in arid Eastern Sierra communities in Inyo Co., including Sagebrush, Creosote Bush, Shadscale, and Alkali Sink scrub, and Joshua Tree Woodland. *Distribution* Ranges from southeastern Oregon and southwestern Idaho through the Great Basin, southern California, the Southwest south to Baja California and interior northern Mexico.

Fig. 11.3.   Mammals. (a) Coyote, *Canis latrans*. (b) American Marten, *Martes americana*. (c) Long-tailed Weasel, *Mustela frenata*. (d) Black Bear, *Ursus americanus*. (e) Common Porcupine, *Erethizon dorsatum*. (f) Striped Skunk, *Mephitis mephitis*. (g) Common Raccoon, *Procyon lotor*

GRAY FOX, *Urocyon cinereoargenteus*   Body: 23.8–29 in (60.5–73.5 cm). Tail: 11–17.3 in (28–44 cm). Black-gray upperparts and tail; white throat and belly. Long, fluffy tail has black stripe down back, black tip. Rusty orange under tail, on sides, along legs, and up neck to the ears. Nocturnal and active year-round. Diet includes rabbits, rats, mice, gophers, berries, insects, and carrion. This fox sometimes climbs trees to forage. Requires denning sites such as hollow logs, rock crevices, caves, and brush piles. Breeds from March to May; 2–7 young per litter. Inhabits both Sierra slopes; in the Eastern Sierra in lower-elevation shrubby areas, including Montane Chaparral, Eastern Sierra Riparian Woodlands, aspen stands, Pinyon-Juniper Woodland, and Sagebrush Scrub.   **Distribution**   Range extends through the United States, except for the northern Great Basin, northern Rocky Mountains, and the Great Plains; ranges south to most of South America.

BEARS: FAMILY URSIDAE

Bears and members of the dog and raccoon families are closely related. They have similar facial features and food habits, and good learning abilities. All bears have nonretractable claws, rudimentary tails, heavy coarse fur, and *plantigrade* feet (the full foot rests on the ground, rather than just the toes). Bears have small eyes and ears but a highly developed sense of smell. Despite their bulk, adult bears climb trees and are good swimmers. Their normal gait is ambling or rolling—that is, they walk by lifting both legs on the same side of the body at the same time. Since the Grizzly Bear *(Ursus arctos)* became extinct in the Sierra around 1925, the Black Bear has been the only species in the region. Adult males occasionally weigh more than 600 lbs (270 kg). Bear signs in an area include biting or claw marks on the trunks of trees, scats, and opened rotten logs. Bears are solitary; individuals maintain separate home ranges and foraging areas. Their diet varies seasonally: grasses and forbs in spring; fruit, insects, and carrion in summer; fruit and seeds in fall. To fatten up for winter dormancy, they wander widely in summer and fall in search of food.

Bears are the largest mammals to become dormant during the winter. Denning sites include tree cavities, caves, or a crude nest on the ground. Dormancy begins anytime from October to January and ends between March and May. The physiological state of dormant bears is quite different from true hibernation. Their body temperature is only a few degrees below active body temperature, whereas in true hibernators, it is much lower—a degree or so above ambient temperature. The metabolic rate is 50–60 percent of the active rate, although the heart rate drops from about 50 beats to as low as 8 beats per minute. Black Bears metabolize body fat during dormancy, often losing 20–27 percent of their fall body weight over a winter. During dormancy, bears do not drink, eat, or eliminate wastes, and they can easily be awakened. In contrast, true hibernators arouse every few days or weeks to eat, drink, urinate, and defecate, but they are not easily awakened while in their torpid state.

Full grown at about four years, Black Bears may live as long as fifteen or twenty years. They become sexually mature between three and seven years; females usually breed every other year. Mating occurs during the summer, when a male and female consort for several weeks. Because ovulation is induced by mating, fertilization is fairly certain. The young are born seven or eight months later, but gestation actually requires less time; the embryos are not implanted until November or December. Cubs are born in the den— hairless, helpless, blind, and only about 8 inches (20 cm) long. They first leave the den in spring with their mother and then den with her a second winter. To learn how to find food and protect themselves, cubs need prolonged association with an adult.

Black Bears rarely bother people. If annoyed, they prefer to escape, but in rare circumstances they will charge. Bears come to campsites for the tasty and calorie-rich food that people eat. "Sophisticated" bears that have learned to open trash cans and steal food carefully hung in trees teach these skills to their cubs. These pests become dependent on human food and are hazardous to tourists. To reduce the risk of encounters between bears and people, problem bears may be transplanted to remote areas, but they often return. A more successful approach has been to eliminate garbage dumps, provide bear-proof trash cans, and educate campers.

BLACK BEAR, *Ursus americanus*   Body: 3.3–6.6 ft (1–2 m). Tail: about 6 in (15 cm). Fig. 11.3d. Large, heavy-bodied, with doglike face. Brown muzzle, black or brown body. Small eyes, round ears, front legs shorter than hind legs. Powerful claws on front and rear feet. Diurnal or nocturnal; usually solitary except for mothers with cubs. Wanders widely in search of food. Omnivorous, consuming grasses, forbs, fruits, seeds, tubers, eggs, honey, insects, larvae, fish, small mammals, carrion, and garbage. Young are born in January and February; commonly 2 per litter, may be 1–3. Remains in den in state of dormancy much of the winter. Roams both Sierra slopes in Subalpine and Montane forest and nearby meadows.   *Distribution*   Range includes the mountainous and coniferous forests of Alaska, Canada, the United States, and northern Mexico.

RACCOONS AND RINGTAILS: FAMILY PROCYONIDAE

The procyonids share several traits with their relatives the bears. For example, Raccoons and Ringtails have five-digit plantigrade feet and are omnivorous; they are clever in finding food and extremely curious. With their remarkably dexterous forefeet, they manipulate food and quickly learn to open door locks and other mechanical devices. Neurological studies indicate that Raccoons have many more touch receptors in their forefeet than in their hind feet. Other studies rank their learning ability between that of cats and primates.

Wherever Raccoons occur, they are abundant; their "handlike" prints are common in the mud along streambanks and ditches. Although they lead solitary lives in separate home ranges, males and females may form a pair bond a month before mating; males may also mate with more than one female. Gestation lasts about 65 days. Several days before giving birth, the female remains in her den. The young animals are born fully furred but helpless, with eyes and ears closed. Movements are awkward until about the seventh week; at eight to twelve weeks they forage with their mother. Raccoons, like bears, become dormant during the colder winter months, living on the fat reserves accumulated in fall. Occasionally, several Raccoons may den together in winter. Ringtails are quicker and more agile than Raccoons; they are swift runners and skilled climbers. Their pair bonds are longer-lasting, and both males and females bring food to the young.

COMMON RACCOON, *Procyon lotor*   Body: 18–25 in (45.5–63.5 cm). Tail: 8–12 in (20.5–30.5 cm). Fig. 11.3g. Robust body; gray-brown above with black-tipped hairs, lighter gray on belly. Long tail with 5 to 7 dark rings alternating with yellow-white rings. Black mask across eyes with light patches above and below; small, rounded ears. Nocturnal. Feeds near streams and lakes; needs water for drinking and washing food (a well-known habit that is still unexplained). Eats fruit, seeds, acorns, crayfish, insects, fish, and amphibians. Den sites are hollow logs or trees, rocks, or burrows; animal retreats to den during spells of cold weather. Breeds from March to August; single litter of 1–8 young. Inhabits both Sierra slopes. On the east slope, from northern Mono Co. north, in Eastern Sierra Riparian Woodlands, and Montane and Subalpine forest.   *Distribution*   Ranges from southern Canada south through the United States (except for parts of the Great Basin and of the Rocky Mountains) and south to Panama.

RINGTAIL, *Bassariscus astutus*   Body: 14–16 in (35.5–40.5 cm). Tail: about 15 in (38 cm). Uncommon. Slender buffy or yellow-gray body, brown-black back. Long tail with 8 incomplete black rings. Delicate face with large eyes and small, pointed ears; narrow black ring around each eye, surrounded by white patch. Strictly nocturnal. Eats small mammals, birds, insects, spiders, and fruit. Breeds from May to October; a single litter of 1–5 young. Both Sierra slopes. In the Eastern Sierra, from southern Mono Co. south, in lower Montane Forest, Eastern Sierra Riparian Woodlands, Pinyon-Juniper Woodland, and Sagebrush Scrub. *Distribution* Ranges from southern Oregon south through most of California and the Southwest, to northern Colorado and Utah, to western Louisiana and Arkansas, and south through Mexico.

MARTENS, FISHERS, WOLVERINES, WEASELS, BADGERS, SKUNKS, AND OTTERS: FAMILY MUSTELIDAE

Mustelids have anal and abdominal scent glands, fine fur, short legs, and five toes on front and hind feet. Like the canids they walk on the toes and front of their feet. Most mustelids lead solitary lives, except for brief periods of pair-bonding during the summer. They are aggressive and use their scent glands to mark their home ranges during the breeding season. Up to twelve months after breeding, the young are born, helpless and sparsely furred. Gestation requires only 30 to 50 days, but delayed implantation of the embryo postpones births until the following spring or summer.

Martens forage on the ground and in trees; they also tunnel under snow to catch prey. During the breeding season, individuals are very aggressive, and females may engage in fierce battles. Marten vocalizations include growls, screams, and chuckles. Courtship between a male and female may last fifteen days, during which time the pair engages in wrestling and play. A female may mate with more than one male.

Fishers, like Martens, climb trees to find food. Fishers regularly eat Porcupines, usually attacking them on the ground. To avoid quills, they attempt to bite the face and soft belly, sometimes requiring half an hour or more to make the kill. Trappers have found that most Fishers carry a few quills, which apparently do not fester or impair their activities.

Wolverines are the largest terrestrial mustelids, weighing up to 60 lbs (27 kg). They alternate three or four hours of sleep with foraging time. Fierce hunters, Wolverines will attack larger animals such as deer. They cache excess food in trees for future use. Male Wolverines are highly territorial but permit one or more females to share part of their home range.

The fine pelts of Martens, Fishers, and Wolverines are extremely valuable. Wolverine fur has been used for lining parka hoods, because it is moisture-resistant and frost is easily brushed off. Since their populations have declined, they cannot be hunted or trapped in California.

The slender bodies of weasels, Ermines, and Minks allow them to squeeze through vegetation, between rocks, and down burrows in pursuit of prey. They are reputed to be bloodthirsty hunters, because they appear to kill more than they need; however, extra food is cached for future use. They usually bite their prey through the back of the skull or through the throat, wrapping their long bodies around the animal to restrain it. Weasels, in particular, must be efficient hunters, since they require 40 percent of their body weight in food each day. Because long bodies have large surface areas, the *Mustela* species lose heat faster than do other mammals of comparable weight.

Courtship in weasels and Minks consists of struggling and fighting. The male mates with the female only after he has grabbed the scruff of her neck and forced her down. All three *Mustela* species are trapped for their luxurious pelts; however, in California trapping is regulated. The white winter pelage of the weasel is sold as "ermine." Although Mink ranching produces a good supply of quality pelts, there is still a market for wild pelts.

Among the mustelids, the Badger is best adapted for digging and burrowing. The body and head are flattened; the legs short and well muscled; the foreleg claws long, thick, and curved; and the ears and eyes small, with the latter protected by a membrane. Badgers hunt mostly at night, digging out the burrows of small rodents. They often plug all but one entrance to a burrow complex and then dig out the remaining tunnel to trap the unlucky prey. During the breeding season, female Badgers will excavate, for their own use, a burrow system that includes main tunnels, side tunnels, and chambers. Although Badgers do not hibernate, they become inactive during food shortages in cold weather. They may remain in their burrows for more than thirty days, until conditions improve.

The bold black-and-white markings of skunks may serve as a warning to their predators: owls, Golden Eagles, Fishers, Badgers, Coyotes, foxes, Bobcats, and Mountain Lions. At the base of the skunk tail is a weapon of last resort—glands that spray a strongly repellent, musky liquid. Voluntary muscles that position a nipplelike opening to each gland control the direction of the spray. Sprayed into a predator's face, the liquid causes nausea and temporary blindness. Skunks warn their attackers before spraying by stamping, shuffling backward, or doing handstands, which may either make the skunk look larger or indicate that it means business. Before spraying, the Striped Skunk raises and fluffs its tail and doubles its body back so that head and rear face the same direction for spraying accuracy. When the Spotted Skunk is ready to spray, it stands on its forelegs with tail held over its head. Resembling an open flower, the long, white tail hairs fall neatly into a circlet, and the skunk backs up toward the attacker. The aim is precise at about 10 ft (3 m) or greater. Skunks have large claws on their forefeet, which are used for digging up insects and other invertebrates. Striped Skunks sometimes excavate burrows and chambers, lining them with grass or leaves. Although skunks are usually solitary, they occasionally den together. Often a female will den with a litter from the previous breeding season.

Although river otters once occurred over much of North America, populations of this graceful carnivore have declined or disappeared in many regions because water diversions, development, farming, and pollution have altered their habitat. In California, they may not be hunted or trapped. Because they are difficult to census accurately, local population sizes are unknown. River otters are highly intelligent animals. Both in captivity and in the wild they show much exploratory and play behavior, such as manipulating small objects like pebbles and shells, wrestling, and sliding down mudbanks or snowbanks. Within their home ranges they have special rubbing and rolling sites, which they use to clean and dry their fur, and long mud or snow slides for repeatedly sliding into water and scrambling out again. Male and female otters establish individual home ranges that overlap; females scent-mark frequently when they are sexually receptive. Breeding usually occurs in the water: the male grips the female by the scruff of her neck and wraps his body around hers. A pair may mate several times over successive days. The young are born helpless but fully furred.

AMERICAN MARTEN, *Martes americana*   Body: 14.2–17.4 in (36–44 cm). Tail: 7–9.4 in (18–24 cm). Fig. 11.3b. Yellow-brown upperparts grading to dark brown on tip of bushy tail and legs. Orange to pale buff patch on throat, and paler undersides. Long body with slightly pointed ears. Active year-round, day and night. Forages in trees, rocks, meadows, and under snow. Eats squirrels, chipmunks, mice, woodrats, rabbits, Pikas, insects, birds, and fruit. Needs talus slopes, snags, or logs for den sites. Breeds from mid-March to mid-April; 1 litter per year of 1–5 young. Locally common to rare on both Sierra slopes in Montane and Subalpine forest. *Distribution*   Ranges through the coniferous forests of Alaska, Canada, the Northeast, and western United States.

FISHER, *Martes pennanti*   Body: 20–25 in (51–63.5 cm). Tail: 13.4–16.6 in (34–42 cm). Dark brown but grayer on head and shoulders, with black legs and black, bushy tail. White spots on neck and belly. Ears more rounded and body heavier than Marten's. Active year-round, day or night. Forages in trees and on the ground. Eats Porcupines, marmots, squirrels, Beavers, woodrats, mice, rabbits, birds, insects, and fruit. Needs dense forest, and logs, snags, or brush for dens. Births peak March to early April; a single litter of 1–4 young. Uncommon on the west slope. Small, scattered populations on the east slope from Plumas through Mono Co. Prefers Montane and Subalpine forest. *Distribution*   Range includes coniferous forest of southern Canada, the Great Lakes region, northeastern United States, the Northwest, and northern Rocky Mountains.

WOLVERINE, *Gulo gulo*   Body: 28–34 in (71–86.5 cm). Tail: 7.5–10.2 in (19–26 cm). Extremely rare species intolerant of human disturbance. Dark brown or nearly black; pale brown or yellow stripe across forehead, and a stripe on each side from shoulder to tail. Except for bushy tail, built like a small bear, with stout, pointed muzzle; small, round ears; sturdy legs; large feet with white claws. Forages on the ground, in trees, burrows, or rock piles. Feeds on gophers, woodrats, mice, marmots, carrion, birds, fruit, and occasionally deer and Porcupines. First breeds at 3 years. Births peak mid-February to mid-March; single litter of 1–5 young. Occurs on both Sierra slopes. On the east slope, from Sierra Co. south through northern Inyo Co. Inhabits Subalpine Forest and Meadow, and Alpine Rock and Meadow Communities. *Distribution*   Populations reduced or extinct in part of range. Occurs in coniferous forest of Alaska, Canada, the Northwest, and Rocky Mountains. Also occurs on other continents.

LONG-TAILED WEASEL, *Mustela frenata*   Body: 8–10.5 in (20.5–26.5 cm). Tail: 3–6 in (7.5–15 cm). Fig. 11.3c. Long, slender body and neck. In summer, brown above with yellow-white underparts and feet, white spot between eyes, sometimes white patch before each ear. In winter, at higher elevations where snow accumulates, all white. Year-round, black-tipped tail with yellow-white underside. Primarily nocturnal. Feeds upon rabbits and small rodents, such as voles, rats, mice, gophers, and squirrels, as well as insects, birds, and fruit. Needs old burrows, logs, or brush for denning sites. Births peak from late April through early May; single litter of 4–9 young. Inhabits both Sierra slopes. The east slope from northern Inyo Co. north in all but the most arid communities: Subalpine and Montane forest, Mountain Meadows, Eastern Sierra Riparian Woodlands, and Pinyon-Juniper Woodland.   *Distribution* Ranges from southern Canada through the United States, except for the Southwest, and south to Bolivia.

ERMINE, *Mustela erminea*   Body: 5–9 in (12.5–23 cm). Tail: 1.6–2.3 in (4–6 cm). A smaller version of *M. frenata,* but less common. In summer, dark brown with white underparts and feet, but no white under tail. In winter, at higher elevations, all white. Black-tipped tail year-round. Active day and night. Diet includes voles, mice, woodrats, chipmunks, birds, amphibians, reptiles, and fruit. Births peak from late April to early May; single litter of 4–9 young. Found on both Sierra slopes; on the east slope from southern Mono Co. north, in dense Subalpine and Montane forest and Mountain Meadows. *Distribution*   Range includes Alaska, Canada, the Great Lakes region, northeastern United States, and coniferous forests of the West. Also on other continents.

AMERICAN MINK, *Mustela vison*   Body: 12–17 in (30.5–43 cm). Tail: 5.9–7.5 in (15–19 cm). Long, narrow body similar to weasel, but more robust. Entire body brown to dark brown year-round except for white chin patch and, in some individuals, irregular spots on belly. Slightly bushy tail. Active day and night. Eats rabbits, mice, birds, and aquatic prey, such as Muskrats, fish, amphibians, and crayfish. Requires permanent water. Forages in dense vegetation and underwater. Dens in tree roots, hollow logs, rocks, or abandoned burrows. Births peak in May; single litter of 4–10 young. Both slopes. In the Eastern Sierra, from northern Inyo Co. north, in Eastern Sierra Riparian Woodlands, Desert Marshlands, and near ponds, rivers, and lakes adjacent to other montane communities. *Distribution*   Range includes Alaska, most of Canada, and the United States, except for the southern Great Basin and the Southwest.

AMERICAN BADGER, *Taxidea taxus*   Body: 18–24 in (45.5–61 cm). Tail: 3.8–6 in (9.5–15 cm). Short-legged, heavy-bodied, powerful animal specialized for burrowing. Gray-brown above with white stripe running from nose, over top of head, and down shoulders. Yellow tinge to belly and short tail. White cheeks and ears, with black patch below the ears, black outline on back of ears. Black feet with long foreclaws. Finds prey by digging out the burrows of chipmunks, ground squirrels, gophers, rats, and mice; also eats snakes, lizards, birds, and eggs. Newly dug oval burrows near ground squirrel burrows indicate Badgers. Active day and night. Births peak from March to early April; single litter of 2–5 young. Uncommon on both Sierra slopes. Prefers open areas with friable soils, including rangeland; grass and forb areas of Sagebrush, Shadscale, and Alkali Sink scrub, and Pinyon-Juniper Woodland; as well as grass and forb areas of Subalpine and Montane forest. *Distribution*   Ranges from southwestern Canada east to Lake Superior and south through the West to central Mexico; the northern Midwest and the Great Lakes region.

STRIPED SKUNK, *Mephitis mephitis*   Body: 13–18 in (33–45.5 cm). Tail: 7.3–15.4 in (18.5–39 cm). Fig. 11.3f. Cat-size, black body with white stripes— a slender stripe down forehead, one broad stripe on both sides from back of head to tail. Strong scent glands at base of tail, for defense. Forages for small rodents, insects, carrion, bird eggs, fruits, and seeds. Usually nocturnal. Births peak from April to June; single litter of 2–10 young. Common on both Sierra slopes. Prefers Montane Chaparral and Sagebrush Scrub. Also occurs in shrub understory in Montane Forest, Pinyon-Juniper Woodland, Montane Meadows, and shrubby Eastern Sierra Riparian Woodlands. *Distribution*   Southern Canada, virtually the entire United States, and northern Mexico.

SPOTTED SKUNK, *Spilogale putorius*   Body: 9–13.5 in (23–34.5 cm). Tail: 4.5–9 in (11.5–23 cm). Small skunk. Black, with white spot on forehead, one spot under each ear, short stripes and spots along back and sides. White-tipped tail. When positioned to spray, stands on forelegs with white-tipped tail spread overhead like parasol. Diet similar to that of Striped Skunk, but the Spotted Skunk will climb shrubs and trees for food. Requires hollow logs, rocks, burrows, or brush for den. Births peak in May; usually a single litter of 2–6 young. Occurs on both Sierra slopes; on the east slope, in open Montane Forest, Montane Chaparral, Sagebrush Scrub, and Pinyon-Juniper Woodland.   *Distribution*   Ranges through much of the United States except the inland Northwest, the Northeast, and Great Lakes region; south to Panama.

NORTHERN RIVER OTTER, *Lontra canadensis*   Body: 23.6–31.5 in (60–80 cm). Tail: 11.8–19.7 in (30–50 cm). Uncommon. Large, long, and slender. Dark brown above with white or silvery neck and throat; short, thick fur. Small ears and eyes, webbed feet, tail thick at base tapering to tip. Large snout, neck as thick as head, and top of head flattened. Requires permanent water with fish: rivers and large streams at lower montane elevations, and lakes at higher elevations. Feeds in the water or on nearby land, taking fish, frogs, aquatic invertebrates, and sometimes birds and small mammals. Usually goes after slower fish, sculpin and squawfish rather than trout. Home range may include more than 15 mi (25 km) of bank or shoreline. Births peak in March and April; single litter of 1–4 young. Family groups of mother and young may remain together as long as 8 months. Found on both slopes of the Sierra: on the west slope, scattered populations; on the east slope from Plumas to Kern Co.   *Distribution*   Range includes Alaska, most of Canada, and the United States except for the southern Great Basin and the arid regions of the Southwest.

MOUNTAIN LIONS AND BOBCATS: FAMILY FELIDAE

The felids are powerful carnivores with well-developed jaw muscles for biting. Hunters by day or night, they have forward-positioned eyes that provide good depth perception; in addition, large pupils that contract to vertical slits enable them to see in dim light. Whiskers provide good tactile ability around the head region. Felids have molars with shearing edges, long canine teeth, and retractable claws; they walk on their toes. Mountain Lions and Bobcats are both asocial species. Some individuals occupy home ranges, while others are transient. Home ranges are marked by piles of feces or by urine. When a female comes into estrus, several males battle for dominance. The mating posture used by cats is unlike that of most mammals. While the female crouches down, the male stands over her with his weight mostly on his front legs, restraining her by biting the back of her neck. Den sites for giving birth are crevices or small caves with little or no bedding material. Mountain Lions become sexually mature at about two and a half years and breed year-round; gestation lasts about three months. Bobcats first breed between one and two years, usually during winter and spring. Gestation lasts two months. Females with cubs do not tolerate males, which may kill the young.

Mountain Lions move with the seasons, following migratory deer, their principal prey. However, where deer are resident, lions do not migrate. They stalk their prey and attack suddenly. A Mountain Lion will jump on the back of a deer, dig in its claws, and kill by biting the back of the neck and severing the spinal cord. Capable of moving a carcass several times its weight, it will drag its prey to cover for feeding.

Bobcats' most common prey are rabbits and hares. Unlike a Mountain Lion, which stalks its prey, a Bobcat usually hides in ambush, quietly waiting for suitable prey, then suddenly pounces. It occasionally tackles a small deer or bedded deer. Often, it caches a carcass for future use, covering it with snow or ground litter.

Because Bobcats are somewhat tolerant of human disturbance, their distribution has not changed greatly from their historic range. Mountain Lions, in contrast, once ranged throughout the United States; they now occur only in the West and southern Florida. However, since they are now protected in some states, including California, Mountain Lions may be expanding their range.

In recent years Mountain Lion sightings have increased in rural and suburban areas, and reports of attacks on people and pets in the mountainous regions of the West have made the headlines and resulted in widespread misinformation. In California the deaths of two women joggers have blown recent incidents out of proportion. The facts are that in California verifiable records of humans being injured by lions indicate that only twelve such incidents have occurred and that since 1909 only two people have been killed. If any wild animal, including a Mountain Lion, becomes dangerous to the public, the California Department of Fish and Game (DFG) has the authority to take immediate action to remove it.

Is there some way to prevent Mountain Lion attacks? Publicizing preventive measures and the need for caution could help immeasurably. Preventive measures are simple, according to a publication of the DFG called *Living with California Mountain Lions:* In lion country, always travel with someone; if you do meet up with a lion, never turn and run but instead shout, scream, wave your arms, and do whatever else you can to look tall and big, and then back slowly away. In lion country, never tie your pet outside where it can neither fight nor run. As more people choose to hike or build homes in remote areas, there are likely to be more encounters between humans and Mountain Lions.

MOUNTAIN LION, COUGAR, OR PUMA, *Felis concolor* Body: 41–60 in (104–152 cm). Tail: 30–36 in (76–91 cm). Large, powerful cat. Long, black-tipped tail and small, rounded black-tipped ears. Occurs in two color phases, both with white underparts: yellow-brown above, sometimes with orange tinge, or dark gray above. Active day and night. Eats mostly deer; also elk, Bighorn Sheep, and small mammals. Uses den sites such as caves, thick brush, and crevices in rocky areas. Births peak April through August; 1 litter of 1–6 cubs every other year. Inhabits both Sierra slopes, all elevations and most communities; less common in High Desert communities. *Distribution* Largest historic range of any North American mammal: western Canada, the United States (now in the West and southern Florida), Baja California, and south to the tip of South America.

BOBCAT, *Lynx rufus* Body: 24–28 in (61–71 cm). Tail: 4–8 in (10–20 cm). Small cat; long-legged with stubby tail, short ear tufts, and long cheek fur framing face. White below and gray, buffy, or reddish above with scattered black spots. Black-tipped tail may be barred. Active day and night. Prey includes deer, squirrels, rats, mice, rabbits, birds, reptiles, and some invertebrates. Needs rock crevice, cave, or brush for den. Births peak in March and April; usually 1 litter a year of 1–7 young. Both Sierra slopes in all communities. Prefers forest edge or shrubby understory in Montane Forest, Pinyon-Juniper Woodland, and Eastern Sierra Riparian Woodlands. *Distribution* Range includes southern Canada, the United States, Baja California, most of Mexico.

ELK, DEER, BIGHORN SHEEP, AND PRONGHORN: ORDER ARTIODACTYLA

Artiodactyls are exclusively herbivorous, feeding on grasses and forbs or foliage. Easy prey while foraging, they have evolved several defenses to reduce their vulnerability. They walk on the tips of their toes, which enables them to run swiftly; two digits have enlarged to form a hoof, while others are reduced or eliminated. Many species are camouflaged by spots, stripes, or disruptive coloration, and their senses of hearing and smell are acute. In addition, the males (and the females in some species) have horns or antlers, which they use to defend themselves against attacks by predators and also for displaying or fighting with other males for access to females.

Artiodactyl teeth are specialized for an herbivorous diet. The upper incisors, important to carnivores for capturing or holding prey, have disappeared. Vegetation is cropped by the lower incisors working against a horny upper lip. In many groups, the canine teeth are also absent. The premolars and molars, which grow continually, have high, wear-resistant grinding surfaces. Deer, elk, sheep, and Pronghorn have a four-chambered stomach in which microbes break down cellulose. Partly digested material returns from the stomach to the mouth for more chewing. These two adaptations efficiently extract nutrients from grass and foliage.

Most artiodactyls are highly social and communicate by stereotyped postures, displays, and vocalizations. The sense of smell is particularly important: They use scent glands to mark home ranges and recognize individuals by their scent. Scent also signals sexual receptiveness. There is usually a hierarchy among males, determined by the outcome of fighting during the *rut* (breeding season). Only the dominant males mate with females.

ELK AND DEER: FAMILY CERVIDAE

Members of the Cervidae have antlers, which are shed each year. Male antler growth is stimulated by testosterone production in spring. The antlers consist of a core of porous bone with an outer layer and base of hard, dense bone. Growing antlers are covered with "velvet" that supplies oxygen and nutrients. By the time the antlers harden and the blood supply diminishes, the male's gonadal development has peaked and he is ready to mate. He then rubs off the velvet, exposing the bare bone. Following the breeding season, when levels of testosterone drop, antlers are shed; new antlers begin to grow about a month later. Since antler size and number of points correspond to age, older males are usually most successful at sparring and, consequently, breeding.

Both subspecies of elk in the Eastern Sierra are introduced. The Rocky Mountain Elk, which occurs from Plumas County north, originated from a small introduced herd near Mount Shasta. The Tule Elk herd, introduced in 1933 to Owens Valley in Inyo County, originated with animals from the San Joaquin Valley. Records from the early nineteenth century indicate that the original range of the Tule Elk in California included most of the Central Valley and the Coast Range from Marin County south to at least Monterey Bay. By 1870 they were virtually extinct, victims of hunters and agricultural development. Around 1875 a pair was found and protected on the Miller and Lux cattle ranch. Supposedly, all of today's Tule Elk are descended from this pair.

After the Tule Elk were transplanted to the Owens Valley, they came in direct conflict with the cattle that had long grazed the valley—bolting through fences, and feeding on and trampling ranchers' alfalfa fields. After years of heated controversy, the problems were resolved by an agreement to keep elk numbers down so that they do not overgraze their range or interfere unduly with ranching; surplus animals are removed to other locations.

The Tule Elk is the smallest subspecies of North American elk, weighing 100–200 lbs (45–90 kg) less than the other subspecies and bearing smaller antlers. It is the most specialized subspecies, adapted to semi-arid habitat. For most of the year, cows and bulls form separate small groups. During spring, the females give birth, usually to a single calf or twins. The rut begins in August and September, at which time the bulls join the cow groups. As the rut progresses, one bull takes command of each cow group, forming a harem and driving out the young bulls. Sparring occurs when two bulls lower their heads, engage antlers, and push. Male aggression during the rut is often accompanied by their musical "bugling."

Two subspecies of Mule Deer naturally occur in the Eastern Sierra: the Rocky Mountain Mule Deer *(Odocoileus hemionus hemionus)* and the Black-tailed Deer *(O. h. columbianus).* The Rocky Mountain subspecies is more common on the eastern slope. Mule Deer occur in small groups in the mountains and foothills, which provide good browse. They are most dispersed in summer, particularly during the fawning season when small groups usually consist of males. Breeding behavior is similar to that described for elk. They remain at higher elevations from spring to fall and migrate to lower elevations for the winter. Mule Deer continually watch for predators, particularly Mountain Lions, using acute hearing and smell. Today, Mule Deer populations are managed, especially by enforcement of hunting restrictions, to achieve some balance between numbers of deer and available forage.

TULE ELK OR WAPITI, *Cervus elaphus nannodes* Body: about 6.5 ft (2 m). Tail: 5.5 in (14 cm). Introduced to the Eastern Sierra. This subspecies is a small, light-colored elk, buffy gray to buffy white over body with white rump patch and red-yellow hue on front of legs; gray-brown mane on throat and neck. The points of the antlers are unbranched. Vestigial toes on either side of hoof. Male sheds antlers in March and April; antlers are fully grown by early August. Eats foliage, aquatic vegetation, grasses, forbs, and crop plants, such as alfalfa. Births peak late May to early June, with 1–3 young. *Distribution* East of the Sierra, in the bottomlands of Owens Valley and on adjacent mountain slopes. Introduced recently to other locations in California west of the Sierra.

A herd of **Rocky Mountain Elk** *(Cervus elaphus nelsoni)* was introduced in the vicinity of Mount Shasta in 1913; wandering individuals are occasionally observed on the east slope, particularly from Plumas Co. north.

MULE DEER, *Odocoileus hemionus*   Body: 3.5–5 ft (1–1.5 m). Tail: 4–9 in (10–23 cm). In summer, yellow-brown or red-brown above; in winter, gray or dark brown above. Underparts and rump patch white with tail either black-tipped or black on top. Face white or gray with dark brown patch on forehead. Fawns are spotted. Antlers on males have forked tines; usually shed in January and February and regrown by late summer. Active day or night, particularly at dawn and dusk. Eats foliage from shrubs, grasses, forbs, and acorns when available. Births peak early April through mid-July; 1–3 fawns. Common on both Sierra slopes; on the east slope, from northern Inyo Co. north. In summer, prefers open Montane and Subalpine forest, Mountain Meadows, Montane Riparian Woodland, and Montane Chaparral. Winters at lower elevations in shrubby habitat, such as Sagebrush Scrub and Pinyon-Juniper Woodland. *Distribution*   Range includes southwestern Canada, the western United States east to Minnesota and Iowa, Baja California, and northern interior Mexico.

BIGHORN SHEEP: FAMILY BOVIDAE

Bighorn Sheep have special adaptations for their precarious lifestyle on rocks and cliffs above treeline. The two toes of each hoof move independently; the back part of each toe has a round, rubbery pad that grips rocky surfaces. Their heavy skull, sturdily supported by the spine, can withstand the fierce, horn-butting fights among males with minimal damage. Acute vision detects predators. Their tall, broad molars enable them to grind up abrasive, dry, dusty grasses and forbs.

Bighorns are very social. Males and females segregate into small bands, and the sexes associate only during the annual rut. Within each band, the individuals establish a hierarchy by fighting; the dominance relationships are usually based on size. Homosexual behavior is common within male bands and is an expression of dominance and submission: A dominant male may treat a subordinate like a female. The rut usually occurs during the late fall on traditional grounds. Most clashes occur between rams of the same size. Before a head-to-head clash, the two rams stand some distance apart and run toward each other. Just before butting heads, they rise on their hind legs to increase the power of the blow. The fight continues until one ram runs away or submits. The dominant male within a band does most of the courting and breeding, although other males may occasionally have opportunities. A female in estrus becomes very aggressive and searches for the largest ram. She may force his attention with various overtures and provoke a chase, which is concluded by the male mounting her. Usually the dominant male guards the female throughout estrus. Gestation lasts about 175 days. On their first day, lambs are capable of following their mothers over steep terrain. At two weeks they begin to nibble grass; at two years they are independent.

Most populations of Bighorn Sheep migrate with the seasons. They remain on their winter range, at lower elevations, for most of the year. Their summer range is above treeline, among high meadows and bare rock. On the winter range, Bighorn Sheep often face competition from livestock, particularly domestic sheep, as well as from deer. Parasites and diseases from domestic sheep have wiped out many Sierra herds. This, along with overhunting and competition for food, has resulted in drastic population declines and extinctions; Bighorn Sheep hunting is now prohibited in California. Medical treatments and management practices are being developed to maintain populations and reintroduce sheep where extinctions have occurred.

BIGHORN SHEEP, *Ovis canadensis*   Body: 3.5–6 ft (1–1.8 m). Tail: 3–6 in (7.5–15 cm). A sturdy, well-muscled sheep; gray-brown to gray with brown to white underparts and creamy white rump. Reduced toes on either side of each hoof. Males have massive horns that spiral from the top of the head, back, down, and forward. Females have small, slender horns that arc backward. Males larger than females. Bighorn Sheep both graze and browse, taking grass, forbs, and the foliage of shrubs and trees. Births peak in May and June; 1 or 2 young. Females lamb on rugged cliffs. On the east slope, in southern Mono Co. and in Inyo Co. Have been reintroduced to Wheeler Crest and Lee Vining Canyon. In summer inhabits open, steep areas of Alpine Rock and Meadow Communities. In winter descends to lower Montane Forest, Pinyon-Juniper Woodland, Sagebrush Scrub, and other High Desert scrub communities. *Distribution*   Scattered populations occur throughout the Canadian Rocky Mountains, the inland mountains and desert ranges of the West, Baja California, and northern Mexico.

PRONGHORN: FAMILY ANTILOCAPRIDAE

Pronghorns are antelopelike grazers of arid grasslands and brush communities. They are not the true antelopes of African plains, but distant relatives. Both males and females grow horns, but in some females the horns are absent or lack prongs. Pronghorns have a distinctive white rump with no tail. If a Pronghorn is startled or frightened, its long, coarse rump hairs stand up, enlarging the size of the rump and serving as an alarm signal for the herd. Pronghorn herds evade their predators by outrunning them; individuals have been clocked as fast as 49–54 mph (79–87 kph). Predators such as Coyotes and Bobcats usually take fawns rather than full-grown animals.

Like other artiodactyls, Pronghorns vary seasonally in herd size and composition. Herds are largest in winter, contain both sexes, and may number hundreds of individuals. In late winter herds break up into smaller groups consisting of females and some males. Females scatter to give birth but form nursery herds when fawns are three to six weeks old. During the rut, older males join female groups or solitary females. Groups of young males may attempt to interfere with the older males.

Although Pronghorns have been protected by law since 1883, poaching and livestock grazing decreased their numbers in Mono County until they became extinct in the 1920s. Similarly, across their range in North America, Pronghorn populations declined because of hunting, poaching, and grazing. However, proper management and reintroductions have increased their numbers in the United States from an endangered low of 13,000 in the 1920s to a healthy 400,000 by the mid-1970s. Today, the Pronghorn remains an important game animal and a symbol of the once vast plains and grasslands of the American West.

*PRONGHORN, *Antilocapra americana*   Body: 4 – 4.8 ft (1.2 – 1.5 m). Pale tan above with white underparts. Prominent, large white rump patch, and 2 broad white bands across neck. Short, dark mane; black patch at angle of jaw on both sides. Both sexes have a pair of horns that are black, hooked back at the tip, with a single, forward-projecting prong. Eats foliage from shrubs, grasses, and forbs. Births peak in May or June; 1–3 young. Absent from the west slope. In the Eastern Sierra, north of Susanville and reintroduced to southern Mono Co. where Pronghorn now inhabit the Bodie Hills, Adobe Valley, and Hammil Valley north of Bishop. Occurs in Sagebrush, Shadscale, and Alkali Sink scrub, and Pinyon-Juniper Woodland.   *Distribution* Ranges across the grasslands and plains of southern Alberta and Saskatchewan, interior western United States east to western Minnesota and Iowa, and south to Texas, Baja California, and northern Mexico.

STAYING PUT

Of all the animals, mammals perhaps have the hardest time of it in the Eastern Sierra. They must endure the same extremes as all other creatures — long summers of withering heat and long winters of bitter cold and scarce food. But unlike the arthropods, fishes, reptiles, and amphibians, active mammals and birds must generate and maintain a high body temperature twenty-four hours a day. To do so, they need food often. But unlike the birds, when food is scarce few mammals can migrate any distance to obtain it. (Bats are the exception; they may migrate hundreds of miles.) Deer may migrate 50 miles (80 km) or more, from their summer range in the mountains to their winter range in the foothills, but most mammals stay put. Their ability to enter daily torpor, hibernate, or estivate enables them to survive times of temperature extremes as well as times of scarce food and water. In fall, many mammals put on fat and then hole up in burrows where their body temperature drops and their metabolic rate may decrease by as much as 50 percent. With fur and fat insulating their bodies and snow insulating their burrows, hibernating mammals can endure the coldest winter months without eating or drinking. Some desert mammals experience a similar physiological state, estivation, to get them through the worst of the desert heat. As they take refuge in cool, underground burrows and their metabolic rate drops, they too can go several months without food or water. The mountain-dwelling Pika copes with scarce food by making "hay" and storing it. The desert-dwelling kangaroo rat copes with scarce water by manufacturing water from its food and by excreting only highly concentrated urine. These and many other specialized behavioral and physiological adaptations enable Eastern Sierra mammals to survive in spite of the harsh conditions at the edge of the Great Basin. They truly experience life "on the edge."

RECOMMENDED READING

Burt, W. H., and R. P. Grossenheider. 1976. *Field Guide to the Mammals.* 3d ed. New York: Houghton Mifflin.

Chapman, J. A., and G. A. Feldhamer. 1982. *Wild Mammals of North America.* Baltimore: Johns Hopkins University Press.

Hall, E. R. 1995. *Mammals of Nevada.* Reno: University of Nevada Press.

Jameson, E. W., Jr., and H. J. Peeters. 1988. *California Mammals.* Berkeley: University of California Press.

CHAPTER TWELVE

WATER

GENNY SMITH

As the prevailing westerlies carry Pacific storms inland, up and over the Sierra Nevada, dumping rain on the foothills and snow on the higher slopes, snowfall totals sometimes reach 25–35 feet (7.6–10.6 m). Sierra snow that accumulates during winter is the major source of water for the plants, animals, and humans that inhabit California and western Nevada. The snowpack on the Sierra's western flanks nourishes more than a dozen rivers and many more creeks, as the melting snow drains into streamlets, fills lakes, and replenishes groundwater (map 3).

Eastern Sierra creatures have far less water available than their kin on the western Sierra slope. Still, the snowpack on the Sierra's eastern flanks gives birth to five rivers: the Susan, Truckee, Carson, Walker, and Owens. None of them reaches the ocean; instead, they terminate in desert lakes and marshes. Today many of these waters are in a sorry state, either dry or drastically shrunken. The large numbers of animals that depended on them historically are no more: some populations are extinct, a few species are near extinction. Winnemucca and Owens lakes are dry. Only a small remnant of Honey Lake remains, as a wildlife refuge. Pyramid Lake has dropped more than 70 feet (21.3 m), Walker Lake over 100 feet (30.5 m), Mono Lake over 40 feet (12.2 m). (During the three winters beginning in 1994/95, extremely heavy rain and melting snow flooded usually dry basins, causing loss of land, homes, and roads. However, unless weather patterns change dramatically, the Eastern Sierra will continue to be an arid region.) Of the Carson Desert's once extensive wetlands, few remain. The Eastern Sierra flyway, once a major migration route for millions of ducks, geese, and other birds, has all but disappeared. The famed Lahontan Cutthroat Trout and Eagle Lake Trout can no longer reproduce naturally but are spawned and reared in fish hatcheries.

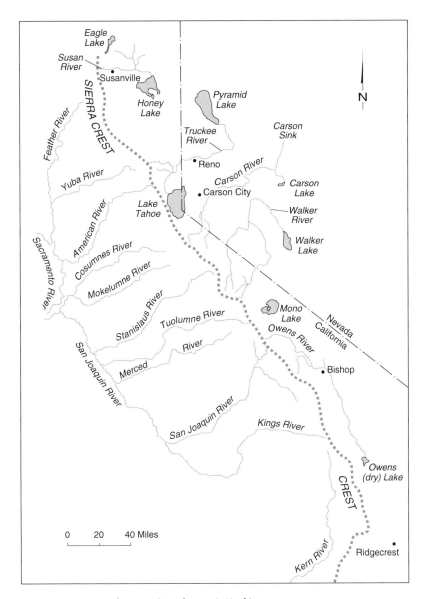

Map 3.   Sierra Nevada rivers. (Map by Annie Kook)

"Bathtub rings" mark exposed shorelines that generate tons of polluting alkali dust. What has happened to the Eastern Sierra's bountiful water?

WHAT HAPPENED TO ALL THE WATER?

It's a long, convoluted story that is believable only if you understand that in
the desert farming is impossible without irrigation and that because water is

scarce, it is as precious as gold. In the desert water is something to covet, to buy and sell, and to get title to by stealing, treachery, or any other means. Something that still generates bitter disputes and lengthy court cases and still bestows power and wealth. Something that in the late nineteenth century induced extravagant dreams of converting thousands of desert acres into green fields with brimming canals. Something that stirred the imagination of dreamers and water engineers. Some of these dreamers were public-spirited; others were speculators, boosters, or at worst con men. Some went bankrupt, borrowing money and losing all of it. Those taken in by the exaggerated claims and promises of cheap water and land often lost their money and their land.

HONEY LAKE VALLEY

In the 1880s Benjamin Leavitt planned and built an irrigation system for the Standish area east of Susanville by diverting the Susan River, the main tributary of Honey Lake, into three reservoirs. His works still irrigate about 6000 acres. (One acre equals 0.4 hectare.) Others in Honey Lake Valley diverted the water of small streams and springs, destroying aquatic habitats. Still other dreamers hatched a plan to pump water out of Honey Lake onto nearby lands. After only one year the lake receded, and the pump could no longer reach the water. Between 1919 and 1937, shallow Honey Lake (its greatest depth was 10 feet, or 3 m) was totally dry. Recently, although barely enough water remains to maintain Honey Lake Wildlife Area, groundwater pumping and export to Reno-Sparks was an issue.

Of all Eastern Sierra irrigation dreams, the least successful involved nearby Eagle Lake. Honey Lake Valley's 150,000 sagebrush acres and the proximity of 20-mile-long (32-km) Eagle Lake sparked imaginations as early as 1872; the mirage of diverting Eagle Lake water out of its watershed and into the valley did not die until 1986. During those years a tragicomic scenario unfolded, as dreamers tried again and again to build tunnels, ditches, and pumping plants to tap what they thought was an inexhaustible supply of water. That Eagle Lake was not drawn down 45 feet (13.7 m), as one project proposed, was not for lack of trying.

Eagle Lake lies 1000 feet (305 m) higher than Honey Lake Valley. Its southeastern shore lies just 1.5 miles (2.4 km) from the head of Willow Creek, a tributary of Honey Lake; between the two is a low 40- to 80-foot (12.2- to 24.4-m) ridge of lava. Tunneling under the ridge, channeling Eagle Lake's water into the natural waterway of Willow Creek, and letting it run downhill to the valley seemed direct and uncomplicated. Captain Charles Merrill, a land speculator and one of the first to promote the idea, filed a claim on Eagle Lake water. Then, with the financial backing and political connections of San Francisco millionaire Alvinza Hayward, Merrill persuaded Congress to pass the Lassen County Desert Land Act in 1875. Under the act, with no residence required, a person could claim 640 acres, irrigate within two years, and buy the land for $1.25 per acre. This act made it easier to obtain desert land than it was under the Homestead Act of 1862, which limited claims to 160 acres. (The Desert Lands Act of 1877 extended the provisions to eleven states and increased the time for purchase to three years.) While the intent of these acts was to benefit family farmers, they proved to be a bonanza for speculators and large land and cattle companies, who thought of many ways—forgery and bribery, among others—to amass large holdings for little or no payment and then sell them (along with promises of water that might never materialize) to unsuspecting farmers. Hayward left the picture after he lost heavily during the Bank of California's debacle in 1875, but Merrill found other investors and hired workers to start a 6000-foot (1829-m) tunnel. When the money ran out, all concerned blamed and sued each other and ended up in court—a pattern that was repeated over and over by Merrill and subsequent entrepreneurs. Parties to the suits often spent years in court, battling over who had rights to what and who should pay the creditors.

In 1891 a new company, the Eagle Lake Land and Irrigation Company, proposed to pump and siphon, rather than tunnel. Without a drop of water yet coming down Willow Creek, the company advertised 150,000 acres of land with water, ready for immediate use, for $1.50 per acre. Its ads included a sketch of Eagle Lake, its depth noted as 20–1000 feet (6.1–305 m). In light of today's engineering standards, the absence of any attempts to determine water supply, rainfall, lake depth, and lake-level fluctuations seems appalling. Locals believed that Eagle Lake's waters were inexhaustible, fed by mysterious springs in the lake bottom. It wasn't until 1915 that anyone sought to measure the lake's depth; it proved to be mainly shallow, with a maximum depth of 105 feet (32 m). Nor did anyone seem to realize that desert lakes fluctuate widely as they respond to above- and below-average rainfall.

In 1915 Leon Bly began the last attempt to bring Eagle Lake water to the valley. He hired a construction company to dig a new tunnel and a channel from the lake to the tunnel's intake. But as the lake dropped, which he had not counted on, he had to lower the cut again and again. By 1932 Eagle Lake had dropped 24 feet (7.3 m). Even in its best years the tunnel irrigated only 5000 acres; in dry years it irrigated none. During the 1930s a rock slide blocked part of the channel and tunnel, the irrigation district went bankrupt, and the State Water Board revoked the water permits. In 1986 all arguments ended when the Bureau of Land Management sealed the tunnel at a cost of $98,000.

The famous Eagle Lake Trout, which weigh up to 11 pounds (5 kg) and are endemic to Eagle Lake, did not fare well in the shrunken lake. Small diversions from tributaries, the low lake level caused by Bly's tunnel, and a series of dry years kept Pine Creek so low the trout could not swim upstream to spawn. By the late 1940s the trout were in danger of extinction; some reports said fish numbers might be as low as just twenty-five trout. From the remaining fish, the Department of Fish and Game initiated a program of artificial spawning and hatchery rearing. The program has been successful, the fishery is productive, but hatchery production is still essential.

THE TRUCKEE AND CARSON RIVERS: THE NEWLANDS PROJECT

The fate of the Truckee River is intimately bound up with that of the Carson River in the Newlands federal reclamation project, in which both rivers were diverted in the early 1900s to irrigate arid lands east of Reno. After Congress passed the Federal Reclamation Act in 1902, Newlands was the first project authorized.

It is important to remember that irrigation, although confined to the West, became a national issue during the late 1800s. Expansion of irrigated agriculture was idealized not only as "a means of economic development but also as the driving wheel of social and spiritual progress" (Kahrl 1982, p. 30). "Making the desert bloom" to benefit family farmers became national policy, which fitted in with the belief, ever since the time of Thomas Jefferson, that a nation of small farmers was the best foundation for the country. At times, however, this benevolent vision became mixed up with boosterism and the quest for power.

Senator Francis G. Newlands of Nevada and other boosters envisioned turning 200,000 acres of desert east of Reno, near Fernley and Fallon, into productive farmland, thus returning prosperity to Nevada after the silver-mining boom collapsed. The project's 32-mile (51-km) canal diverted about half of the Truckee River out of its watershed into the Carson's watershed. A dam 162 feet (49.4 m) high contained the waters from both rivers in Lahontan Reservoir west of Fallon. Lake Tahoe also served as a storage reservoir for the Truckee. Although a small dam controlled only the upper 6 feet (1.8 m) of the lake, on so large a lake that amounted to usable storage of 744,000 acre-feet, or more than twice the storage of Lahontan Reservoir. (One acre-foot is the amount of water needed to cover 1 acre of land to a depth of 1 foot.) The Truckee and its tributaries also supplied irrigation water to about 29,000 acres in Truckee Meadows, municipal water to Reno-Sparks and, in more recent years, to the growing resort towns at Lake Tahoe. Beginning in 1910 the Truckee's terminal lakes, Winnemucca and Pyramid, dropped precipitously, and by 1939 Winnemucca was dry. The native Pyra-

Squaw Creek, tributary to the Truckee River. High Sierra snowbanks nourish hundreds of small streams that meander downslope and coalesce, forming the Eastern Sierra's five rivers. (Photo © Tom Lippert)

mid strain of Lahontan Cutthroat Trout became extinct in the 1930s, partly from overfishing but mostly because the trout could no longer swim upstream to spawn. After diversion, not only were Truckee streamflows too low and too warm, but also as Pyramid dropped, the delta at the river mouth blocked the trout and the Cui-ui from swimming to their spawning grounds.

Before diversion, cutthroat swam as far as Tahoe tributaries to spawn. Cutthroat from another lake have been introduced to Pyramid and are maintained by hatchery production.

The West and East forks of the Carson River rise in the Sierra snowpack between Carson and Sonora passes, flow down and join near Genoa in the Carson Valley, and then flow east. The waters naturally come to rest in the broad, low portion of Lahontan Valley known as the Carson Desert or Carson Sink, which contains a series of shallow lakes and marshes, some ephemeral and some not. Farmland irrigation has always been the major user of Carson River water. Irrigated lands total about 47,000 acres in the Carson Valley, less acreage in Diamond Valley, and small acreages in Dayton, Churchill, and Eagle valleys. The largest user today is the Newlands Project, which irrigates about 57,000 acres near Fallon—primarily alfalfa and pasture, with some grain and truck crops, notably cantaloupes. The 200,000 acres touted in 1903 proved to be unrealistic because of poor drainage and saline and alkaline soils, among other problems that were not recognized or planned for. The soils and the high groundwater level required an extensive drainage system to remove water applied to flush out the salts and to remove excess water lest crop roots become waterlogged. As river waters were diverted to the farmlands, many of the Carson Desert marshes and ponds dried up.

Litigation over many aspects of the Newlands Project went on for years—over control of the Lake Tahoe dam and over rights to Truckee and Carson water. Litigants in the various suits included power companies, Tahoe lakeshore property owners, the federal Bureau of Reclamation, Truckee-Carson Irrigation District, and upper Carson and Truckee Meadows ranchers. Conflicts became bitter in the late 1920s and 1930s during a series of dry years when frustrated farmers were not receiving the water they expected. The Bureau of Reclamation had designed the Newlands Project during years of wetter-than-average precipitation, without planning for the dry years that are a recurring weather pattern in the Eastern Sierra. With the increased demand upon the rivers from Newlands, there was not enough water to go around. When Lake Tahoe fell below its natural rim in 1930, the power company, the Truckee-Carson Irrigation District, and other water users thought up various plots to put Tahoe water down the almost-dry Truckee. One plan revived the idea of pumping Tahoe water over the dam. Another involved a steam shovel sent with a police guard to the Tahoe dam to dig a diversion ditch to the rim; there were rumors of dynamiting the dam. A California sheriff's posse and a court injunction stopped the digging.

The Bureau of Reclamation, seeking to establish water rights, sued existing water users on the Truckee in 1913. The Truckee River Agreement of 1935 finally settled such issues as Tahoe lake levels, pumping, and some water rights. The bureau in 1925 sued water users on the Carson; the major conflict was between ranchers on the upper Carson and Newlands ranchers on the lower river. The suit was not settled until 1980, making it one of the longest-running water cases in the federal courts.

WALKER RIVER

The West and East Walker rivers rise in the Sierra south of Sonora Pass, flow northward into Nevada, join in Mason Valley south of Yerington, make a horseshoe bend, and flow south to empty into Walker Lake. Ranchers divert most of their waters. The West Walker irrigates lands in Antelope, Little Antelope, Slinkard, and Smith valleys; water is stored in Topaz Reservoir. The East Walker irrigates lands in Bridgeport and Mason valleys; water is stored in Bridgeport Reservoir. To store water for irrigating Walker River Indian Reservation lands that are downstream from the valleys mentioned, Weber Reservoir was completed in 1935. Between 1882 and the 1990s, the terminal lake of both rivers, Walker Lake, has dropped 126 feet (38.4 m). Lahontan cutthroat once migrated upstream as far as Robinson Creek above Bridgeport Valley. Today the trout cannot spawn naturally but are maintained by hatchery plantings.

WATER FOR THE BIG CITIES

California coastal cities realized long ago that their local water supplies were limited and, because of recurring dry years, unpredictable. In the 1860s a San Francisco engineer, Alexis Von Schmidt, formed a company to supply water to San Francisco (and some Mother Lode mining camps on the route of the proposed aqueduct) and constructed a small dam at Lake Tahoe. The Donner Lumber and Boom Company immediately challenged Von Schmidt over land and water rights. When financial difficulties ended the plan, San Francisco turned to Hetch Hetchy on the western Sierra slope.

Los Angeles engineers were also aware of their limited and unpredictable supply. Foremost among them were Fred Eaton and William Mulholland. Eaton at age twenty had become superintending engineer of the Los Angeles City Water Company. Self-taught, he combined his expertise in water and his devotion to the growth of Los Angeles in the effort to find additional water sources. Two surveys made in 1885 and 1891 showed it was possible to transport Owens River water to Los Angeles by gravity alone. Eaton tried to interest others in such a project, but no one would listen. However, holding on to his dream, he hit upon municipal financing as the only way to raise enough money for a 235-mile aqueduct and combined that with a scheme to establish a perpetual source of wealth for himself by controlling the water

Kearsarge Peak and Mount Mary Austin above Owens Valley. The Sierra snowpack is the major source of water for the Eastern Sierra and all its creatures—plant, animal, and human. (Photo by Tom Ross)

source. Eaton was elected city engineer and, in 1898, mayor of Los Angeles. William Mulholland, also self-taught, started out as a ditch-tender for the City Water Company, became a protégé of Eaton, and succeeded him as superintendent. When the city bought the water company, it also bought Mulholland. By 1904 Los Angeles was growing rapidly; both Mulholland and Eaton realized that local water supplies could not support a continually growing city.

Unlike the water wars mentioned above, in Owens Valley the farmers battled to *keep* their water. But the city 250 miles away, with money and powerful connections, was too much for them. The intrigue, the secrecy, the politics, the broken promises, and the strong personalities of the key players in the drama are the subjects of nonfiction works as well as several novels— tragic, outrageous, or humorous reading (perhaps all three), depending on your sympathies.

OWENS VALLEY AND THE MONO BASIN

Knowledgeable water engineers knew that Owens Valley was a prime candidate for an irrigation project under the 1902 Federal Reclamation Act. The

Reclamation Service chose Joseph Lippincott, an educated, well-respected consulting water engineer in Los Angeles, to be its supervising engineer for California. Lippincott accepted the position with the proviso that he retain his consulting practice—a proviso whose meaning was later debated. In 1903 Lippincott, inspecting Owens Valley, was so impressed with its potential for irrigation that he withdrew over 550,000 acres from entry under the Homestead and other federal land acts.

The next year, as Eaton realized that the Owens Valley project could end his cherished dream, he hastened to visit Lippincott in Owens Valley. Los Angeles had long been one of Lippincott's major clients; Eaton had worked closely with him and Mulholland on various projects for years. All three men knew the city's water supply was limited. After learning much about the project's details, Eaton returned to the valley for a month, this time with Mulholland, who had been looking for (but not finding) new water sources for the growing city. It was not difficult to persuade Mulholland that here was a water supply that could support a population of 2 million, ten times the number of people then in the city. At the same time, Owens Valley farmers had become enthusiastic about the irrigation project and were urging the government to proceed.

By September 1904 Lippincott was aware that Eaton and Mulholland were intensely interested in the Owens River as a water source for Los Angeles. He and Frederick H. Newell, the director of the Reclamation Service, met with them and their colleagues to assess the city's intentions. Lippincott proposed that the city receive a copy of a just completed report, with all streamflow data, which stated that on average the project could increase irrigated lands by more than 100,000 acres. At about this time Lippincott concluded that the value of water for the growing city far outweighed the value of irrigation for a small number of farmers, that using the water in southern California was "the greatest public necessity" (Kahrl 1982, p. 58). He began actively working with Eaton to persuade the city to fund the aqueduct and signed a contract (against Newell's instructions, for he was still working for the Reclamation Service) to work with the city on the aqueduct.

Eaton began buying up options on Owens Valley lands needed for the aqueduct, being careful not to say whom he was working for. Assuming that he was working with or for Lippincott on the irrigation project, many farmers sold willingly. Eaton also bought options in Long Valley, resold half of the acres to the city for close to the full purchase price, and kept half himself, including the key site for a reservoir in Long Valley—a profitable deal for Eaton that led later to a bitter rift with his old friend Mulholland over the price Eaton demanded for the reservoir site.

Up to this point Los Angeles plans and dealings were still secret, but

Eaton's purchases were raising suspicions. Realizing the necessity to act quickly before aqueduct news became public and land prices soared and others established water rights to the Owens, the city secured all the options necessary for the aqueduct before the *Los Angeles Times* broke the story in late July 1905. Valley farmers were outraged. In a scare campaign to get city voters to approve a $1.5 million bond issue to pay for land and water rights purchases, Mulholland made speeches and the *Times* printed articles about a nonexistent drought between 1892 and 1904 and predicted dire consequences if the aqueduct were not built. The aqueduct bond issue, the first of many, passed in September 1905 by a margin of 14 to 1. During this year accusations against Lippincott mounted. He was charged with a conflict of interest because he had been working for both Los Angeles and the Reclamation Service. The following year he resigned and accepted a better paying job with the city. He later said he had no regrets whatsoever over his dual role, emphasizing that his goal had always been the greatest public good, never personal gain.

Building the largest aqueduct in the Western Hemisphere was an engineering feat second only to construction of the Panama Canal. Without a single pump, gravity carried the water through canals, pipes, siphons, tunnels, and reservoirs all the way to Los Angeles. As the city continued to buy land, some farmers sold willingly; others were tricked or frightened into selling. During the hard times of the 1920s—with many farms mortgaged and with widespread hopelessness and bitterness over the city's trickery and broken promises—valley feelings erupted, and peaceful farmers turned to violence. Between 1924 and 1927 valley farmers blew up the aqueduct seventeen times; at one point they seized the Alabama Gates and turned all water out of the aqueduct for five days.

But the city prevailed. The Department of Water and Power (DWP) continued to buy land and water rights, extending its holdings north to Long Valley and the upper Owens in Mono County, and also pumping groundwater. To pick up the waters north of the Owens drainage, DWP filed claims on several streams, bought ranches and their water rights around Mono Lake, and in 1941 began diverting Mono's streams through the Mono Craters tunnel into the upper Owens. With ever more people to serve, in 1970 DWP completed a second aqueduct, increasing total capacity by 48 percent, and eventually expanded its groundwater pumping.

Little by little the extensive marshes and wetlands dried up, farmlands returned to brush, and both lakes shrank. Owens Lake at the southern end of Owens Valley, the natural terminus of the Owens River, became completely dry. Between 1941 and 1982 Mono Lake dropped 45 feet (13.7 m). With few nesting and resting areas remaining, numbers of migrating waterfowl declined drastically. The Owens Pupfish and Owens Tui Chub declined as shallow habitats along the streams disappeared and as Brown Trout, Largemouth Bass, and other predatory fish were introduced for sport fishing. The pupfish and chub were close to extinction when they were placed on the state and federal lists of endangered species.

CHANGING VALUES

Recent conflicts have taken a different direction, reflecting today's concern with values other than subsidized irrigation and urban growth. In the fervor of idealizing the family farm, no thought was given to the potential impacts of the diversions—the effects upon Indian water rights, migrating and nesting birds, native fish, scenery, or upon the clouds of alkali dust the winds would pick up from newly exposed lakeshores.

By 1959 ideas had changed in Lassen County. Support for diverting Eagle Lake water had declined, and local groups were opposing a new diversion plan, testifying that the lake's recreational and scenic values far outweighed the costs and benefits of irrigation. The lake's famous trout had become a major attraction; their near extinction in the 1940s only emphasized the value of the lake's extraordinary fishery. In 1962 the State Water Board denied any diversions from Eagle Lake.

Then, as impacts of the Newlands and Los Angeles massive water diversions accumulated and became obvious, injured parties filed lawsuit after lawsuit. The Pyramid Lake Paiutes, whose reservation surrounds the lake, saw the lake dropping and becoming more saline, their fishery declining, and Truckee River water dwindling to a trickle by the time it reached the lake. Two fish in Pyramid Lake—the Cui-ui and the Lahontan cutthroat—are on the federal list of threatened and endangered species. The Fallon Paiute-Shoshone tribe saw promises of irrigated land broken by other Newlands commitments. Fears are mounting that Walker Lake may soon become nonproductive as its water becomes more saline and alkaline and the level of total dissolved solids increases. If these trends continue and its water becomes more concentrated, fish will not be able to survive in Walker Lake. At Mono Lake, California Gulls lost one of their major nesting areas when the low lake level exposed a land bridge that allowed predators to reach a former island.

With wetland habitats drastically diminished in the Eastern Sierra as well as in Mexico and Central America, the once important Eastern Sierra flyway no longer supported many millions of migrating and nesting waterfowl, shorebirds, and other birds. Phalaropes, pelicans, Bald Eagles, White-faced Ibis, Long-billed Dowitchers, and Snowy Plovers, to name a few, are among those that lost habitats important to their survival. At Mono Lake alone, the number of waterfowl that frequent what is left of the freshwater marshes has dwindled from 1 million to 15,000 birds. Alkali dust kicked up by winds from Owens (dry) Lake and Mono's exposed lakeshore exceeds the federal air quality standard for particulate matter. In Owens Valley populations of some native plants disappeared as groundwater pumping lowered the water table, and the Owens Pupfish (1967) and the Owens Tui Chub (1985) were listed as endangered.

The lawsuits were complex, with many litigants. Some are yet to be resolved; others have reached settlements. Public Law 101-618, passed by Congress in 1990, reauthorized the Newlands Project to serve additional purposes: recreation, fish and wildlife, and municipal water for Fallon. It addressed the issues raised by the Pyramid and Fallon Paiutes, and the issues of endangered fish at Pyramid Lake and declining wetlands in Lahontan Valley. It will be many years before details are worked out and the legislation implemented, although the purchase of water rights for Stillwater Wildlife Refuge in Lahontan Valley has begun. Concern over Walker Lake and its fishery is just beginning to mount. Inyo County and Los Angeles have a tentative agreement that restores certain wetlands and limits water export when valley groundwater level drops below the root zone of specified native plants. After fifteen years of lawsuits, Mono Lake in 1994 received a decision from the California State Water Board that will restore its level to 6392 feet (1948 m) above sea level, about 17 feet (5.2 m) above its 1994 level; Los Angeles diversions were significantly reduced.

BALANCING VALUES

If, early on, politicians and engineers had recognized the enormous variation in Eastern Sierra rainfall that limits water supplies, and if they had recognized the multiple values affected by diverting water, we would not today be undoing some of the major mistakes made a hundred years ago. Today's laws would not give anyone the leeway Isaac Roop had in the 1850s, when he recorded, "I, the undersigned, claim the privilege to take all the water of Smith Creek . . ." (Amesbury 1967). As it has turned out, many of the irrigated desert farms were marginal, and today the small family farm is largely a romantic myth. With few exceptions, small is not economically feasible; small farmers often must have other jobs to make a decent living. Some farms are viable enterprises only because they are subsidized with cheap water and government crop purchases. Today large individual holdings and very large company farms are the norm.

As public values change, balancing the values inherent in wildlife, scenery, irrigation, family farms, municipal use, and power production will be a never-ending exercise. Today the balance is swinging toward a recognition that desert marshes and lakes are valuable and beautiful, and that the wild animals and plants that inhabit them have a worth of their own, whether we use and enjoy them or not. A recognition that in arid lands such as the Eastern Sierra, what happens to the rivers and streams eventually affects every living thing, including us humans.

RECOMMENDED READING

Amesbury, Robert H. 1967. *The Search for Water in the Honey Lake Valley.* Bulletin 17. Lassen County Historical Society.

California Department of Water Resources. 1991. *Carson River Atlas.* Sacramento: DWR.

————. 1991. *Truckee River Atlas.* Sacramento: DWR.

————. 1992. *Walker River Atlas.* Sacramento: DWR.

Kahrl, William L. 1982. *Water and Power.* Berkeley: University of California Press.

Purdy, Tim I. 1988. *Eagle Lake.* Susanville: Lahontan Images.

CHAPTER THIRTEEN

WONDERS AND SECRETS AT
THE EDGE OF THE GREAT BASIN

GENNY SMITH

In these pages we have introduced you to the Eastern Sierra, a region where desert valleys of long summers and snow-spangled mountains of long winters lie side by side, a land of immense views and overarching sky. Where 500-million-year-old folded rocks and the 100-million-year-old granites that shouldered them aside are on dramatic display. Where Earth's crust has moved and continues to move, up, down, and horizontally, lifting the mountains and dropping the valley floors. Where red cinder cones, pale gray pumice, and steaming hot springs speak of long-past violent eruptions. Where the Sierra Crest towers a mile—even two miles—above its adjacent desert valleys. An arid land where plants and animals must survive intense sun, blazing heat, freezing cold, and little water. Yet despite its ruggedness and seeming austerity, the Land of Little Rain (as Mary Austin lovingly named it) is friendly and easily accessible. Hundreds of miles of good roads traverse the valleys and wind up the canyons to passes at 9000 or 10,000 feet (2743 or 3048 m).

To understand this complex region, so different from most of California, we must look eastward, for it is akin to the Great Basin—an arid region of gray-green sagebrush that stretches from the base of the Sierra across Nevada and half of Utah. If the land seems bleak and lifeless, that is only because we have not yet learned that desert lands do not flaunt their wonders or give up their secrets easily. We must seek them out, at the right season and at the right times of day and night. The Eastern Sierra teems with life; we need only to know when and where to look. We may never see valley wildflowers, great swaths of yellow and white, if we come only in July and August. Ground squirrels frequent campgrounds at any time of day, but most other mammals are secretive, leaving their hiding places only at dawn or dusk. Some animals

The Eastern Sierra near Bishop. Desert valleys of long summers and snow-spangled mountains of long winters lie side by side. (Photo by David Hamren)

are active only at night, others only at certain seasons. Desert mammals know that the best place to be on hot summer days is in the shade or underground. We may never see them at all, unless we learn something about their habits. When do they feed and hunt, what is their food, and where do they find it? Where do they go for water? When and where do flowers bloom? Where do birds nest, and when do they migrate? Where are the springs and ponds and wet meadows? For life of all kinds thrives in wetlands. It takes time to ferret out the secrets of this land and its creatures—maybe a lifetime.

Each season brings its wonders. Spring and fall days are a joy, the air crisp and the sun warming. Summer's crowds are absent, the pick of the campgrounds is yours. Winter too has many glorious sunny days. Never are the mountains more dramatic than after a winter storm, when snow banners fly from the high ridges. Many animals will have migrated away or are hibernating underground, but not all of them! The valleys will have surprising numbers of birds, such as hawks, that have migrated south from their breeding grounds.

Much of the Eastern Sierra belongs to all of us: It consists of federal land managed by the U.S. Forest Service or the Bureau of Land Management. These lands support a little mining (the large, productive mines were mined out years ago), some logging (although trees such as lodgepole and pinyon are not suitable for timber), and grazing that dates back well over a hundred years. Much of the land is managed for recreation—camping, hiking, fishing, sightseeing, cycling, skiing, and snowboarding—and a long, narrow strip of magnificent wilderness straddles the Sierra Crest.

What is the future of this dramatic, colorful, starkly beautiful Eastern Sierra? When our great-grandchildren come camping and hiking, will there still be mountain meadows? Native fishes, eagles, bluebirds, wild sheep? Will we pay attention to scientific studies, such as the 1996 *Sierra Nevada Ecosystem Project: Final Report to Congress,* that tell us what is happening to our lands and how to care for them?

There are encouraging signs that more and more of us do care about our lands and the wild things that live there. Encouraging signs that we are learning to walk softly, to think of meadows and forests as gardens that we would no more trample than we would our own front yard. Encouraging signs that more and more of us are aware that the world is a very small planet—a wondrous creation inhabited by plants and animals so numerous that we don't yet know them all. Aware that since we alone have the power to annihilate our fellow creatures, we have all the more reason to assume responsibility for the welfare of the wild things that are here on Earth along with us— throughout the world as well as here at the edge of the Great Basin.

INDEX

Species preceded by an asterisk (*) are found only east of the Sierra Crest, not west of the Crest. Arthropods found mostly east of the Sierra Crest are indicated by *E;* species found only east of the Crest or at high altitudes are indicated by *ES.*

467

Buteo
    jamaicensis, 197, 290, 295, 306–7
    lagopus, 307
    regalis, 307
    swainsoni, 306
Buttercup, Alpine, 147, pl 10a

Cactus,
    Beavertail, 204–5, pl 15h
    *Hedgehog, 205
    *Mojave Mound, 205
    *Old Man, 205
California Coffeeberry, 157, 159
California Snow Scorpionfly^E, 242, 243
Callipepla californica, 311, 313
*Callisaurus draconoides, 278
Calocedrus decurrens, 29, 31, 128, 129
Calypte
    anna, 334
    costae, 334
Calyptridium umbellatum, 108, 114, 115
Camas Lily, 32, 33, 150–51, pl 10b
Camassia quamash, 150–51, pl 10b
Camponotus modoc, 215, 216
Canis latrans, 31, 430, 431, 432
Canvasback, 302
Capnia species, 242, 243
carbon dioxide, at Horseshoe Lake, 58
Cardiophorus tenebrosus, 242, 243
Carduelis
    pinus, 386–87
    psaltria, 387
    tristis, 387
carnivores, 429–42
Carpodacus
    cassinii, 375, 385
    mexicanus, 385–86
    purpureus, 384–85
Carson Desert, destruction of, 450, 456
Carson Range, 25, 27, 28, 61, 71, 86
Carson River, 25, 30, 251, 255, 257, 258; im-
    pact of diversions, 450. See also New-
    lands federal reclamation project
Carson Valley, 25–26, pl 1
Casa Diablo, 22
Cassiope mertensiana, 110, 111, 143
Castilleja
    *angustifolia, 172
    lemmonii, 150

miniata, 150, pl 11d
*minor, 188, pl 14h
nana, 108, 113, 114
Castor canadensis, 22, 393, 418–19
Cathartes aura, 295, 297–98
Catharus
    guttatus, 355, 365
    ustulatus, 364
Catherpes mexicanus, 359
Catoptrophorus semipalmatus, 318
Catostomus
    *fumeiventris, 250, 261
    *platyrhynchus lahontan, 253, 261–62
    *tahoensis, 34, 261, 262
Ceanothus
    cordulatus, 140, 141
    velutinus, 140–41, pl 12b
*Centrocercus urophasianus, 311, 312
Cercocarpus ledifolius, 138, 139
Certhia americana, 355, 357–58
Cervus
    elaphus nannodes, 444–45
    elaphus nelsoni, 444, 445
Ceryle alcyon, 290, 327, 335
*Chaetodipus formosus, 417
Chaetura vauxi, 331
Charadrius
    alexandrinus, 315, 461
    semipalmatus, 316
    vociferus, 312, 316
Charina bottae, 282, 283
*Chasmistes cujus, 254, 255, 262, 455, 461
*Cheesebush, Burrobrush, 205, pl 15e
Chickadee, Mountain, 353–54
Chickaree. See Squirrels, Douglas'
*Chionactis occipitalis, 287
Chionea species, 242, 243
Chipmunks, 405, 407, 409–10
    Allen's, 409
    Alpine, 407
    Least, 19, 407, 408
    Lodgepole, 409
    Long-eared, 409
    Merriam's, 409–10
    *Panamint, 410
    *Uinta, 410
    Yellow-pine, 409
Chlidonias niger, 33, 322
Chondestes grammacus, 373

Text:	10/13.5 Minion
Display:	Franklin Gothic Book and Demi
Design:	Barbara Jellow
Composition:	G&S Typesetters, Inc.
Printing and binding:	Malloy Lithographing, Inc.